**Preston L. Moore**

With chapters by
GEORGE S. ORMSBY
ROBERT D. GRACE
CHARLES C. PATTON
DWIGHT K. SMITH

The Petroleum Publishing Co.
Tulsa, 1974

DEDICATION

This book is dedicated to my wife. She has given support when support was needed, she has helped when help was needed, and she has become an integral part of all my activities including the preparation and writing of this book. Our partnership has grown and strengthened in our twenty-five years together, and in truth the authors should have been listed as Preston and Mary Jo Moore.

Contents

Preface

THIS book includes ideas and concepts that have been accumulated and observed during my twenty-five years of association with the oil business. It would be difficult to delineate the new from the old. In most ways the drilling business has followed an evolutionary trend and most improvements are simply changes and advancements in existing technology.

It is not uncommon in general conversations to hear the remark that a specific practice was introduced by a specific person. While this may heap glory on the individual, it is doubtful if any of us can claim exclusive credit for any practice. The drilling business includes many pioneers, and very often changes and improvements that were introduced simultaneously from many different sources.

Contributions to technology in drilling have come from basic research and trial-and-error field tests. There are no sure-fire ideas that can claim success before field applications. Actually, success is dependent on: (1) Will it work? (2) Is it safe? (3) Will costs be reduced?

Conclusive results from field tests are at times difficult to obtain because the applications are always subject to the available equipment and personnel. This simply means that a practice one group considers successful may be classified a failure by another group. One fact is relatively clear: Operating personnel consider their practices good or they would change. Thus, any advancement program attempted by staff personnel must by necessity be introduced carefully and with the full cooperation of those in operations.

The contents of this book are opinions that show the general level of proven technology as of January 1, 1974.

With more time, improvements could be made. Some will immediately criticize certain approaches made, and their criticism may be justified. In fact probably the most critical reviewer is the author himself. While reading the manuscript, the temptation was always present to alter substantially some portion of the book. If the temptation is allowed to flourish the book will never be written and publication will never be possible.

DRILLING PRACTICES MANUAL

1

Introduction to
Drilling Practices

ROTARY drilling began in 1900 and was expanded and developed through a combination of art and technology. Art has been at least an equal partner with technology. Many of the advances in drilling have been introduced by courageous men who refused to accept the phrase, "It cannot be done." Evidence of this attitude is demonstrated by many of the past accomplishments of drilling personnel. The most pronounced achievements in recent years have been jungle locations accessible only by helicopter, offshore operations in extremely rough waters and locations close to the North Pole, almost inaccessible to man only a few years ago. Science has played a part in these operations; however, a large measure of the credit must belong to the rig operators who were willing to be the pioneers.

The use of technology in drilling has been accelerated during the 1970's, although to some degree injections of technology began with the inception of rotary drilling in 1900. Technology is often applied as separate packages rather than total concepts. It is not uncommon to consider improvements in muds, bits and other accessory equipment separately. Some of the first efforts towards injections of technology in drilling were introduced by improvements in drilling muds. Before 1920, there were many efforts to improve drilling muds and these efforts have been accelerated in recent years. Many believe that any problem in drilling can be related to the drilling mud. As a general answer, mud cannot be blamed with all the problems; however, mud has been properly termed the heart of the drilling operation.

A major problem in the development and treatment of drilling muds is the lack of general agreement on requirements. Also in many cases, there

3

are difficulties in equating easily obtained surface measurements with down-hold performance. Special laboratory measurements under high temperatures and pressures are frequently made; however, it has been difficult to equate even qualitatively some of these measurements to measurements made in the field. There is no question that measurement procedures can be developed for field use that are accurate. The primary problem has been the economic justification for more expensive equipment and personnel. The economic hurdle has been overcome in recent years because of the daily operating costs of some of the offshore rigs, which exceed $60,000.00 per day. Time will still be required for equipment developments and qualified personnel to overtake demand.

Technology has played significant roles in well control. Operators can determine formation pressures and fracture gradients accurately. Equipment available for well control meets almost any conceivable requirement. If a piece of equipment fails, most rigs are equipped with multiple back-up systems to maintain control of the well. Occasionally a purely human mistake will be made and even this part of the operation is under attack through multiple training programs.

There is a continuous search for new methods to drill wells. Some of the procedures that have been tried include, (1) Flame drilling, (2) Electric-arc drilling, (3) Plasma-jet, (4) Laser beams, and (5) Erosion drilling. Erosion drilling appears to have special applications for use. The other procedures were workable, however, the costs of generating energy made them uneconomical when compared with methods in general use. Special techniques such as down-hold turbines to turn the bit have been used on a limited basis and may be used more in the future. Percussion drilling has in general been confined to air-hammers which are used sparingly in air drilling.

Technology as an applied science is changing continuously. Most changes are simply extensions of old ideas; however, on a limited basis, some old concepts are discarded and directions reversed. There appears to be no real substitute for actual applications to prove the worth of a practice proven by mathematical analysis and laboratory tests. Very often the statement is made that the drilling industry has changed very little since the introduction of rotary drilling. In truth the drilling business has changed substantially but the process has been evolutionary rather than revolutionary.

This book emphasizes drilling practices. Hopefully the philosophy is developed that the well talks to the operator. Once the operator learns the well language, the books contents are meant to help the operator develop meaningful answers to warnings of potential trouble. No set of instructions are ever complete and many times success or failure is based on preplanned actions which are not possible unless the operator is well versed in both the technology and the art.

Cost Control

TH E primary objective in cost control for drilling operations is to minimize the total well cost. Methods proposed to accomplish this objective include; (1) Speed drilling introduced in the 1950 to 1960 decade, (2) Minimum cost drilling concepts introduced in the same decade, and (3) Special programs aided by rig site or off-site computers introduced during the decade from 1960 to 1970. The use of optimized drilling programs aided by computers to assemble drilling data and to determine optimum programs is continuing into the decade from 1970 to 1980. There is no question that science and technology has invaded the drilling business and will continue to play a substantial part in cost control.

Actually the methods being used to minimize drilling costs are not new. All of the current procedures were introduced many years ago. The difference in current operations is that more operators are aware of cost control procedures and more willing to pay the cost of implementing cost control programs.

The first pre-requisite for any type of cost control program in drilling is good well planning. Second the drilling operation must be supervised closely and third the operator should follow-up the well completion with a thorough analysis of the drilling operation.

Well planning details will not be considered in this discussion because there is a separate chapter on well planning. It should be mentioned, however, that a review of separate cost functions for past wells should be reviewed carefully. Table 2–1 lists one such analysis for a low cost 7500 foot well in Northwestern Oklahoma.

Wells in this same area have been drilled and completed for total costs in the range of $70,000.00. Thus before drilling in this area an operator

TABLE 2–1
Separate Well Cost Functions

Item		Cost
I. Drilling		
1. Footage		$40,000
2. Day work		5,000
3. Other		2,000
	Total	$ 47,000
II. Intangible		
1. Location and roads		$ 5,000
2. Coring		3,600
3. Logging		3,000
4. Formation testing		1,800
5. Fuel		–
6. Water		1,500
7. Drilling fluid		9,000
8. Cementing		3,200
9. Transportation		1,600
10. Perforating		1,000
11. Stimulation		7,500
12. Bits		–
13. Rental Equipment		1,500
14. Miscellaneous		3,200
	Total	$ 41,900
III. Tangible		
1. Casing – Surface		5,000
Production		15,000
2. Tubing		9,400
3. Christmas tree & Surface connections		5,400
4. Other equipment		2,500
		$ 37,300
	Total Well Cost	$126,200

should obtain a cost breakdown such as that shown in Table 2–1 for as many wells as possible. From the cost breakdown add the minimum cost for each function from wells drilled on a similar basis. This total well cost represents the minimum cost well with no changes in normal drilling practices. In one case a program of this type resulted in an estimated cost of less than one-half of any previous well. The projected cost was never reached; however, the improvements on the next well were substantial. Improvements are many times made when the routine is changed by establishing new objectives that are potentially possible.

All operators want to drill minimum cost wells to the desired objectives. Their drilling philosophies may vary substantially on how to drill the minimum cost well. Sometimes the operator is influenced substantially by the last problem he encountered and as a result his drilling program for the next well reflects his desire to avoid the problem rather than minimizing cost. One operator may feel drilling a straight hole is more important than drilling a fast hole. Another may be concerned more with pressure control than drilling fast with low mud weights.

Hole enlargement many times may prompt the operator to increase well costs to prevent the problem. The specific hole problems are discussed in another chapter and will not be discussed in this chapter; however, hole problems are common and do account for a big percentage of deep well drilling costs. While hole problems should be considered in planning the well, it is believed that the cheapest wells will in general be those that reach their objective in the shortest period of time.

In addition, drilling with this philosophy of speed will probably result in more wells reaching their desired objective. Opponents of the fast drilling concept can use specific examples to show where problems have occurred. In fact, it is possible to prove almost anything by selecting specific field examples. General trends offer the best evidence of drilling progress and the fast drilling concepts developed along the Louisiana coast during the latter part of the 1950–1960 decade reduced substantially drilling times and costs in that area.

As in the case of any operation, the concept of fast drilling has to be tempered by other objectives or drilling problems encountered in a specific well. For example, drilling may have to be slowed in very soft surface hole sections to prevent the mud from becoming so heavy that circulation is lost. Gumbo clay sections may completely plug flow lines and stick the drill string if drilling rates are not controlled in such sections. In pressure transition zones, the primary objective is to select a protective casing seat and this takes precedence over drilling fast. These exceptions are given just to illustrate that no philosophy or concept can cover all the contingencies encountered in drilling.

Fast drilling promotes the idea of attacking the hole problems and withdrawing slightly if necessary rather than establishing a program based on holding a position which to that point has never been very strong. This simply says drilling in problem areas should never proceed from well to well using the same practices that resulted in hole problems. Sometimes this is done because it is more comfortable to remain in the main-stream of activities than to make changes which may or may not result in improvements and which are hard to explain if problems arise.

There are certainly many factors that have to be considered in well costs rather than those related directly to drilling time. Casing programs are of prime importance in deep high pressure wells. Primary cementing may hold the key to reasonable completion costs. Formation evaluation methods may substantially affect well costs. Mud programs may be the key to minimizing specific hole problems. All these considerations and many others may have to be considered in planning a specific well. The first consideration in this section will be a study of factors that affect penetration rate and thus the total well time.

FACTORS THAT AFFECT PENETRATION RATE

Variables that affect penetration rate are enumerated as follows:
1. Drill bit
2. Bit weight
3. Rotary speed

4. Bottom-hole cleaning
5. Mud properties

Fixed factors that affect drilling rate such as rock hardness and type and formation pore pressure cannot be changed; however operator recognition is very important.

The affect of any of the variables on well costs has to be determined by a daily record of drilling costs, generally expressed in dollars per foot. Cost per foot determinations may be made using Equation 1.

$$C_T = \frac{B + C_r(t + T)}{F} \tag{1}$$

In this equation, C_T represents the total drilling costs in dollars per foot. The bit costs are shown by B and C_r represents the rig costs in dollars per hour. Rotating time is t and round trip time is T, both expressed in hours. Total footage per bit run is F. Example 1 shows the use of Equation 1.

■ EXAMPLE 1:

Well depth = 10,000 ft.
Bit costs = $1,400.00
Rig costs = $100.00 per hour
Rotating time = 60 hrs.
Round trip time = 6 hours
Footage per bit = 600 ft.
Determine the cost in dollars per foot
Solution:
$$C_T = \frac{1400 + (60 + 6)100}{600} = \$13.33/\text{ft.}$$

The drilling costs in dollars per foot should be maintained on a routine basis. This record of costs will provide a quantitative method of evaluating the variables that affect penetration rate.

Bit Selection

Bit selection should be based on past bit records, geologic predictions, well logs from offset wells and drilling costs in dollars per foot. In recent years, bit selection has been complicated by many changes in bit design. The most significant change has been the sealed friction bearing insert bits. The bits are expensive but run longer than regular roller bearing bits, thus a true picture of their value can only be obtained by a record of drilling costs in dollars per foot. Diamond bits have also been used more extensively in recent years and bottom-hole rotating drives have made the use of diamond bits more attractive and also substantially more expensive.

Even with increased bit costs, drilling times and costs are being substantially reduced by improved bit selection procedures. There is a chapter on bits thus no more detail will be given on bit design and type.

Bit Weight and Rotary Speed

An increase in bit weight and rotary speed will increase drilling rate. However, these increases will also accelerate bit wear. Field tests have shown that drilling rate increases in direct proportion to bit weight. Figure 2–1 shows typical field results.

Note in Figure 2–1 that the drilling rates in the hard formation are not proportional to bit weight at weights below 2000 pounds per inch. The bit regardless of type has to impose enough pressure on the rock to overcome the compressive strength of the formation rock. In this case about 1500 pounds per inch of bit weight was required to fracture the rock. It is noted that in both hard and soft formations that drilling rate increased in direct proportion to bit weight as long as bottom-hole cleaning was adequate. This is generally true with any type bit including the insert bits. The upper limit of the proportional relationship between drilling rate and bit weight occurs when the matrix of the bit comes in contact with the formation.

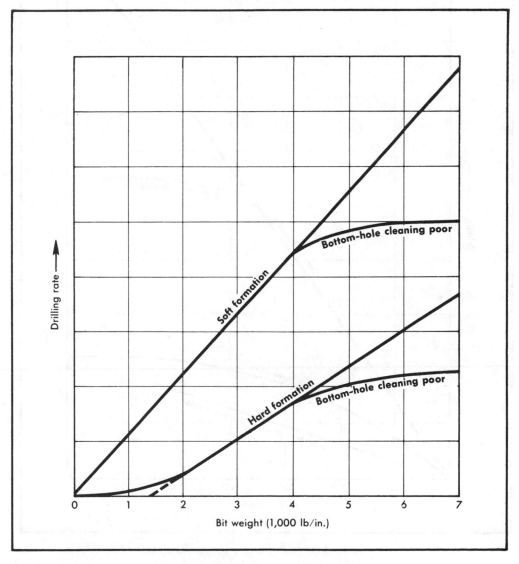

FIG. 2–1. *Drilling rate vs. bit weight*

The relationship between drilling rate and rotary speed is shown in Figure 2–2. It is noted from Figure 2–2 that the drilling rate is directly proportional to rotary speed in soft formations. In hard formations the rate of drilling rate increàse, decreases with increases in rotary speed. This is the primary reason that high rotary speeds (150–250) are used in soft forma-

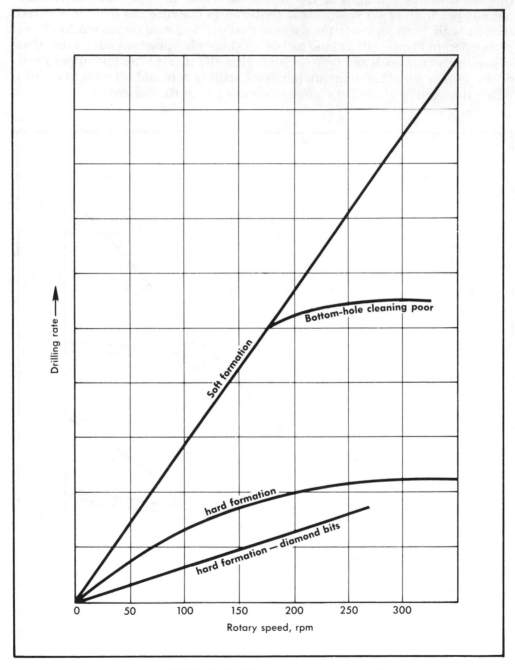

FIG. 2–2. *Drilling rate vs. rotary speed*

tions and low rotary speeds (40–75) are used in hard formations. Diamond bits are an exception as shown on Figure 2-2. Drilling rates increase in close proportion to rotary speed providing bottom-hole cleaning is adequate. This relationship is the reason bottom-hole drives that will rotate diamond bits at several hundred RPM's are being used by many operators.

Figures 2–1 and 2–2 do not give any indication of the best bit weight or rotary speed to use. They simply indicate the effects of bit weight and rotary speed on drilling rates. The objective is to use the bit weight and rotary speed that will provide the minimum cost in dollars per foot.

The effort to select bit weights and rotary speeds to provide minimum cost in dollars per foot was termed at one time, "minimum cost drilling." The increase in drilling rates due to increases in bit weight or rotary speed were combined with the reduced bit life to predict the best operating limits for bits. Equation 2 can be written for drilling rate versus bit weight and rotary speed from Figures 2–1 and 2–2.

$$R = KWN^a \tag{2}$$

Equation 2 is the instantaneous drilling rate at any point in time or can be used as an average penetration rate for a complete bit run. In actual fact for tooth type bits there is a dulling trend and Equation 2 has been written as shown in Equation 3.

$$R = \frac{KWN^a}{1 + K'(D)} \tag{3}$$

In this equation K' is a constant and D is normalized tooth wear. The constants K', D, and a would have to be determined from field operations.

The relationship between bit life and bearing life for roller bearing bits is given in Equation 4.

$$L = \frac{K''}{NW^b} \quad \text{Life (Bearing)} \tag{4}$$

In this equation L is the bit life in hours and K'' is a constant depending primarily on the type of drilling fluid being used. The exponent b is a function of fluid type and will vary between 1.0 and 3.0 depending on the abrasive characteristics of the fluid in contact with the bearings. The best way to determine a constant such as b is to keep an accurate record of bit life versus bit weight.

A plot of the data on log paper is the best method to obtain b. A typical plot of this type is shown in Figure 2–3. It is noted that, b, is the slope of the straight line in Figure 2–3. Field data should plot a reasonably straight line; however, this will depend on the basis used for pulling bits and the technique used for grading bits that are not worn-out. Specific bit grading procedures are included in the chapter on bits. The primary methods used to determine when to pull bits include; (1) a low rate of penetration compared with expected normal rates, (2) a sudden increase in rotary torque and (3) an economic analysis based on cost per foot calculations.

Drilling rate has been one method used to pull bits since the inception of rotary drilling. Many times a combination of drilling rate and time are

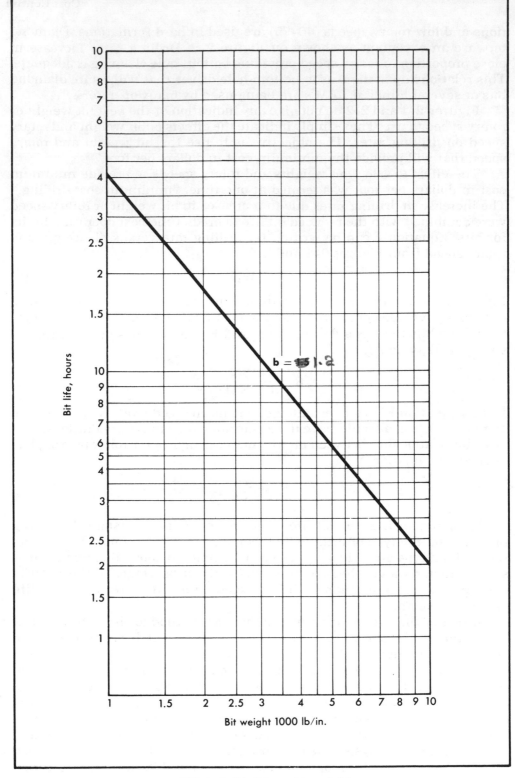

FIG. 2-3. *Bit life vs. bit weight*

used. This procedure is not very precise and places a premium on bit grading procedures.

Drilling torque has become a common method to determine the need to pull a bit. This procedure is subject to having a good torque indicator and to the operator's recognition that the increase in torque is due to locked bearings and not a change in formations. Even when a bit is pulled because of the increase in drill torque some operators have not graded the bit, B-8. This is a mistake. If the bit bearing locked enough to increase torque the bit bearing should be graded B-8. If grading procedures follow a consistent pattern of using torque properly, then the data in Figure 2-3 should be reliable.

The third method used to determine when to pull bits is the economic analysis. This procedure involves a frequent determination of cost per foot while drilling. The bit is pulled when the cost per foot versus drilling time reaches a minimum. This procedure is shown in Example 2.

■ EXAMPLE 2:

Well depth = 8000 ft.
Bit costs = $200.00
Rig costs = $100.00 per hr.
Round trip time = $\frac{1}{2}$ hr. per 1000 ft.
Rate of penetration = 30 − t
Note: a relationship between the rate of penetration and rotating time has been assumed, in field operations this would not be necessary.
Determine: When the bit should be pulled
Solution:
Table 2–2 lists the cost data versus time using Equation 1 and the relationship between penetration rate and rotating time.

TABLE 2–2
Cost per Foot Data for Example 2

t, hrs	R, ft/hr	F, ft	Total Cost	C_T $/ft
5	25	137.50	1100.00	8.00
10	20	250.00	1625.00	6.50
12	18	288.00	1829.00	6.35
14	16	322.00	2032.00	6.30
16	14	352.00	2235.00	6.35
18	12	378.00	2438.00	6.45
20	10	400.00	2640.00	6.60

The data from Table 2-2 have been plotted in Figure 2-4. As shown in Table 2-2 and Figure 2-4 the drilling costs in dollars per foot were a minimum at the end of 14 hours and the normal time to pull the bit would be at the 15 or 16 hour time. At the end of 16 hours of rotating time the bit is still drilling at the rate of 14 fph; however according to these data the drilling costs in dollars per foot have begun to increase. If the bits are being pulled on an economic basis it would not be possible to obtain meaningful bit wear information as shown in Figure 2-3. Almost any grading system

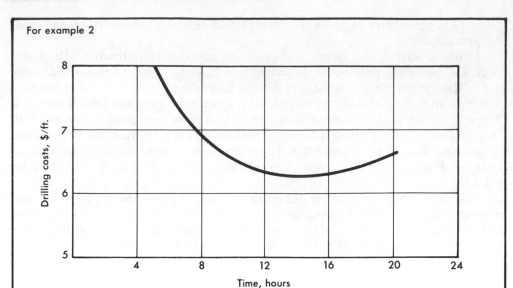

FIG. 2–4. *Drilling costs vs. rotating time (for example 2)*

based on an interpretation of bearing wear at some level before the bearings lock would not be accurate enough to use as shown in Figure 2–3.

It will be noted from Equation 4, that bit life is inversely proportional to rotary speed. Thus in soft formations where drilling rate is directly proportional to rotary speed this would indicate that rotary speeds should be very high. Actually other limitations have to be considered such as the horsepower available for rotation, pipe wear and the danger of twist-offs and the proper cooling of bits.

There are methods in use to predict the optimum bit weight and rotary speeds for drilling at a minimum cost in dollars per foot. Many techniques have been introduced. There are variable bit weight and rotary speed programs, variable bit weight and constant rotary speed programs and constant optimum weight and rotary speed programs. Most of the methods depend very heavily on the proper interpretation of bit wear.

All of the methods used to optimize bit weight or rotary speed will not be discussed. The basic procedure for optimizing is to begin with a cost per foot equation such as Equation 1, then write expressions for drilling rate as a function of bit weight and rotary speed and finally express bit life as a function of bit weight and rotary speed. This has been done in this chapter in Equations 1, 2, and 4 and an optimum bit weight can be determined using Equation 5.

$$\text{Wopt.} = \left[\frac{C_r \, K''}{(b-1)N(B + C_r T)} \right]^{1/b} \tag{5}$$

The use of this equation is shown in Example 3.

■ EXAMPLE 3:

Well depth = 10,000 ft.
Bit costs = $200.00
Rig costs = $100.00 per hr.
Round trip time = $\frac{1}{2}$ hr. per 1000 ft.
Bit weight = 40,000 pound
Rotary speed = 150 RPM
Bit wear, b = 1.5
Bit life = 10 hours

Determine the optimum bit weight

Solution:

Using Equation 4, $\dfrac{K''}{N} = (L)W^b = (10)(40,000)^{1.5}$

$\text{Wopt.} = \left[\dfrac{(100)(10)(40,000)^{1.5}}{(1.5-1)[200+(100)(5)]}\right]^{1/1.5}$

$\text{Wopt.} = \left[\dfrac{(100)(10)}{(0.5)(700)}\right]^{0.667} \; 40,000 = 80,000 \text{ pounds}$

To show the effect of b consider a change in the bit wear function from 1.5 to 2.0.

$$\text{Wopt.} = \left[\dfrac{(100)(10)}{700}\right]^{.5} \; 40,000 - 17,480 \text{ pounds}$$

To further emphasize the effect of b, consider a doubling of the rig operating cost from $100 to $200 per hour.

$$\text{Wopt.} = \left[\dfrac{(200)(10)}{(.5)(1200)}\right]^{.667} \; 40,000 = 90,400 \text{ pounds}$$

It is noted in Example 3 that an increase in the bit wear function of 33 per cent reduced the optimum bit weight from 80,000 to 47,840 pounds while a doubling of the rig operating costs resulted in an increase in the optimum bit weight from 80,000 to 90,400 pounds. The primary purpose of this example is to illustrate the absolute necessity of obtaining accurate bit wear data for any program of optimizing bit weight or rotary speed.

It should also be mentioned that the rig operating cost, C_r, is dependent on many variables such as drill string maintenance, pump maintenance, mud costs and resulting hole problems. High rotary speeds may not effect the rig operating costs enough to cause concern, however, if a twist-off occurs the resulting hole problem might be significant. This simply means that regardless of calculated results, judgment has to be used in the application of any of the optimizing techniques.

It is difficult to apply optimizing methods with the friction bearing insert bits. Bit wear data as a function of bit weight show an irregular pattern. It is common to use a maximum bit weight of 5000 pounds per inch on these bits. Many operators use less and a few operators have reported using up to 6500 pounds per inch. Some field tests indicate a substantial reduction in bit life when the bit weight is allowed to exceed 5000 pounds per inch. The close tolerances in the bearing areas of these bits make it difficult to cool the bits properly.

Bit overheating is also the reason that rotary speeds are generally maintained in the range of 40 to 60 Rpm's. These ranges of bit weights and rotary speeds are simply reported as common practice. Laboratory and field tests should be maintained in order to optimize future operating levels. It has been noted in many field tests that drilling rates with the insert bits are directly proportional to bit weight as shown in Figure 2–1. The relationships between drilling rate and rotary speed follow the pattern shown in Figure 2–2 for hard formations; where the curve generally flattens at about 50 to 60 Rpm's.

Diamond bits fall in a different category than either roller cone mill tooth bits or insert friction bearing bits. They have no moving parts, diamond sizes are variable and bit designs may be substantially different. Cooling of the diamond bits is of prime importance. Generally an increase in bit weight will produce a proportional increase in drilling rate with diamond bits. The limitation of course occurs when the matrix contacts the formation and this will be determined by bit design, diamond size and formation hardness.

The normal range of bit weight has been from 3000 to 5000 pounds per inch. Drilling rates with diamond bits are directly proportional to rotary speed. As a result high rotary speeds are desirable. The limitations imposed on rotating speed is surface horsepower available, drill string limitations, and the cooling permitted by bit design and the fluid hydraulics program. It is common to rotate diamond bits at speeds above 500 Rpm's using bottom-hole rotating drives. Again common practice has been related, future practices are subject to laboratory and field test programs. It is probable that bottom-hole drives will become more common, thus higher rotary speeds with diamond bits will also become common. No prediction will be made on future bit weight levels.

Bottom-Hole Cleaning

There is a chapter on hydraulics which describes the need for bottom-hole cleaning. There has been a tendency in recent years to ignore bottom-hole cleaning in deep wells with abnormally high pore pressures. This has been done because in many cases a very fine balance is reached between a well kicking and lost circulation. It should be emphasized that with no adverse effect on hole conditions bottom-hole cleaning can many times be increased substantially and the beneficial effect of reduced drilling time should increase the possibilities of reaching the desired objective.

Mud Property Effects

Mud property effects on drilling rate have been an object of research for more than 20 years. At various times there have been papers proving almost any property of the mud affects drilling rate. Eckel[1] combined the mud properties into a dimensionless quantity, he referred to as Reynold's number. Eckel's[1] basic relationship is shown as Equation 6.

$$R = \left(\frac{296\,\rho Q}{\mu D}\right)^{0.5}$$ (6)

Laboratory results using Equation 6 were favorable, field results in general have failed to confirm this relationship. The primary problem is the density term ρ. In most field tests an increase in mud density has the most pronounced detrimental effect on drilling rate and this would not conform to that shown in Equation 6.

Mud properties that have been claimed to affect drilling rate include the following:

1. Mud weight
2. Solids content and type solids
3. Mud viscosity, laminar and turbulent
4. Water-loss and also spurt loss
5. Liquid phase, water or oil

Mud Weight

It is known that increases in mud weight will decrease drilling rate. The quantitative relationship between mud weight and drilling rate has not been established for general use. In calculating the d exponent for pore pressure prediction methods, the log of drilling rate is considered inversely proportional to the log of mud weight, and this has proved to be a reasonable relationship. However, this may be substantially in error as shown by Figures 2–5 and 2–6.

Figure 2–5 represents a field test in western Canada by Murray where the increase in pump pressure and the noted back pressure were imposed by a hand adjustable choke in the discharge flow line. It was reported that this test was run at a depth of 2500 feet using a 9.0 ppg mud, thus the increase in back pressure of 250 psi represents an effective mud weight increase of 2.0 ppg. Note the decrease in drilling rate is from about 95 to 30 fph. and this is a common result at these mud weight levels. Figure 2–6 shows a laboratory test of drilling rate versus imposed hydrostatic pressure. Of significance in this test is the fact that the increase in hydrostatic pressure from 0 to 2000 psi resulted in a decrease in penetration rate from 18 to 5 fph. The increase in hydrostatic pressure from 2000 to 4000 psi resulted in a decrease of only 5.0 to 3.5 fph. Results such as those shown in Figure 2–6 are typical of those in actual field operations. The first unit of mud weight in excess of formation pore pressure will reduce drilling rate more than each subsequent unit of mud weight increase.

Figure 2–7 shows the effect of mud weight on rotating time in two South Mississippi wells. Well A was drilled with 10.4 ppg mud and typified the normal practice in the field before drilling well B. Using a thinner mud weighing 9.6 ppg in well B resulted in a rotating time of 210 hours or less than one-half of the 500 plus rotating hours required to drill well A. More

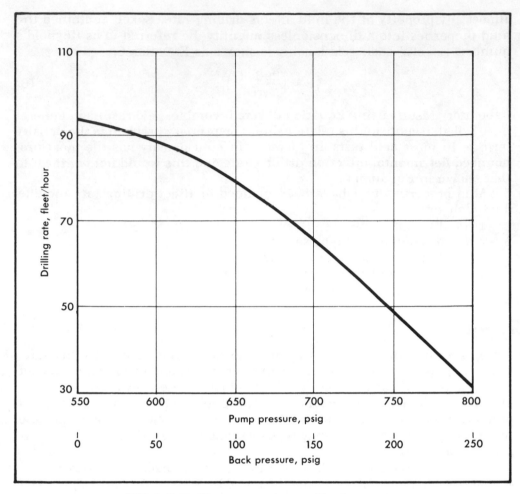

FIG. 2-5. *Drilling rate vs. imposed back pressure*

than just the mud weight was changed and this test alone does not prove that all of the improvement was due to the reduction in mud weight. However, the reduction in mud weight made the other improvements in mud properties possible.

The effect of mud weight or the additional hydrostatic pressure on drilling rate is believed to be caused by the chip hold-down effect of the differential between hydrostatic and formation pore pressure. This theory gains support when drilling with air or gas as shown in Figures 2-8 and 2-9.

In Figure 2-8, a composite of wells drilled with mud are compared with a composite of wells drilled with gas. The results show 14 days drilling time required for the interval between 1500 and 9800 feet with gas as compared with 90 days for the same interval with mud.

Figure 2-9 shows a comparison between drilling with air and water in 2600 foot wells in the West Texas area. The rotating time with air was about one-half the rotating time with fresh water. In both of the series of field tests in Figures 2-8 and 2-9 the degree of improved performance occurred

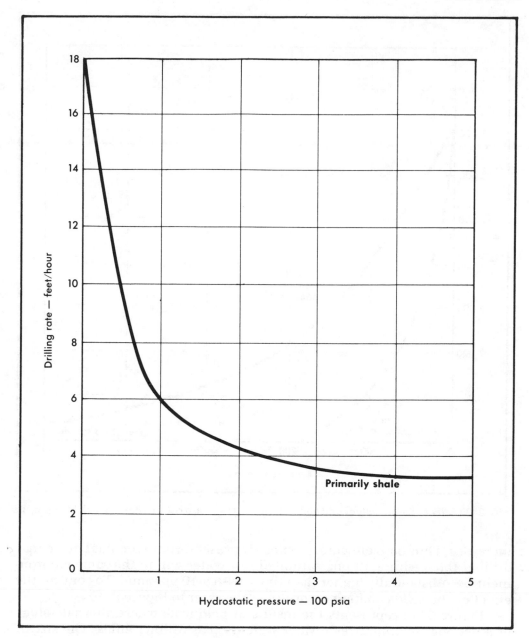

FIG. 2-6. *Drilling rate vs. hydrostatic pressure*

primarily because the hydrostatic pressure was less than the formation pore pressure.

Solids Content and Type Solids

It is difficult to separate the solids content effect on drilling rate from the mud weight effect. As the solids content increases the mud weight also

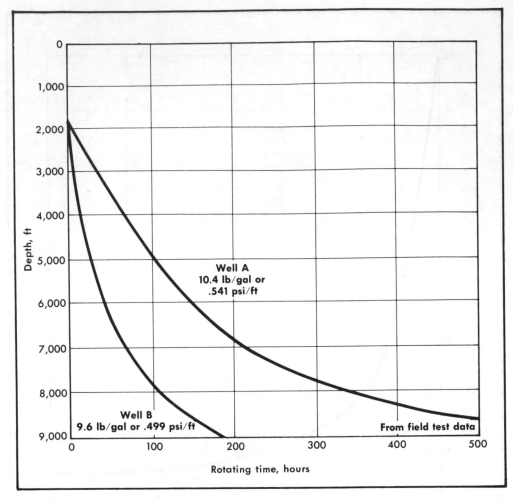

FIG. 2-7. *Effect of mud weight and hydrostatic pressure on drilling time — South Mississippi*

increases. Thus both effects are generally present. However, it is known that drilling rates with a 10 ppg saturated salt water will in the same environment be substantially higher than those with a 10 ppg mud. To confirm the effect of the solids content on drilling rate, refer to Figure 2-10.

Figure 2-10 represents the results of field tests where alternate slugs of a 9.2 ppg mud and fresh water were used as drilling fluids. The size of the slugs of each were selected to provide a constant hydrostatic pressure in the annulus. Drilling rates were measured as the water and mud passed through the bit. The drilling rates with water are shown as the 100 per cent drilling rate curves on Figure 2-10. The drilling rates measured when mud passed through the bit are shown as 60 per cent of those with water using jet bits and 40 per cent of those with water using conventional bits. These tests were conducted primarily in the West Texas area and confirm that the solids content of the mud affects drilling rate. Further proof of the effect of the solids content on drilling rate is shown in Figure 2-11, which is a laboratory test.

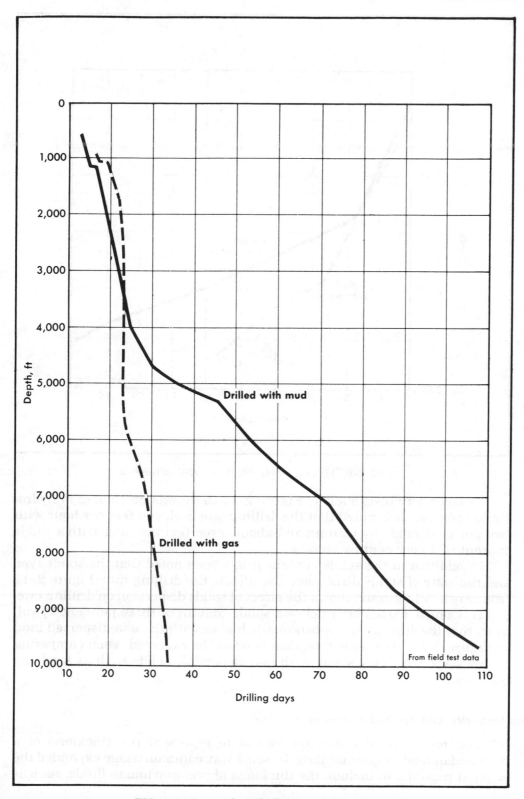

FIG. 2–8. *Gas and mud effect on drilling time*

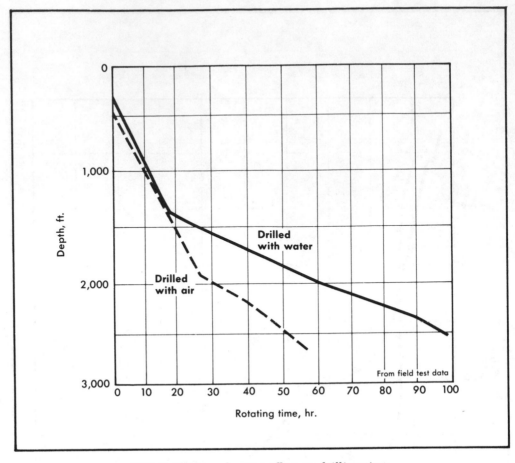

FIG. 2-9. *Air and water effect on drilling time*

In laboratory tests such as Figure 2-11 there was no change in hydrostatic pressure. It is noted that the drilling rate is eleven feet per hour with two per cent solids by volume and about three feet per hour with a solids content of 12 per cent by volume.

In addition to the solids content it has been noted that the solids type and the state of solids dispersion also affects the drilling rate. Figure 2-12 from work by Lummus[2] shows the effect of solids dispersion on drilling rate.

It is noted, particularly below a solids content of three per cent by volume, that the drilling was substantially higher with the non-dispersed mud in Figure 2-12. This same type results would be expected when comparing highly dispersible clay solids with non-dispersible solids such as sand and limestone.

Mud Viscosity, Laminar and Turbulent

The term viscosity was introduced to represent the thickness of a Newtonian fluid in laminar flow. In some way common usage expanded the original meaning to include the thickness of non-newtonian fluids, such as

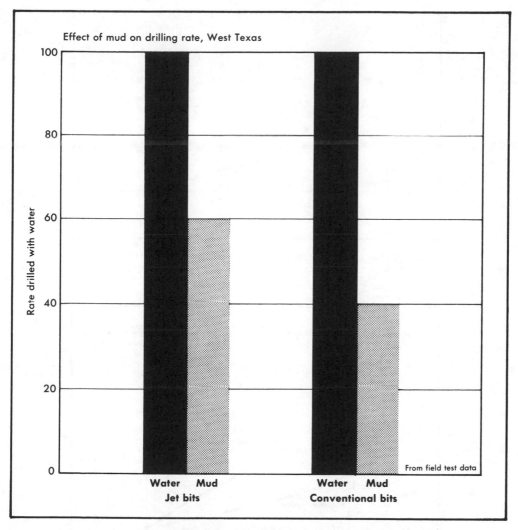

FIG. 2–10. *Relative drilling rate, per cent*

muds. Later the concepts of thickness and viscosity became synonymous for drilling fluids regardless of flow pattern and it has become common practice to refer to mud thickness as viscosity. In laminar flow the mud thickness is defined by two terms, either plastic viscosity and yield point or from the power-law concept by n and K. These relationships are discussed in detail in the chapter on drilling mud. In turbulent flow thickness is primarily a function of solids content, and the laminar flow viscosity terms have no meaning for thickness in turbulent flow.

The laminar flow viscous effects would affect drilling rate only by the additional circulating pressure loss imposed in the annulus as a mud thickens. This additional pressure would increase the effective hydrostatic pressure head in the annulus and affect drilling rate as shown in the section on mud weight effects.

The flow pattern of the drilling mud will be turbulent as it passes

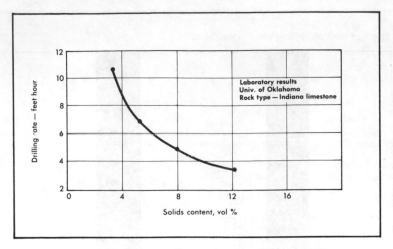

FIG. 2-11. *Quantitative effects of mud solids on drilling rate*

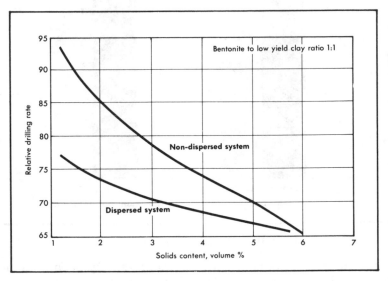

FIG. 2-12. *Effect of solids dispersion on drilling rate*

through the bit and the laminar flow viscous properties of the mud cannot be used to determine the mud thickness. However, because plastic viscosity is used as a means to determine the relative solids content of a mud, the plastic viscosity can also be used as an indicator of the mud thickness in turbulent flow. It has been generally determined that a thin mud at high shear rates will result in higher penetration rates than a thick mud if no other change in mud properties occur. This was shown in Eckel's work and is represented in his drilling rate equation.

Figure 2–13, taken from Eckel, shows five muds with the same apparent laminar flow viscosity and with different ratios of yield point to plastic viscosity. As the ratio of yield point to plastic viscosity increases the shear thinning characteristic of the mud increases. It is noted from Figure 2–13 that the mud with a yield point to plastic viscosity ratio of 4.0 sheared thinner than the other muds at high shear rates. Muds of this type will generally be those with a low and flocculated solids content or some type of polymer for thickening. As noted in Figure 2–13 most muds are shear thinning, the low solids thick muds are in general the ones that thin the most at high shear rates.

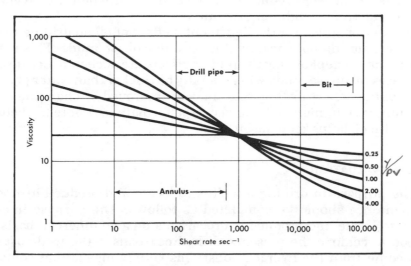

FIG. 2–13. *Viscosity vs. shear rate*

Filtration Rate and Spurt Loss

The concept that drilling rates increased with increases in filtration rate was introduced by operating personnel. Drillers noted that drilling rates were decreased when the filtration rates were decreased. In all probability this reduction in drilling rate was due more because of the materials added to reduce filtration rate than because of the filtration rate reduction.

When low solids muds were introduced it was not uncommon to have an API filtration rate of 10cc's and a high initial loss of fluid, called spurt loss. The spurt loss was simply the loss of fluid necessary to form a filter cake of solids. As a result, drilling rates with the low solids mud with an API water loss of 10cc's may be substantially higher than a higher solids mud having an API water loss of 20cc's. Drilling rate may or may not be affected by changes in filtration rate.

Fluid Phase

The effect of the fluid phase on drilling rate may be substantial. Air or gas is light in weight and the effects of these two fluids were shown in Figures 2–8 and 2–9. A continuing point of discussion is water versus oil. Drilling rates with a continuous oil phase will generally be about 20 to 30 per cent less than the same weight mud using water as the continuous liquid phase. There will be exceptions. In a deep deviated well using cone type bits there may be no noticeable difference between drilling rates with oil base and water base muds. This will probably be true because the reduction in friction between the drill string and the hole is more with oil, which results in an increase in actual bottom-hole weight on the bit for a given surface indication of bit weight. By contrast in deep wells while drilling in limestone formations with diamond bits, drilling rates with a given weight water base mud were reduced from 12 to 4 fph when a conversion was made to the same weight oil base mud.

One reason given for the detrimental effect of oil on drilling rate is that oil phase muds do not shear as thin as water phase muds. This reason has merit if the concept of shear thinning effects are accepted, because at high shear rates oil phase muds will generally be thicker than water phase muds. One situation where drilling rates may be improved by changing to oil base muds might be in high yielding clay type formations where the better lubrication with oil helps keep the bit clean.

FORMATION EVALUATION

The purpose in drilling a well is to locate and produce hydrocarbons. This sentence should be completed as follows: The purpose in drilling a well is to locate and produce hydrocarbons on a commercial basis. It does no good to confirm the presence of hydrocarbons if the well cost exceeds the income from the hydrocarbons. This simply means that some type of compromise needs to be arranged when well costs become excessive primarily because of extended formation evaluation programs.

One problem in the so-called compromise is that no department or individual will ever admit that their current programs can be trimmed without sacrificing critical information. For this reason, some of the primary formation evaluation methods, such as sample analysis, coring, logging and testing, will be discussed.

Sample analyses are a significant part of the search for hydrocarbons. Flourescence from hydrocarbon stained samples may be the first indication that a potential productive horizon is present. In general the drilling operation may be affected very little. There has been some debate relative to the use of oil in the mud and this led to the use of essentially non-flourescent oils.

Actually this problem is being minimized because ecology has also frequently eliminated the use of oil, particularly in offshore exploratory operations. In any event the elimination of oil emulsified in water-base muds may or may not affect drilling operations. Thus this is one area where very little argument arises between those concerned with drilling and those concerned with formation evaluation.

Conventional, wireline and side-wall coring represent desirable methods of evaluating potential productive horizons, particularly in exploratory wells. The only fault that should be emphasized relative to the frequency of coring are practices that continue extensive coring operations in development areas. Often coring in the development area is justified as the only means of obtaining pertinent information relative to porosity, water saturation and permeability. Actually information may be obtained at a much lower cost from logs and pressure build-up or draw-down data.

While coring will slow down drilling operations and increase cost, the coring operations would have a minimum effect if many special drilling precautions were not introduced. It is not uncommon to include in most drilling programs the statement, "before commencement of coring get the drilling mud in shape." Getting the mud in shape generally includes lowering water-loss, either lowering or increasing viscosity and an improvement in the general appearance of the mud. These practices are habitual and in general have no beneficial effect either from the standpoint of core recovery or core condition.

It is more or less obvious that fluid property changes would have very little effect on core recovery. Specific cases may arise where it is desirable to use special coring fluids. A desire to look closer at water saturation may prompt the use of oil base drilling fluids. In addition solids plugging of permeability may introduce a search for special muds. Many factors would have to be considered in the special cases; however in general oil base fluids offer very little improvement in any case and frequently complicate evaluation.

Special fluids to prevent solids plugging of cores may be very expensive and probably one of the best fluids for this purpose would be bentonite and fresh water, which is cheap and easy to prepare. Arguments against the bentonite and water mixture emphasize that bentonite is a very small particle and will be very detrimental. The truth is that bentonite is a small particle which adsorbs large quantities of water and will plaster on the core face in most cases rather than entering the pore spaces. Thus many expensive procedures that have been associated with coring could be eliminated with absolutely no loss in the information obtained.

Logging operations have become extensive in recent years in both exploratory and development wells. Some logs are run specifically to determine formation pore pressure in wells where abnormally high pore pressures have been detected. Thus logs are both for formation evaluation and drilling operations.

Emphasis has been placed on selecting only logs required for evaluating the formation and this aspect of the business has been improved substantially in recent years. Some duplication still exists. Self-potential and gamma-ray logs may both be run to determine lithology, where only one has utility. Several porosity logs may be run and different answers may increase the confusion rather than providing clarification.

One bad habit that has developed in logging operations is the short trip, run back into casing and back to bottom before pulling the drill string to log. This has become a common and expensive practice in Coastal operations and can become a prohibitive cost in offshore floating drilling. This practice has been evaluated by several companies. Some studies have observed operations for more than two years.

The result in all cases revealed that the short trip before logging was of no value as insurance in getting the log to bottom. In other words, the logs failed to go to bottom with the same frequency regardless of whether the short trip was made or not. Operators should check this practice carefully rather than letting failures to get logs to bottom result in expensive, non-beneficial programs of prevention.

Logging as well as coring may introduce many changes in mud properties. The primary changes in mud properties are generally directed towards reducing the invasion of filtrate into the formation. Also at times the composition of the liquid phase may be changed. Unfortunately the quantity of filtrate invasion has been equated to the level of the API water-loss. While it may be true that a reduction from 20 to 10 cc's may reduce the short term invasion it is probable that the total invasion over a period of several days would probably be affected very little.

A reduction in the API water loss from 10 to 5 cc's might actually increase the short term invasion because of an increase in dynamic fluid loss. A more detailed explanation is given in the chapter on drilling muds. It is suspected that the total invasion of filtrate and the depth of filtrate invasion may be related more to the pressure differential and the time of exposure than to the magnitude of the API water-loss. This conclusion is an opinion only; however, the logic should be considered and when the opportunity presents itself operators should make meaningful evaluations.

Formation testing may be done using wire line testers, conventional drill stem tests or regular production tests. Reasons for testing may include an evaluation of fluid type, formation pore pressures, and formation productivity. An open-hole conventional drill stem test is often run to determine productivity before spending money on a production casing string. However, it may also be run even though the operator has already decided to run casing. It is not uncommon for operators to test formations shortly after they are exposed rather than wait until total depth is reached. Reasons given for this practice include the following: to minimize exposure time for the formation of interest, to determine fluid contents and to check pore pressures.

Problems with the practice are typified by a 12,000 foot well in Wyoming where 14 open-hole drill stem tests were run. The well record showed a total of 17 days of making open-hole drill stem tests. Actually out of 154 days, 75 days were required to run the tests and correct the subsequent hole problems. It would have been substantially cheaper to drill to total depth and set a small production string of casing.

In exploratory wells of this type some consideration should be given to cheaper and better alternatives of formation testing than open-hole drill stem tests that more than double the well time, increase costs substantially and are inferior to better and cheaper methods.

Another pattern of conventional drill stem testing that should be examined is the situation where samples, cores and logs indicate the presence of producible hydrocarbons and a confirmation test is run before setting casing. If the confirmation test is run to determine whether to set casing or not and the test will be used to make this decision, then the test is probably justified. If casing is to be set regardless of the drill stem tests results, a common practice in some areas, then running the test is an expensive exercise in futility.

The drilling man is interested in finding hydrocarbons as well as drilling cheaply and the formation evaluation man wants to drill cheaply as well as find hydrocarbons. The only conflict comes in the interpretation of requirements. While improvements in communication will help, the best method to improve is to study procedures and practices on the basis of results versus dollars spent.

COMPLETION COSTS

A company may or may not include completion costs in drilling costs, these costs are certainly a part of total well costs. No attempt will be made in this section to enter into a discussion on suggested completion practices; however, problems that happen during or after the well completion are often blamed on drilling practices and this point will be discussed.

For many years, practices were promoted that slowed drilling and increased drilling costs in the Rocky Mountains because of collapsed casing caused by running salt. Although the problem may never be cured completely, cementing between the casing and salt section and heavier casing opposite the salt section has minimized the problem substantially. Fast drilling and thin fluid which were considered causes of the problem, were actually beneficial to minimizing the problem.

Very often a bad primary cement job is blamed on drilling practices, when during the cement job the casing is not moved or properly centered in the hole. As a result drilling costs are increased and completion costs remain the same.

SUMMARY

The secret to cost control is to consider the total well costs. This will include location, drilling, formation evaluation, casing and completion costs. Cost control in drilling starts with the rig selection after the well is planned. Too often operators begin a deep expensive well with an inadequate rig. Rigs should be examined carefully. Pumps are important relative to size and normal operating procedures followed by the rig operator. Careful attention should be focused on the drill string and the operating efficiency of the rig on other wells. It is clear that the success of any cost control program will be dependent on the rig.

Cost control or a minimum cost well is sold as a service by many organizations. In most cases some type of computerized operation is used and an accurate record of current and past operations is essential.

In general all of the factors which affect drilling rate should be considered and cost per foot determinations should be used on a continuous basis. It should be emphasized that all organizations may not agree with all the conclusions in this discussion. Also any cost control program has to be tempered by the potential hole problems.

In most cases, an increase in emphasis and supervision will result in reduced drilling costs. Expertise is an essential ingredient in the increased supervision. The high cost drilling operations create the necessity of having competent personnel at the rig during the drilling operation.

Formation evaluation practices should be studied carefully. Extensive coring and logging programs do not necessarily increase the ratio of successful wells to dry holes. These programs do reduce the number of wells drilled. Formation testing programs should be analyzed. It may be cheaper to set small casing than conduct multiple open-hole drill stem tests. Also be certain the tests results are used to make the right decisions.

Completion practices may be affected by drilling practices; however make sure that suggested changes that raise costs are not actually excuses for a dry hole.

A total cost control program between geologists, engineers and operating personnel is possible if they recognize the importance of being on the same team. Competition between these groups will result in face saving excuses that indicate a cause for problems that in many cases are not related to the actual problem.

NOMENCLATURE

Cost Control

1. C_T = Drilling cost $/ft
2. B = Bit costs, $/bit
3. C_r = Rig operating costs, $/hr
4. t = Rotating time, hours
5. T = Round trip time, hours
6. F = Footage per bit, feet
7. R = Drilling rate, fph
8. K = Drillability constant
9. W = Weight on bit, pounds/inch
10. N = Rotary speed, Rpm
11. a = Drilling exponent for rotary speed
12. K^1 = Drillability constant related to bit wear
13. K^{11} = Drillability constant
14. b = Drilling exponent related to the effect of bit weight on bearing wear
15. L = Bit life hours
16. ρ = Mud weight, ppg
17. Q = Circulation rate, gpm
18. μ = Mud viscosity, centipoise
19. D = Diameter, inches

REFERENCES

1. Eckel, John R., "How Mud and Hydraulics Affect Drilling Rate," The Oil and Gas Journal, June 17, 1968.
2. Lummus, J. L., and Field, L. J., "Non-dispersed Mud: A New Drilling Concept," Petroleum Engineer, March, 1968.

Hole Problems

$\mathbf{D}$R I L L I N G a small hole into underground formations that may or may not be well consolidated introduces the possibility of some type of hole problem. The potential of having hole problems may be shown by geology used in well planning or by past experience which shows a specific hole problem may occur. The well plan should be designed to attack the potential hole problems. Drilling programs based on caution and containment generally result in expensive wells and very often wells that fail to reach the desired objective.

In one sense drilling operations are unique, in that an operator is almost sure to encounter some type of hole problem in any area. Many hole problems are unexplained. Drilling practices may be standard and yet the hole sloughs, where in several years of previous work no similar problems have existed. One obvious answer is that the underground formations are not homogeneous and new problems may occur at any time in carefully planned wells. Unfortunately many times hole problems in one or two wells will dictate practices for an area where no such problems have existed previously. As a result, well times are extended, the hole problems continue, costs increase and the good old days are forgotten.

Major hole problems that will be discussed include lost circulation, hole sloughing and pipe sticking. Crooked hole problems will be discussed in a separate chapter.

LOST CIRCULATION

Lost circulation is the most common problem in drilling. The normal range of lost circulation problems begins in the shallow unconsolidated sands and extends to the well consolidated formations that are fractured by the hydrostatic pressure imposed by the drilling fluid.

Shallow Unconsolidated Formations

In the shallow unconsolidated surface formations the drilling fluid may flow freely into the formation because of its high permeability. Drilling may continue without circulation or the mud may be thickened to slow the rate of loss.

Drilling without circulation may be dangerous in shallow unconsolidated surface formations and before following this procedure the operator should be familiar with the area. Lost fluid into such formations may result in underground cavities caused by washing and the result could be a surface cavity, the loss of a rig and possible injury to rig personnel.

The most common method of handling lost circulation in shallow surface formations is to thicken the mud. This may be done in fresh water muds by adding flocculating agents such as lime or cement. In areas where only water is being used, the water may be thickened using polymers. Also, lost circulation material may be used and because strength is not a requirement the best material would be cheapest bulk material available, such as cotton-seed hulls or sawdust.

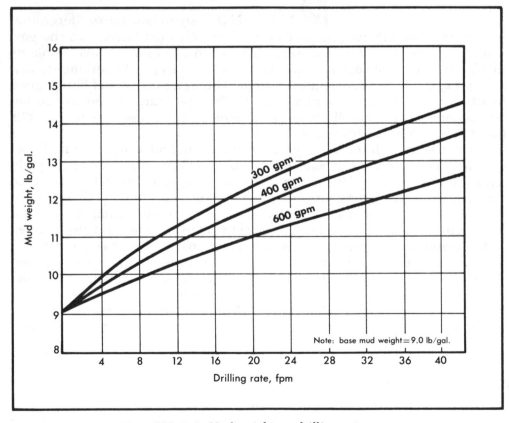

FIG. 3–1. *Mud weight vs. drilling rate*

In addition, fast drilling rates in soft surface hole formations may result in high annular mud weights. Figure 3-1 shows the increase in annular mud weight experienced while drilling a 12¼ inch hole with a 9.0 ppg fluid.

It is noted in Figure 3-1 that the mud weight can exceed 12.0 ppg in the annulus with very high penetration rates and a relatively low circulation rate of 300 gpm. Increasing the circulation rate to 600 gpm reduced the mud weight substantially and the higher circulation rate would have a very small effect on circulating pressure losses. If trouble continues it may also be necessary to limit the rate of penetration to control mud weight. Problems such as this are common in surface formations of the Gulf Coast areas of the United States.

Below Surface Casing in Normal Pressure Formations

It would be difficult to name all of the causes of lost circulation in normal pressure formations below surface casing. Natural occurring fractures are common. These fractured formations may or may not have subnormal pore pressures. Also many of the unfractured formations may fracture easily. The operator may not know the precise cause of lost circulation or even the exact location where circulation is lost.

The zone of loss may be determined if desired by running a temperature log. If the temperature log is not definitive on the first run, pump an additional 50 barrels of fluid and rerun the temperature log. More expensive procedures are not generally necessary; however, radio-active tracers may be used if the temperature log is not definitive.

A general solution to lost circulation in this part of the hole is to drill without fluid returns to the surface. This practice requires large volumes of water and close supervision. In principle the generated cuttings are removed from bottom and deposited in the lost circulation zone. Thus the fluid velocity has to be sufficient to lift the cuttings to the lost circulation zone. The drilling operation has to be watched closely for evidence of drill string torque or drag. With close supervision the operator can detect potential hole problems and correct the operation before disaster strikes.

Specific problems in this part of the hole that may require special attention include seepage losses, a complete loss of circulation where there is a requirement for obtaining cuttings back to the surface for formation evaluation purposes, and an area where there is a serious limitation on the available water supply.

Seepage losses are common in areas such as the Anadarko Basin. In such cases the operator may use fine or course lost circulation material to reduce the rate of loss. The fine material may be beneficial for this type loss and has the advantage of permitting the continued use of a shale shaker to remove formation cuttings. In general the lost circulation material is used in the entire mud system for this type loss.

The requirement for getting cuttings back to the surface or the lack of water may preclude drilling without fluid returns to the surface. Methods that have been used to correct these problems include:

1. The use of medium and coarse lost circulation material, batch-mixed

to a concentration of 20 to 30 ppb in 25 to 100 barrels of mud. The amount of mud used will depend on the specific problem.

2. The use of solid plugs such as cement or deisel oil-bentonite. Usually a cement plug would consist of a light weight cement slurry spotted opposite the zone of loss in stages. A deisel oil-bentonite plug is difficult to use. All water in tanks and the pipe must be removed, then from 300 to 350 ppb of bentonite is mixed in the deisel oil. This slurry is displaced and spotted in the zone of loss. The zone of loss must contain water to hydrate the bentonite or this procedure has no chance of success.

3. A third method of getting circulation is to use an aerated fluid column or a viscous foam. This procedure has been used often in areas of Libya and the Rocky Mountain areas of the United States. The amount of air required to obtain a specific hydrostatic head is shown by Example 3 in the chapter on Drilling Muds. The viscous foams can be used, however, some compromise has to be made so the foams can be broken on their return to the surface. Aeration or foams increase the costs of the operation substantially and are generally last resort methods.

Many lost circulation problems have been cured simply by keeping the drilling fluid weight close to the weight of water and using very thin fluids. Over 50 per cent of the initial lost circulation problems in Libya were cured in this manner.

Lost Circulation While Using Weighted Muds

Lost circulation while using weighted muds generally occurs as a result of induced fractures caused by the hydrostatic pressure imposed by the drilling mud and associated activities such as fluid circulation and pipe movement. Lost circulation of this type may occur in normal pressure formations above a pressure transition zone or below protective casing in the abnormal pressure section of the hole. The pressure transition section of the hole is defined as that section of the hole where the pore pressure begins to increase above normal to the point that protective casing is set to prevent the loss of fluid.

The loss of fluid in normal pressure formations above the pressure transition zone may be at any point from the bottom of the surface casing to the top of the pressure transition zone. The danger in a case such as this is that when circulation is lost, fluid will begin to flow from a formation in the pressure transition zone. This danger places a premium on knowing the surface pressure that can be contained if the blowout preventers and surface choke lines are closed.

Procedures to be followed in closing-in wells are discussed in the chapter on Pressure Control. As a precaution the operator should always test the cement job around the surface casing when abnormal pore pressures are expected. This may be done by drilling 15 to 20 feet below the surface casing shoe and applying pressure to the level of the total pressure expected at this point. If for some reason, hole conditions at a later time make it appear the total pressure will exceed the test pressure, the pressure test should be repeated before increasing the mud weight above the previous test level.

It should be remembered that a pressure test at the surface casing shoe does not indicate that a mud weight of the test magnitude can be used, because a lower formation may fracture at a lower pressure. For this reason it is not uncommon to pressure test open-hole sections. Such tests of open hole sections should always be performed with the drill string in the casing to prevent sticking. Specific casing seat testing procedures are given in the chapter on pressure control.

Along the coastal areas of Texas and Louisiana it has been noted in recent years, that the test pressure that can be contained below surface casing may increase with open-hole exposure time. For example, it is not uncommon to find the test pressure below surface casing to be equal to an equivalent mud weight of 14.0 ppg when testing immediately after the cement job and on a subsequent test after several days of drilling find the formation at the same point will hold a 15.5 ppg mud.

This increase in formation strength may be due to the plugging of the formation permeability with drill solids, which raises pressure required to fracture the formation. Regardless of the cause, the phenomena does exist frequently and tests several days after the cement job may permit the operator more freedom in selecting the setting depth of protective casing.

In coastal area formations, the accidental loss of circulation in this part of the hole should not occur. In some areas such as the deep Anadarko and Delaware Basins where highly fractured low pressure formations may be encountered in many zones below surface casing the problem is more complicated and testing programs are less definitive. However, it should be emphasized that pressure testing programs are still desirable in areas such as the Anadarko and Delaware Basins.

If circulation is lost in the normal pressure section of the hole in these areas, a batch treatment with lost circulation material may be used or a solid plug. Both procedures were discussed in the previous section. However, in this case, if and when circulation is established consideration should be given to setting protective casing, because the operator is faced with the potential of an underground blowout.

Consider next the use of weighted muds below protective casing or protective liners. In the abnormal pressure section of the hole below protective casing the following steps should be taken to avoid the accidental loss of circulation:

1. Pressure test 15 to 20 feet below the casing shoe to equivalent pressure level expected or to the leak-off pressure. Leak-off pressure is the pressure at which the formation begins to take fluid. A specific explanation of leak-off pressure is given in the chapter on pressure control.

2. After drilling the first sand below the protective pipe, retest to the same levels given in Step 1.

3. If equipment limitations prevented reaching the leak-off pressure, retest after increasing mud weight.

4. If the test pressure used equaled only the anticipated pressure requirements and subsequent drilling developments have raised these requirements, retest to leak-off. All of the testing should be performed with the drill string inside the protective pipe. Specific testing procedures for this section of the hole are given in the chapter on pressure control.

If circulation is lost in the hole section below protective pipe, two conditions may exist: (1) circulation is lost but there is no evidence of a well kick, and (2) circulation is lost and an underground blowout is taking place.

In the first case, if circulation is lost and there is no evidence of a well kick, one of the first decisions that must be made is whether to pull the drill string off bottom into the protective pipe. This is a desirable procedure because the drill string may stick if left in open-hole. This is an undesirable procedure because if the well kicks when trying to pull the pipe it may be very difficult to control the well properly.

For these two reasons the question of pulling the pipe cannot be answered in a general discussion of this type. It is a field decision and drilling information, formation pressure data, formation types, control equipment and crew excellence would be some of the factors that must be considered. However, the operator should be aware of the potential problem and establish procedures that would help in making the decision if the problem occurs.

To cure the lost circulation problem, the general procedure would be to reduce the hydrostatic pressure either by reducing mud weight or by thinning the mud. The use of lost circulation material at this point without reducing the hydrostatic pressure may be more detrimental than beneficial. If circulation material were displaced into the zone of loss, it would act as a propping agent and most probably would prolong and perpetuate the problem.

In the past, recommendations have been made to wait a given number of hours; however, in recent years this procedure does not appear to be necessary. Some waiting time is built into the basic procedure for a determination of the problem. When circulation is lost, the first requirement is to determine if the well is kicking. Second, the operator must make the decision whether to pull the drill string back into the protective pipe. Third, an attempt should be made to fill the annulus with water, measuring the quantity of water required to fill the annulus carefully. Fourth, determine from the mud and water levels the actual pressure the formation will support. Fifth, begin to displace fluid slowly.

In this latter case lift the pipe as circulation is initiated to minimize any surge of additional pressure on the formation. The time required to fill the annulus with water, measure the quantity of water and initiate circulation is all the waiting time necessary. Example 1 shows how to determine the mud weight that can be supported by the formation and also the mud weight that will control the subsurface pressure.

■ EXAMPLE 1:

Well depth = 16,000 ft.
Protective casing seat = 12,500 ft.
Mud weight = 17.0 ppg
Drill pipe size = 4½ inch
Hole size, Csg. I.D. = 8.5 inches
Annulus volume = .05 Bpf
Water required to fill hole = 20 bbls.
Determine the effective hydrostatic head and mud weight in ppg.
Solution:

$\dfrac{20}{.05}$ = 400 ft. of water

Pressure imposed at total depth:
400 ft. of water × .433 = 173 psi
15600 ft. of mud × .884 = **13,800 psi**
Total pressure at 16,000 ft. = 13,973 psi

Effective mud wt. = $\dfrac{13,973}{(16,000)(.052)}$ = 16.8 ppg

Pressure imposed at the casing seat:
400 ft. of water × .433 = 173 psi
12,100 ft. of mud × .884 = **10,700 psi**
Total pressure at 12,500 ft. = 10,873 psi

Effective mud wt. = $\dfrac{10,873}{(12,500)(.052)}$ = 16.75 ppg

It will be noted that there is a slight difference in the effective mud weight at the casing seat, which in this case is in the vicinity of the assumed zone of loss, and at the bottom of the hole. Trouble may be experienced when trying to regain circulation. It is difficult to determine all the operating conditions such as the amount of additional hole required, the specific conditions that resulted in losing circulation and specific mud weight required to contain the high pressure formations. However, if difficulty is experienced in regaining circulation in the manner prescribed and particularly if any gas or water cutting of the mud is noted during any fluid circulation the operator should consider another protective casing string or liner.

The second condition where the well kicked when circulation was lost is one of the most dangerous conditions experienced in drilling. Surface control of the well kick has been lost. There is a general problem of not only losing the hole but also having a dangerous blowout. The problem in a case such as this is to first control the kicking formation and then attempt to cure or seal-off the lost circulation zone.

Methods that may be used to control the kicking formation include: (1) displacing a heavier mud in the open-hole below the lost circulation zone and decreasing the mud weight above the lost circulation zone; (2) setting a barite plug in the open-hole below the lost circulation zone; and (3) setting a cement plug in the open-hole below the lost circulation zone. Experience in an area will help determine what should be used. Again, the final decision should be made by a knowledgeable operator at the rig site.

Displacement of a mud that is heavier than that being used to drill the hole in the open-hole section below the zone of loss may be used when the productivity of the kicking formation is low. No attempt will be made to define low and if there is a substantial doubt, method one should probably not be attempted. Example 2 illustrates the basic procedure.

■   EXAMPLE 2:

Well depth = 18,000 ft.
Last casing seat = 14,500 ft.
Mud weight = 18.5 ppg when well kicks
Hole size = $6^1/_8$ inches
Determine the feet of 22.0 ppg and 18.0 ppg mud required to equal a total column
    of 18.6 ppg mud.
Solution:
Let X = length of column of 22.0 ppg mud in feet
18,000 − X = length of column of 18.0 ppg mud in feet
18,000 ft. of 18.6 ppg mud = 18,000 × 0.967 = 17,406 psi
Thus: 1.144X + .936(18,000 − X) = 17,406
        1.144X + 16848 − .936X = 19,406
        .208X = 558
            X = 2680 feet of 22.0 ppg mud
        18,000 − 2680 = 15,320 feet of 18.0 ppg mud
Volume of 6.125 inch hole = .0364 Bpf
Bbls. of 22.0 ppg mud = (2680)(.0364) = 97.5 bbls.
This shows at least 97.5 barrels of 22.0 ppg mud would be required to fill 2680 feet of $6^1/_8$ inch hole. It is suggested that about 120 barrels of 22.0 ppg mud be used because some hole enlargement could be expected. If the hole were exactly to gauge, 120 barrels of mud would fill 3300 feet of hole and this would still be below the casing seat at 14,500 feet.

After the determination of mud weight and volume requirements, the 120 barrels of 22.0 ppg mud should be displaced with 18.0 ppg mud until the height of the 22.0 ppg mud is about 100 feet higher inside the drill string than outside. Then pull the drill string back to the casing seat and continue to displace slowly with a thin 18.0 ppg mud. Do not attempt to pull the drill string off bottom if the well is still kicking after displacing the 22.0 ppg mud. Proceed to setting a barite or cement plug.

The second procedure for controlling the formation kick is the setting of a barite plug. Two procedures may be used to set a barite plug: (1) the barite plug may be displaced and the drill string pulled out of the barite; or (2) the barite plug may be displaced and the drill string left in place. Again the procedure used will depend on the formation productivity and experience in the area or in similar areas.

The procedure for mixing the barite plug will be the same regardless of the setting method used. Essentially the procedure should be as follows:

1. Raise the pH of the mix water to about 10.0.
2. Add 6 ppb of chrome lignosulfonate. Stir vigorously.
3. Add barite to a maximum weight of about 21.0 ppg; weight may be less.

It should be noted that some barite at weights above 19.0 ppg may not settle readily when left quiescent and this should be checked prior to the need for a barite plug. In critical areas where the danger of lost circulation is associated with a potential well kick, if the barite available does not settle readily then a special order of barite that will settle should be kept available. The potential problem is too great not to be prepared.

If the drill string is to be pulled out of the barite plug, the top of the barite inside the drill string should be left about 100 feet above the barite top in the annulus. This procedure will help ensure that the barite inside the pipe is cleared as the drill string is pulled. If the barite plug is successful, the barite will settle into a solid plug and that portion of the hole is lost.

If the drill string is to be left in the barite, the barite should be over-displaced out of the drill string by about five barrels. This procedure will clear the bit and permit if necessary a second barite plug or a cement plug.

The advantage of a barite plug over the cement plug is: (1) the plug weight is higher, and (2) more than one plug can be set if the first is unsuccessful. A cement plug is a last resort procedure which leaves the operator in a very poor position if it is not successful, because he can no longer circulate fluid. This is not meant to imply that a cement plug should never be used. If the formation fluid is gas, it is in general suggested that at least the first attempt to shut-off the well kick should be with a barite plug. If the formation fluid is water and the flow is strong the operator may decide to use a cement plug. Again specific recommendations on a general basis are not very useable, decisions of this type have to be made at the rig site where all the potential problems can be evaluated.

Assuming the well kick is controlled by setting a plug of some type, the next step will be dependent on the method used to control the kick. If the placement of the 22.0 ppg mud were successful, then a liner should be set if further operations in the well are considered desirable. If a solid plug were set and the kick controlled, the lost circulation problem can be generally solved by reducing mud weight. Further side-tracking operations would be conducted in the normal manner. However before reaching the high pressure permeable sand that resulted in the problem initially, a liner will have to be set, cemented and the cement job tested as discussed previously.

It will be noted that no mention is made relative to using lost circulation material or setting a solid plug to cure the lost circulation problem when it is below the protective pipe and the well kicks. There may be areas where practices of using solid plugs have successfully solved the problem as defined, if so there is no argument against the procedure. The ultimate objective is to cure the problem and any solution that has proved successful should be considered. The general suggestions in this discussion are based on industry wide experience but certainly cannot include every specific operating area.

As mentioned, specific rules are difficult to establish for controlling lost circulation; however, some procedures have been tried repeatedly, acclaimed by some and discarded later and these should be discussed briefly.

First, should lost circulation material be used on a precautionary basis? For seepage losses that occur with low solids, low weight muds, fine lost circulation materials carried in the mud have been claimed to be successful. Other than cost, these materials would do very little harm so their need may be determined experimentally. The use of course lost circulation materials on a precautionary basis with low-weight mud is considered a poor practice.

The use of any lost circulation material on a precautionary basis in weighted drilling muds is considered a very poor practice. The practice continues to flourish in some areas. In the early part of the decade from 1960 to 1970 the practice of using about six ppb of a fine lost circulation material on a precautionary basis in weighted muds was a common practice in coastal areas of the United States. Currently the practice has, in general, been discontinued in the coastal areas. However, the practice of using lost circulation material on a precautionary basis in weighted muds is a common practice in the Rocky Mountain areas of the United States. Many claims of improvement using these procedures have been made. The practice is considered poor as it was many years ago in the coastal areas and with time it will be discontinued.

The use of lost circulation material in weighted muds before losing circulation is believed to be detrimental to the problem, not beneficial. Lost circulation material is a solid, circulating pressure losses are increased as the solids content increases, and any increase in circulating pressure would increase the danger of lost circulation.

There continues to be an introduction of new lost circulation material into the drilling business. New claims are made and on specific occasions successful results can be shown for any material. Laboratory tests of materials are not reliable, thus field testing becomes the only reliable procedure and field tests are difficult to evaluate.

PIPE STICKING

Many reasons have been advanced for the sticking of the drill string. In 1937, Hayward[1] reported results from laboratory and field research on why drill pipe became stuck; included was a series of discussions by prominent industry personnel. This paper was followed by another by Warren[2] in 1940.

Both of these papers concluded that pipe sticking was due to keyseating, an accumulation of cuttings around the pipe or balling up of the bit. In their discussions they also discussed the effect of sticking pipe in filter cake. The first paper on wall or differential pressure sticking was presented by Helmick and Longley[3] in 1957. This paper was followed by many others,[4,5,6,7,8,9,10] which also discussed the mechanics of wall-sticking.

Keyseating is still recognized as a primary reason for sticking pipe. Figure 3–2 illustrates how keyseating might occur. Noted is the fact that circulation will be maintained even though the pipe is stuck. One recommendation for preventing this problem has been to keep the hole straight. Today this is recognized as a possible solution, but emphasis is placed on controlling rate of deviation and sudden changes in hole direction. Methods that can be used to free pipe from a key seat include: (1) a jarring action, (2) spotting of oil to reduce friction, and (3) pipe rotation if possible.

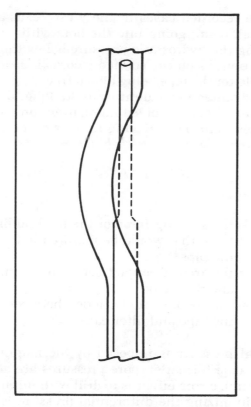

FIG. 3–2. *Keyseating drill collars in crooked hole*

Hole cavings and cuttings that accumulate in cavities offer potential hazards to sticking pipe. These problems have been discussed, and the manner in which they might stick drill pipe is obvious. When pipe is stuck in this manner, free fluid circulation is generally shut off or the pressure required for circulation increased substantially.

Suggested methods for freeing pipe include:
1. If circulation is not possible, shut-off the pump and release pressure. Work pipe slowly.
2. If circulation is possible but limited, circulate clear water to help remove cuttings from around pipe.
3. If Step 2 is unsuccessful spot oil around the pipe to reduce friction.

Differential Pressure Sticking

The term, "Differential pressure sticking," was introduced in about 1955 to identify the sticking of a portion of the drill string in the filter cake. Actually, operating personnel on drilling rigs had identified this type problem as wall sticking many years before introducing the term differential pressure sticking. In 1937, Hayward[1] recognized that the pipe could become stuck in wall cake and suggested that one method of solving the problem was to stop circulating in order to reduce hydrostatic pressure. Also in 1938 and 1939 there are recorded cases in the West Texas area of backing-off pipe above the stuck point, going into the hole with a packer, setting the packer and releasing the hydrostatic pressure below the packer to free the pipe. The specific remarks on the service records indicated that the procedure was responsible for the pipe being blown free.

The primary advantage of calling wall sticking, differential pressure sticking, was the latter term received more attention and engineers began to emphasize the problem. The sticking force is described mathematically in Equation 1.

$$F_s = \Delta PAf \tag{1}$$

Where:

F_s = the sticking force or the total pulling force that would be required to free the pipe.

A = the area of contact between the pipe and filter cake.

f = the coefficient of friction between the pipe and filter cake.

The differential pressure is imposed by the magnitude of the hydrostatic pressure because formation pore pressures are at fixed levels. Thus one method to minimize this effect is to drill with minimum mud weights. The problem of minimizing the differential pressure is often complicated by long sections of open-hole, where the formation pore pressures are substantially different. For this reason a given mud weight may be necessary to control the pore pressure in one open formation and this will impose a large pressure differential across another open formation.

The area of contact represents the total area of the pipe covered by filter cake across which the pressure differential is effective. Thus the area of contact will be affected by the following:

1. The length of the permeable zone where the pipe contacts the filter cake.
2. The hole size and pipe size.
3. The pipe shape, whether externally flush or pipe with raised shoulders.
4. The thickness of the filter cake.
5. External stabilization of the pipe.

The length of the permeable zone is a fixed parameter thus this factor cannot be changed by the operator. Even the hole and pipe size may be something that cannot be changed or at least the reason the size combinations are used become more important than any special effort to minimize differential pressure sticking.

The pipe shape is a very important parameter and is one that can be changed easily. Large externally flush drill collars represent the ideal type equipment for differential pressure sticking. In recent years special drill collar configurations have been used, some of these are: (1) spiral collars, with circulation grooves in the external surface of the drill collars; (2) square drill collars; (3) shouldered drill collars; and (4) heavy-weight drill pipe, which could be called turned-down drill collars.

The effect of the special drill collars is to reduce substantially the area of wall contact and the manner in which this is accomplished can be easily recognized. More discussions on special collars are given in the chapter on straight hole drilling. The heavy weight drill pipe has reduced substantially the differential pressure sticking problem, particularly in directional wells. This pipe, some of which is made by turning down the O.D. of regular drill collars, has upsets in the middle and on each end of each 30 foot joint. This configuration reduces substantially the area of contact. There are other advantages to heavy weight pipe, the reduction in stiffness reduces some of the dangers of keyseating and has substantially reduced twist-offs at the connection point between the drill collars and the drill pipe.

The potential effect of the filter cake on contact area is shown in Figure 3–3.

It is noted that the area of contact may be more than doubled by a thickening of the filter cake. This has been the primary reason for con-

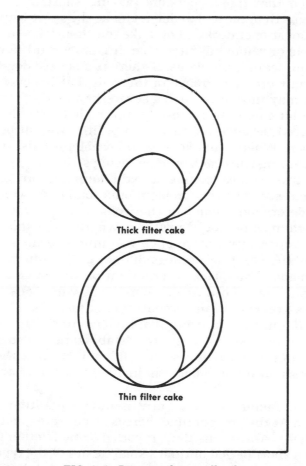

Thick filter cake

Thin filter cake

FIG. 3–3. *Pipe stuck in wall cake*

trolling the high temperature high pressure filtration rate. Actually the rate of filtrate loss was of no concern to the operator, his primary concern was the filter cake-thickness. This will be discussed in more detail in the chapter on drilling muds.

It should be emphasized that it is generally assumed that filter cakes during normal drilling reach an equilibrium thickness. This simply means that the rate of erosion by the circulating fluid equals the rate of deposition of new solids in the filter cake. This concept of cake erosion was used to show that short trips were not necessary during long bit runs or for the purpose of removing filter cake before wire-line logging.

When diamond bits were first introduced in deep drilling in the coastal areas of the United States, it became common practice to pull the bit back into the last casing string every 24 hours, commonly called a short trip. The primary purpose of this short trip was to remove excessive filter cake and ensure an open hole. Pipe-sticking problems were already common in the coastal areas and if the pipe stuck while using a diamond bit the first question asked was, how long was the bit in the hole? This placed pressure on operators that did not believe in the short trips. However, studies of the problem showed that those operators making short trips had more pipe sticking problems than those that did not make the short trips.

Currently most operators do not make the short trips for the purpose of opening the hole or removing filter cake. It is important to watch carefully for potential key-seating problems. If there is a severe dog-leg in the hole, the drill pipe may create a groove in the wall that is deepened with time. A deep groove may introduce such a severe key-seating problem that drill collars and the bit cannot be removed from the hole. Problems such as this can be recognized because the increased pulling weight occurs while the pipe is in motion. Where such key-seating problems exist it may be necessary to make reaming trips back into the casing string.

The short trips before logging are still common, but largely unnecessary. A complete study of this problem on a total of 50 wells, where short trips were made on about one-half the wells, revealed the success in running logs to bottom could not be related to whether the short trip was made or not made. Apparently the practice continues because most operators remember only the cases where logs did not go to bottom when the short trip was not made. This places a premium on investigating all practices that require extra time, because off-shore wells where costs exceed $30.00 a minute places a real premium on conserving time.

External stabilization of drill collars is a common practice to minimize crooked hole problems; however, this stabilization also minimizes the area of wall contact and reduces the tendency to stick the pipe. A more complete discussion on stabilizers is included in the chapter on straight hole drilling.

The friction factor in Equation 2 is never quantitatively defined in field operations. At the present time there is no accepted method to measure the friction factor. Annis,[11] in 1962, reported some friction factor measurements and for a short period of time a small tester of the type used by Annis was tried in operations. Later the tester disappeared. In 1970, Mondshine[12] published results concerning Drilling-mud lubricity. Mondshine emphasized his work was related to lubricity and not to friction factors for differential pressure sticking.

Specific numbers for friction factors will not be given from previous tests because in general there is no agreement among investigators. In general it is agreed that bentonite helps reduce the friction factor. Oils are known to be good lubricants and the most common way of freeing stuck drill pipe is to spot oil or an oil-base mud around the stuck portion of the pipe. Other special additives that have been used include the following:

1. Asphalt
2. Graphite
3. Detergents
4. Walnut hulls
5. Special lubricants, generally a type of fatty acid

Various claims have been made for these materials. Detergents are commonly used in weighted and unweighted muds. In some areas the detergents help prevent bit balling; however, there is no evidence that they have been beneficial where pipe sticking is a problem. Walnut hulls have been used to reduce rotating torque. At times they have been successful; however, to maintain the benefits the walnut hulls must be added almost continuously. Special lubricants have been successful in reducing friction factors and have been beneficial. These lubricants are expensive and it is sometimes difficult to evaluate results in the field. Asphalts and graphite are generally materials that are used more because they appear to have lubrication qualities than any reported results.

The best way to minimize the friction factor is to use an oil base mud. With water-base muds, the first prerequisite for low friction factors is a good bentonite content for the filter cake. Oil in almost any concentration is beneficial and special lubricants have been beneficial. Materials that are considered detrimental include most drill solids and barite.

In field applications, the operator should watch for wall sticking indications. Example 3 is one illustration of watching the problem:

■ EXAMPLE 3:

Problem sand at 12,000 feet
Pore pressure of sand = 8730 psi
Hydrostatic pressure of 14.8 ppg mud = 9250 psi
Force required to pull collars after stopping opposite the sand = 60,000 pounds
Maximum pull permitted on the drill string = 100,000 pounds
Determine the maximum permissable mud weight if no other changes are made.
Solution:
$F = \Delta PAf$
$60,000 = 520\, Af$
$Af = \dfrac{60,000}{520}$

$\Delta P = \dfrac{(100,000)(520)}{60,000} = 867$ psi
Permissible hydrostatic pressure = 8730 + 867 = 9597 psi.

$$\text{Permissible mud weight} = \frac{9567}{(12,000)(.052)} = 15.4 \text{ ppg.}$$

Also the use of barite to increase mud weight will increase the friction factor thus to even reach a mud weight of 15.4 ppg the operator would have to change the drill string, thin the filter cake, or add a lubricant.

SLOUGHING HOLE

Hole sloughing is generally a problem associated with shale. Each drilling area of the world will have specific names for the sloughing shale zones in that area. This simply means that it is common for shale to slough into the hole and when the problem occurs repeatedly operators begin to plan drilling programs to combat the shale problems.

It is very difficult to define the specific reason shale sloughs into the hole. It is known that shales in general have a strong affinity for water and when water is adsorbed by the shale the stability of the shale section is reduced. This does not mean that the shale would necessarily slough or cause trouble. Other associated hole and drilling conditions tend to further promote the problem of sloughing hole. Some of these are: (1) a high percentage of high yielding clays, such as sodium montmorillonite; (2) steeply dipping shale beds; (3) pressured shale sections, where the pore pressure exceeds the hydrostatic pressure; and (4) turbulent flow in the annulus which helps to promote erosion.

It is known that almost any shale will adsorb some water if it is available. When the water is adsorbed the yield strength of the shale will be reduced and the shale will generally expand in a direction perpendicular to its bedding plane. The amount of water a shale will adsorb will depend on its hydration state and composition. Shales containing sodium montmorillonite are likely to expand more than those containing a clay such as kaolinite. However, it has been shown that all shales tend to adsorb some water.

Some shales adsorb water faster than others and in actual drilling operations may cause some trouble very quickly, others adsorb water slowly and several days may elapse before any sloughing is initiated. Because most of the shale expansion is in a direction perpendicular to the bedding plane, horizontal shale beds are not likely to slough as a result of this expansion. In fact, an increase in bed dip, increases the probability that a given shale will slough when exposed to water.

It has been common practice to reduce water-loss when sloughing shale becomes a problem or when drilling through known sloughing shale zones. Based on the discussion concerning the effect of water on shale this practice sounds reasonable. However, most of the wetting of shales occurs by the imbibition of water into the shale, not by a filter process associated with permeability.

The actual magnitude of the hydrational stress or pressure that forces the water into the shale may be on the order of magnitude of 50,000 psi. Actually it is determined from the ratio of the aqueous vapor pressure of shale to the aqueous vapor pressure of water. This simply means that if there is water present it will wet the shale even if the measured API water-loss is reduced to zero.

One argument for reducing filtration rate has been that shales had interbedded sand streaks or fractures which are invaded by filtrate and the shale is permitted more contact with the water. If the filtration takes place because of the sand or fractures in the shale the argument about greater contact with water would be true; however, the shale would be wet in either case and the surface wetting in the vicinity of the borehole is the most detrimental.

Methods to prevent shale sloughing and hole enlargement have been studied almost since the beginning of rotary drilling. The concept of controlling water-loss to stop sloughing was introduced in the decade from 1920–1930. Lime muds were introduced for this purpose and others in the 1930–1940 decade. Special calcium type muds were introduced in the 1950–1960 decade to inhibit shales to prevent hydration. Oil base muds already used for the purpose of inhibition of shales changed to include salt in the water phase during the decade 1960–1970. Current practices include potassium chloride-ploymer muds which are claimed to prevent clay swelling by inhibition and encapsulation.

The question might be asked, why do certain practices endure if they are not beneficial? First, it should be remembered that any practice may appear successful at times because formations are not homogeneous. Second, the operator feels he must do something even if it is not successful. Consider water-loss control, many papers and talks were given on the beneficial effects of drilling with a low water-loss to prevent shale sloughing. It would be impossible to determine the specific effects of this practice at the time; however, it is suspected that the control of water-loss helped improve filter cake qualities and many times increased lifting capacity of the muds, which reduced problems of bridges in the hole and pipe sticking. This, of course, made the practice worthwhile at the time even if the diagnosed effect was not correct. Caliper logs helped show that hole enlargement could not be correlated with the API water-loss.

Calcium treated systems for the control of sloughing flourished during the decade of 1950 to 1960. Hole enlargement continued with calcium treated muds although some papers were written showing less hole enlargement in specific areas. Beginning in about 1960 claims were made that shale swelling could be prevented using high concentrations of a chemical thinner called, Chrome Lignosulfonate. Papers were written showing that bentonite would not hydrate if the concentration of chrome lignosulfonate was 12 ppb or higher. Thus began the era of highly treated lignosulfonate muds, concentrations ran up to more than 20 ppb. Unfortunately the hole enlargement problem continued.

The first so-called breakthrough on controlling shale sloughing came with the oil base muds when the water phase of these muds was saturated with salt. The use of calcium chloride in the water phase of oil base muds was first introduced by Mondshine[13] in 1966. He showed that water could be pulled into the shale from oil base muds if no calcium chloride were used and that water could be pulled out of the shale if an excess amount of calcium chloride were used. The desired result would be to ensure no movement of water into or out of the shale. To do this required a balance between the vapor pressure of the water phase in the oil and the vapor pressure of the water of hydration in the shale.

In 1969 Chenevert[14] showed that the ratio of vapor pressures could be determined very easily by using a correlation based on the relative humidity. However current practice is based primarily on using calcium chloride in the water phase of oil base muds in a concentration known to be high enough to prevent water invasion into the shale. No specific effort is generally made to prevent pulling water out of the shale.

These oil base muds with salt in the water phase have been very successful in preventing hole enlargement. It is claimed that the continuous oil phase provides a membrane on the wall of the hole across which water in the oil or formation must move if the shale is to be wet by water or dehydrated. If the vapor pressure of the waters are the same there is no force to cause any movement of water. Equation 1, a thermodynamic relationship, provides an insight into the reason for water movement.

$$P_h = \frac{RT}{\overline{V}} \ln \frac{P}{P_o} \qquad (1)$$

Where:

P_h = hydrational stress
R = units conversion constant
T = absolute temperature
$\overline{V}$ = partial molar volume of pure water
P = aqueous vapor pressure of shale and
      water of hydration
P_o = aqueous vapor pressure of water in oil base mud

Note if $P = P_o$ in Equation 1 there is no force to cause water movement.

The primary question an operator has to answer on the use of oil base muds of this type is whether the additional hole stability is worth the added cost of using oil. Again this is not a question that can be answered in a general discussion.

The potassium chloride-polymer muds of the current type were first used in Western Canada. However, potassium chloride as an inhibitor has been used for 15 years in the Rocky Mountains and California. Specific results have shown that the use of potassium chloride and polymers claimed to encapsulate the shales have helped reduce substantially hole enlargement in some areas of Western Canada. These muds are also expensive and there general application is still to be determined. Thus the use of such muds at this time should be considered experimental and economic comparisons should be watched closely.

These discussions have been concerned primarily with methods to prevent expansion of the shales. Another cause of sloughing is where the pore pressure in the shale exceeds the hydrostatic pressure. At times the operator may drill without excess shale problems even if the hydrostatic pressure is less than the pore pressure of the shale. If the dip of the shale beds is low, say less than 5 degrees, some under-balance of the hydrostatic pressure can be tolerated because the primary direction of shale expansion is perpendicular to the bedding plane. If water can be prevented from entering shale, as discussed above, then the yield strength of the shale may be high enough to prevent excessive sloughing even though the hydrostatic pressure is less than the pore pressure. Evidence of this latter claim has been demonstrated in the Delaware basin of West Texas where 11.0 ppg oil base muds have been used as compared with 13.5 ppg water base muds. The primary problem in using the lighter oil base mud is if a permeable zone is encountered the mud weight is too low to prevent a well kick, which may result in more problems and expense than using the heavier water base mud. Again this is an economic decision that should be based on experience for the specific area.

In general, the method used to prevent sloughing in pressured shales is to maintain a mud weight that is equal to or greater than the pore pressure of the shale. An alternative is to simply permit some sloughing as drilling progresses.

There are frequent debates over whether high annular velocities will cause hole erosion and if so what is the magnitude of this effect. If the annular flow pattern is laminar, the fluid velocity would probably have no effect on hole erosion. If the flow pattern is turbulent, the fluid velocity would probably substantially affect hole erosion. Thus in high angle fractured shale zones that have a history of severe sloughing, it is suggested that the mud be thick enough to remove cuttings instead of using thin fluids at high annular velocities that would create turbulent flow patterns. In all probability the hole will slough in either case, a turbulent flow pattern would simply accelerate the problem and once enlargement begins the thin fluid would not clean the hole.

An operator must decide why he wants to prevent shale sloughing before he can justify expensive programs of prevention, which also may not always be successful. If the shale zone lies between two sands that must be isolated from each other for completion and production reasons, the reason is valid. If on the other hand the desire to prevent sloughing is just because of frequent bridges in the hole, fill on bottom or delayed drilling because of the excessive solids build-up then the prevention program becomes an economic decision.

One secret to shale problems is to clean the hole by raising the lifting capacity of the mud. Lifting capacity can be increased by flocculating clays, adding more bentonite or by using polymers. One frequent problem encountered is that the operator wants to increase lifting capacity and at the same time control water-loss. Any flocculation of the clays to thicken mud would increase the water-loss. The addition of bentonite will increase lifting capacity; however, unless a thinner is also added the mud is very likely to flocculate.

If a thinner is added, the lifting capacity is reduced and the shale problem continues. Polymers may be added for thickening; however, again cost has to be considered. A very important point to remember is that if the hole is not being cleaned, absolutely nothing can be gained by adding a thinner to lower water-loss. If the filtration rate must be reduced in this case, use a filtration control agent such as starch or CMC not lignosulfonates or lignites.

One other problem that may arise in thickening mud for hole cleaning is the possibility of losing circulation. If lost circulation problems are associated with hole cleaning problems, a compromise may be necessary. Thicken 25 to 50 barrels of mud, by one of the methods mentioned. The amount of thickening will depend on the specific problem; however, a yield point of 60 to 100 should be achieved easily. Pump this batch of thick mud with the thinner mud. The thick mud should clean the hole with a minimum effect on total annulus pressure.

The frequency of use for the thick mud will depend on the specific problem. There is no question that 75 per cent of the serious shale problems are allowed to be serious problems. They could be corrected by cleaning the hole.

REFERENCES

1. Hayward, J. T., "Cause and Cure of Frozen Drill Pipe and Casing": Drilling and Production Practices, p. 8, 1937.
2. Warren, J. E., "Causes, Preventions, and Recovery of Stuck Drill Pipe": Drilling and Production Practices, p. 30, 1940.
3. Helmick, W. E. and Longley, A. J., "Pressure Differential Sticking of Drill Pipe": The Oil and Gas Journal, June 17, 1957, p. 132.
4. Outmans, H. D., "Mechanics of Differential Pressure Sticking of Drill Collars": Petroleum Transactions, Vol. 213, p. 265, 1958.
5. Tschirley, N. K. and Tanner, K. D., "Wetting Agent Reduces Pipe Sticking": The Oil and Gas Journal, Nov. 17, 1958, p. 165.
6. Tavan, Jr., E. V., "Casing, Pipe Freed During Fishing Jobs with Oil-Base Mud": Drilling, Feb., 1959.
7. Sartain, B. J., "Drill-Stem Tester Frees Stuck Pipe": Petroleum Engineer, Oct. 1960.
8. Fox, F. K., "New Pipe Configuration Reduces Wall Sticking": World Oil, Dec. 1960.
9. McGhee, Ed., "Gulf Coast Drillers Whip the Wall-Sticking Problem": The Oil and Gas Journal, Feb. 27, 1961.
10. Haden, E. L. and Welch, G. R., "Techniques for Preventing Sticking of Drill Pipe": Drilling and Production Practice, 1961.
11. Annis, M. R. and Monoghan, P. H., "Differential Pressure Sticking – Laboratory Studies of Friction Between Steel and Mud Filter Cake": Petroleum Transactions, May 1962.
12. Mondshine, T. C., "Drilling-mud Lubricity," The Oil and Gas Journal, December 7, 1970.
13. Mondshine, T. C. and Kercheville, J. D., "Successful Gumbo Shale Drilling," The Oil and Gas Journal, March 28, 1966.
14. Chenevert, M. E., "Shale Alteration by Water Absorption," presented at Fourth Conference on Drilling and Rock Mechanics, Austin, Texas, January 14–15, 1969.

4

Planning a Well

ROBERT D. GRACE
Consulting Engineer
Amarillo, Texas

$\mathbb{P}$ROPER planning of any drilling venture is the key to optimizing operations and minimizing expenditures. Companies are in business to find and develop oil and gas and it is the ultimate responsibility of the Drilling Engineer to establish the geologic objectives and accomplish those objectives at a minimum cost.

In order to function in this capacity, the Drilling Engineer must be intimately familiar with the drilling practices in the area of immediate interest. However, familiarity with common practices is not enough. In addition, it is necessary for the Drilling Engineer to develop an expertise in every phase of drilling in order that improvements can be made and savings can be accomplished.

The first step in planning a well should be the gathering of all available data on past wells. In this respect it is important to be completely familiar with all sources of information, the availability of the sources, and the information normally associated with each source.

As outlined, a major responsibility of the Drilling Engineer is to establish the objectives of the well. With respect to this responsibility, the Drilling Engineer must obtain a Geologic Prognosis similar to the one illustrated as Figure 4–1. It is imperative that the prognosis be obtained and is normally provided on request from the Geology Department within major oil companies. Independent Drilling Engineers or Drilling Engineers representing drilling contractors can generally obtain the prognosis by making a formal request of the operating management.

Of primary importance is the fact that the prognosis represents a firm commitment as to the Drilling Engineers ultimate responsibility. The loca-

GEOLOGICAL OUTLINE

Name and Location:
  Dry Hole No. 1-"A," 700' FNL & 660' FEL Section 82.
  Block B-1, H&GN RR Survey, Northwest Mendota Field, Roberts County, Texas

Objective Horizon and Contract Depth:
  Base of Upper Morrow Sand plus 100'; Approved depth 11,350'

Estimated Formation Tops:
Estimated Elevation K. B.	2,857'
Top Wichita-Albany Anhydrite	2,950'
Top Wolfcamp Dolomite	4,150'
Top Possible Lost Circ.	4,300'
Top Douglas Sands	7,100'
Top Granite Wash	9,950'
Top 13 Finger Lime	10,910'
Top Morrow "Formation"	11,000'
Top Morrow Sand	11,165'

Possible Producing Zones:
Douglas Sand	7,100–7,200'
Stray Douglas Sand	7,400'
Des Moines	9,050–9,900'
Granite Wash	9,950–10,800'
Upper Morrow Sand	11,165'

Samples:
  Catch 10' samples from 6,800' to TD. Wash thoroughly, air dry, and tie in 100' bundles. 10' drilling time from 3,350' to TD.

Coring:
  One 50. oriented core of Upper Morrow Sand 11,165 to 11,215', approximately. (Need core for dipmeter study and environmental analysis.)

Drill Stem Testing:
  Possibly one test in Granite Wash

Surveys:
  Schlumberger IES & Compensated Density log

Remarks:
  Set surface casing at 3,350'; set intermediate casing at 10,950' ($5\frac{1}{2}$"). Possible string of $2\frac{1}{2}$" tubing to be set outside of $5\frac{1}{2}$" casing in order to test Granite Wash.

FIG. 4–1

tion and spot are provided along with the anticipated total depth. The depth to fresh water is included which in many instances establishes the surface casing requirements.

Potentially productive objectives are listed which provides insight into hole size and production casing requirements as well as ultimate completion requirements. Geologic intervals from the surface to total depth are included along with their anticipated tops. The geology department will normally outline sample requirements, logging requirements, and any anticipated drill stem tests and cores.

This information is fundamental to the successful planning of any well and should be considered basic knowledge by management. It is expensive to "tight hole" those responsible for accomplishing the established objectives.

It is equally important to the Drilling Engineer to be intimately familiar with the basic geology in the area of interest. It is vital then that geologic maps be obtained. These maps establish structure information and outline regional and local dips and abnormalities. This information allows the Drilling Engineer to establish control wells as wells geologically similar in every respect to the prospect.

It must be stressed that control wells should be geologically identical in every respect to the prospect. That is, intervals should be correlative, structural position should be similar, and the prospect should be on trend with the control wells. When information is not abundant, it may be necessary to improvise a control well from a composite of several wells recognizing the limitations of each. Seismic data is generally available and should always be included to establish control.

Proper control is vital and must be established. The old, over used statement, "every well is a wildcat" is nonsense. Drilling problems and characteristics are normally a result of formation characteristics and can be mapped and correlated just as a particular formation can be mapped.

Ironically, drilling characteristics are in many instances more reliable and revealing than geologic markers. That is, establishing structural position relative to control wells while drilling is often best done by analyzing bit records and drilling characteristics as opposed to plotting drilling time or analyzing samples. Given adequate information, a good drilling man is never lost.

Land maps are essential in establishing control wells and provide fundamental orientation. As Figure 4–2 illustrates, land maps which are readily available within most companies as well as readily available commercially, provide offset operators, total depth, formation tops and development dates. The land map is essential in establishing orientation and control.

Once control wells have been established from geologic considerations, it is imperative that the Drilling Engineer obtain bit records and electric logs on those control wells. Bit records such as illustrated in Figure 4–3 are available free of charge from the major bit manufactures. Electric logs are not as readily available; however, logs can normally be obtained commercially.

Bit records contain a wealth of information essential to the Drilling Engineer. As illustrated, the contractor, contractors rig number, and contractors tool pusher are listed along with all pertinent casing point dates. Included with the drill pipe and drill collar descriptions are complete pump descriptions including power and liners.

The drilling rig and power are adequately described along with the number, sizes, and types of bits required. Casing sizes and depths are listed. Pump pressures, bit jet sizes, and pump strokes per minute are furnished. Bit weight, and rotary speed are included with bit hours and condition-information essential for sophisticated minimum cost drilling programs. Drilling problems such as vertical deviation and fishing jobs are often included. Finally, a basic insight into the mud system used is obtained.

It is difficult to appreciate or elaborate on all the information provided

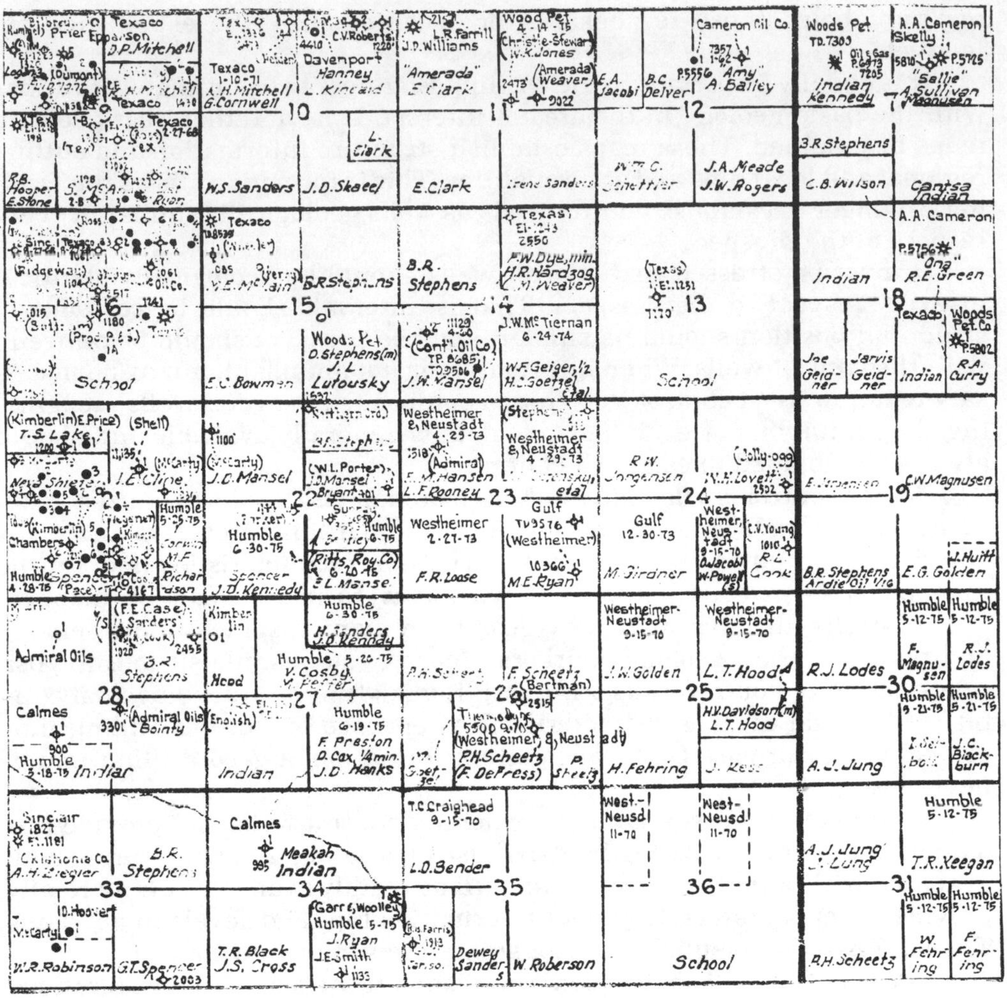

FIG. 4-2. *Typical lease map*

by a bit record. Adequate data for evaluation hydraulics and minimum cost drilling are but a part of the information available. The bit gradings are an integral part of minimum cost drilling and should be viewed with a skeptical eye. In this respect the Drilling Engineer should become intimately familiar with bit gradings and should compare numerous bit records for consistency when planning a well. Minimum cost drilling programs based on a single bit record should be subject to suspicion.

Electric logs on the control wells are beneficial in well planning. Bit records from control wells plotted on their electric logs provide essential insight in planned bit selection, Figure 4-4. In fact, experienced personnel with a thorough knowledge and understanding of rig costs and bit performance under various weights and speeds have combined this knowledge with the electric log-bit record concept to consistently out perform the best of minimum cost drilling programs. Sophisticated minimum cost drilling programs of the future will incorporate the bit record-electric log concept.

COUNTY	FIELD	STATE	SECTION	TOWNSHIP	RANGE	LOCATION	WELL NO.
LEA	W/C	N. MEX.	13	11-S	34-E	BOGLE FARMS	1

CONTRACTOR	RIG NO.	OPERATOR	TOOLPUSHER	SALESMAN
MORAN OIL PROD. & DRLG CORP.	3	EARL T. SMITH & ASSOCIATES	MILO PENFIELD	TOMMY EVERHART

SPUD	UNDER SURF.	UNDER INTER.	SET SAND ST.	REACHED T.D.	O.D.	PUMP NO. 1	PUMP NO. 2	LINER	LINER	PUMP POWER	TYPE MUD
2-11-69	2-12-69	2-17-69				GD	NAT	C-250	5½	COMPOUND	

DRILL PIPE	TOOL JOINTS	SIZE	TYPE	NUMBER	DRILL COLLARS	O.D.	I.D.	LENGTH	DRAWWORKS POWER
4½		4	H-90	28	28	8"		86	MUD CONT. U-712-A 2-LRZ

CUMU-LATIVE $/FT.	NO.	SIZE	MAKE	TYPE	JET 3Lnd IN	SERIAL	DEPTH OUT	FEET	HOURS	FT/HR	ACCUM DRLG. HRS.	WT. 1000 LBS	RPM	VERT. DEV.	PUMP PRESS	COST $/FT.	SPM 1	SPM 2	MUD WT VIS WL	T	B	G	OTHER	FORMATION REMARKS
1.77	1	17-1/2	HUGHES	OSC-3A		RT	396	396	6-1/2	61	6.5		110	1/2°	1800	1.77			SPUD MUD	N	D			RUN 13-3/8" CSG. 2-12-69
1.11	2	11-1/2	HUGHES	OSC-3A	3-10	JB592	2135	1736	21-3/4	80	28	50	100	1/2°	1800	0.16	52		NATIVE	N	D			
1.40	3	11-1/2	HUGHES	OSCJ	3-10	HL403	3001	866	24	36	52	50	80	-1/2°	1800	2.11	52			N	D			
1.64	4	11-1/2	HUGHES	OWV	3-10	HT329	3597	596	17-1/4	35	70	50	80	1°	1800	2.42	52			N	D			
1.89	5	11-1/2	HUGHES	OWV	3-10	FF923	4130	533	24-1/4	22	94	70	80	1/2°	1800	3.32	52			N	D			RUN 8-5/8" CSG. 2-17-69
2.07	6	7-7/8	HUGHES	XWR	240-9	HR792	4490	360	18	20	112	60	60	1/4°	1800	3.52	52			N	D			2-18-69 4137
2.23	7	7-7/8	HUGHES	XWR	241-9	HR784	4857	367	17-3/4	21	130	60	60	1°	1800	4.12	52		4.20	6	7			
2.35	8	7-7/8	HUGHES	XWR	3-10	HE753	5498	641	27	24	157	60	60	1/2°	1800	3.25	52		4.20	8	7			
2.49	9	7-7/8	HUGHES	XWR	3-10	FP598	5801	303	12-1/4	25	169	60	60	1/2°	1800	3.99	52		4.20	8	8			
2.36	10	7-7/8	HUGHES	XWR	3-10	HE924	5991	190	15-1/2	33	174	60	60	1/4°	1800	4.26	52		4.20	7	7			
2.66	11	7-7/8	HUGHES	XWR	3-10	HY914	6497	506	20-1/4	25	195	60	60	1/4°	1800	3.60	52		4.20	6	8		C D	
2.75	12	7-7/8	HUGHES	XWR	3-10	JTW223	6790	293	15	20	210	60	60	1/4°	1800	4.79	52		4.20	6	8			
2.84	13	7-7/8	HUGHES	XWR	3-10	RR#10	7062	272	14-1/4	19	224	60	60	1/4°	1800	4.62	52		4.20	6	7			
2.94	14	7-7/8	HUGHES	XWR	3-10	HN909	7424	362	19-1/2	19	244	60	60	1/4°	1800	4.67	52		4.20	6	7			2-26-69 7570
3.04	15	7-7/8	HUGHES	XWR	240-11	JTW217	7662	238	14-3/4	16	258	60	60	3/4°	1800	5.94	52		MUD UP	7	7			
3.12	16	7-7/8	HUGHES	XWR	3-10	JW217	7910	258	15-1/2	16	274	60	60	1/2°	1800	5.75	52		98 40	8	8			
3.21	17	7-7/8	HUGHES	XWR	3-10	HW208	8228	318	18-1/2	17	293	60	60	1/2°	1800	5.2	52		98 40	8	8			
3.31	18	7-7/8	HUGHES	XWR	3-10	HS233	8372	144	10-3/4	13	303	60	60	3-1/4°	1800	8.30	52		98 40	8	8			
3.40	19	7-7/8	HUGHES	XWR	3-10	JTW214	8597	225	18	13	321	60	60	2°	1800	7.28	52		98 37	6	7			
3.76	20	7-7/8	HUGHES	XWR	240-11	HS618	8835	238	16-1/4	15	338	60	60	1/2°	1800	6.47	52		98 37	8	8			
4.12	21	7-7/8	R	SCM-J		580	9267	432	35-1/4	8.3	343	35	40	-3/4°	1900	11.09	56		98 37	3	6			3-4-69 8970
4.43	22	7-7/8	R	SCM	3-10	977	9733	466	62-1/4	7.5	435	40	40		1900	11.82	56		98 36	4	9			3-7-69 9457
4.52	23	7-7/8	R	SCM	3-10	581	10,093	360	52	6.9	507	40	40	-3-1/4°	1900	12.84	56		98 36	4	7			3-11-69 10093
4.79	24	7-7/8	HTC	WRT	3-10	AJF253	10,165	72	11-1/2	6.3	519	40	40		1900	17.78	56		94 47	7	7			
5.07	25	7-7/8	R	SCM	3-10	984	10,476	311	44-1/2	7.0	563	60	60	1°	1800	13.43	56		97 44	6	6			
5.31	26	7-7/8	R	SCM	3-10	597	10,831	375	55	6.8	618	40	40	1/2°	1900	12.86	56		96 36	N	D			
5.54	27	7-7/8	R	SCM	3-10	617	11,165	314	45-3/4	6.9	664	40	40	1°	1900	13.63	56		96 32	8	P			
5.63	28	7-7/8	R	SCM	3-10	JTW224	11,367	202	37	5.5	701	40	40	1/4°	1900	18.42	56		96 32	4	8			
5.92	29	7-7/8	HTC	XWR	3-11	442	11,415	48	9-3/4	4.8	711	60	60		1900	25.57	56		96 32	N	D			
6.05	30	7-7/8	R	FRCSJ	3-10	AV162	11,612	197	31-3/4	6.2	742	40	40		1900	17.53	56		96 32	7	D			3-24-69 11571
6.23	31	7-7/8	HTC	WRJ	3-10	JR082	11,649	37	6-3/4	5.5	749	50	50	1/2°	1800	28.47	56		97 38	7	7			
6.43	32	7-7/8	HTC	X55R	3-10	JR082	11,765	116	34	3.4	783	40	40		1900	30.97	56		96 38	N	D			3-26-69 11697
6.53	33	7-7/8	HTC	XWR	3-10	KD488	11,817	52	12	4.3	795	40	40		1900	26.43	56		95 37	N	D			
6.85	34	7-7/8		SCM1	3-10	RR	11,936	119	27-3/4	4.3	823	40	40		1900	20.08	56		95 37	4	8			
7.10	35	7-7/8	HTC	RG7XJ	3-10	JTW123	12,114	178	58-1/4	3.1	881	40	40		2.00	28.42	56		95 37	3	8		BTM	
7.32	36	7-7/8	HTC	RG7XJ	3-10	LM444	12,411	297	58-1/2	5.1	940	40	40		2.00	17.5	56		96 37	3	8		BTM	
7.50	37	7-7/8	HTC	LD407	3-10	LD407	12,645	234	48-3/4	4.8	888	40	40		2.00	9.25	56		96 38	4	7		BTM+	
7.57	38	7-7/8	HTC	RG7XJ	3-10	JR088	12,880	235	40	5.9	1028	40	40		2.00	16.96	56		96 38	4	7		CR	
7.65	39	7-7/8	HTC	WRJ	3-10	AJ	12,940	60	12-1/4	4.8	1041	60	60		2.00	23.72	56		96 39	4	5			
7.84	40	7-7/8	HTC	XWR	3-10	JF898	12,990	50	10-1/4	4.8	1051	60	60		2.00	26.39	56		96 38	8	6		C D	4-11-69 13081
7.92	41	7-7/8	R	SCMJ	3-10	RR	13,098	108	29-3/4	3.6	1081	60	60		2.00	31.28	56		95 41	4	8			
8.02	42	7-7/8	HTC	XWR	3-10	6N905	13,147	49	19-1/2	3.8	1084	60	60		2.00	30.09	56		95 42	N	D			
8.09	43	7-7/8	HTC	XWR	3-10	HS243	13,220	73	19-1/2	3.8	1113	60	60		2.00	25.36	56		96 42	8	8			4-14-69 13223
8.15	44	7-7/8	HTC	XWR	3-10	HC799	13,280	60	12	5.0	1125	60	60		2.00	23.64	56		96 43	7	7			
8.23	45	7-7/8	HTC	XWR	3-10	LU285	13,315	35	6-3/4	5.2	1132	60	60		2.00	37.56	56		96 42	8	8		DST 1	
8.23	46	7-7/8	HTC	XWR	3-10	LU840	13,385	70	15-1/4	4.6	1147	60	60		2.00	23.09	56		96 45	7	7			
8.28	47	7-7/8	HTC	XWR	3-10	LJ288	13,396	11	1-3/4	11.2	1149	60	60		2.00	73.33			95 46	1	1		DST 2	4-17-69

LOGGING 4-17-69   PLUGGED & ABANDON 4-20-69

FIG. 4-3. Typical bit record

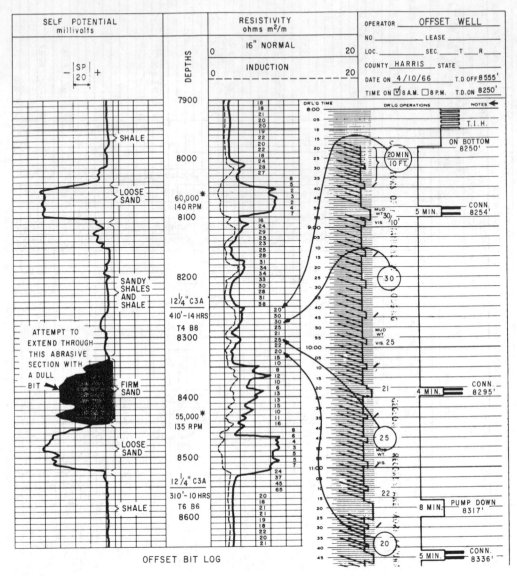

After John Alterman, World Oil Mar. 1969

FIG. 4–4. *Bit records plotted on electric logs*

The use of electric logs to determine frac gradients and pore pressures is well established. In some areas the techniques are quantitative; in virtually all areas logs qualitatively describe pore pressures. Similar relationships should exist for defining abnormally low pore pressure; however, to date no substantial investigation describing zones of severe lost circulation has been reported.

Drilling mud recaps from the control wells are generally difficult to obtain, but they are available from the mud companies. These recaps contain data significant to proper planning. As illustrated in Figure 4–5, total

DRILLING MUD RECAP

COMPANY _APACHE OIL_   WELL _HONEY #1_

CONTRACTOR _UNIT DRLG._   FIELD _N.W. FT. SUPPLY_

LOCATION _35-25N-23W_   COUNTY _HARPER_   STATE _OKLAHOMA_

MUD COST	5,745.97
SALES TAX	114.93
DRAYAGE	262.92
TOTAL COST	6,123.82
COST PER FOOT	.76

MATERIALS USED

ITEMS		UNITS	COST	ITEMS		UNITS	COST
Mil-Bar	Bag			Ligco	Bag		
Milgel	Bag			Ligcon	Bag		
Super-Col	Bag	81	182.25	Lime	Bag	5	5.50
Salt Water Gel	Bag	13	33.80	M-D	Can		
Aplosol S	DM	28	1,411.20	Mil CMC	Bag		
Ben-Ex	Bag			Mil-Flo	Bag		
Bicarb	Bag			Mil-Starch	Bag		
Caustic Soda	Bag	12	63.60	Milmica	Bag		
Flo-Sal	Bag	100	702.00	Pres.	DM		
LD-7	Can			Synergic	Can	32	165.12

MUD AND CHEMICAL COST

ITEMS		UNITS	COST	
Uni-Cal	Bag	9	123.75	
G.S. Hulls	Bag	393	1,179.00	
Kwik Seal	Bag	7	61.25	
Mil-Cak-Dl	Bag	4	16.00	
Mil-Fiber	Bag			
Mil-Flakes	Bag			
Milmica	Bag			
Drispac	DM	30	1,782.00	
Chek Loss	Can	2	20.50	

ENGINEER'S REPORT

SURF PIPE 8-5/8" FT. 825 DATE 6-23-67

	Ft.	Date
Intermediate		
Casing 5-1/2	7400	7-9-67
Mud Up Depth At 3500	DR 6000	
Total Depth and Formation	7535	
Days on Mud AT 4	DR 9	
Date Completed	7-9-67	
D.S.T. Depth and Formation		

LOST CIRCULATION ZONES
(1)	2900-50 Bbl.
(2)	
(3)	
(4)	
(5)	
(6)	Cores
(1)	
(2)	7256-7272 Morrow
(3)	7287 Chester

DRILLING MUD RECORD

DATE 1967	DEPTH FT.	WEIGHT LB/GAL.	SEC./QT.	PLASTIC VISC. CPE	YIELD POINT CPE	GEL IN.	GEL 10 MIN.	C.C.	CAKE 32ND	SAND %	% OIL	% SOLIDS	pH	SALT PPM	CALCIUM PPM	SULPHATES PPM	PF	LCM #/BBL	PRES. lbs/bbl	REMARK NUMBER	TREATMENT
6-22	1065	10.5	34							3.6			6.6	100,000	Hvy	Hvy	0	1			
6-23			Ran 82.5'	8-5/8 WOC																	
6-24	2233	9.8	29	3	0	0	0	NC		.5			9.0	95,000	Hvy	Hvy	0	0			
6-25	2888	9.9	30	4	2	0	1	NC		.6			6.7	80,000	Hvy	Hvy	0	0	AT	
6-26	3517	9.7	35	8	10	7	9	100	4	3.5	2.5	9	6.6	68,000	Hvy	Hvy	0	2	4	
6-27	4403	9.6	35	9	10	8	9	82	4	2.9	4.5	9.5	6.6	55,000	Hvy	Hvy	0	3	3	
6-28	5040	9.5	39	11	14	11	13	74	4	5	5	9.5	6.6	44,000	Hvy	Hvy	0	3	3	
6-29	5655	9.5	36	12.5	12	6	8	80	4	2	4	9	6.7	36,000	2210	Hvy	0	4	2	
6-30	6100	9.3	39	13	18	8	9	86	4	3	3.5	7.5	6.6	35,000	2000	Hvy	0	4	1	2	DP SA UC CS
7-1	6687	9.5	40	15	10	8	12	12	1	2.9	1	6.5	8.5	31,000	300	Med.	.15	2	2	14 10 8 8
7-2	6873	9.7	43	11	8	8	9	12	1	3	4	9.5	7.1	45,000	800	Med	0	4	2	3	4 2
7-3	7072	9.7	45	13	9	8	9	10	1	2.6	4	10	6.8	46,000	1200	Med	0	4		5 4
7-4	7300	9.8	50	18	12	8	14	10	1	3	5	8	8.0	44,000	600	Med	0	6		2 5 4
7-5	7373	9.8	54	22	16	12	16	9.9	1	4	4	9	6.9	44,000	700	Med	0	4	3	3 3 7
7-6	7535		Ran Logs 2nd DST																3	
7-7	7535	9.7	32	19	12	8	11	11.4	1	3	5	10	6.7	50,000	1000	Med	0	4	4	
7-8	7535	Ran DST #4 3100 laid down drill pipe																		
7-9	7535	Ran 5-1/2" pipe																		

REMARKS: (1) DOWN 12 HR. MOTOR REPAIR

(2) RAN CONE OFF BIT – LOST ALMOST 30 HRS.

(3) RAN DST – NO PROBLEM.

(4) RAN LOGS – HIT ONE BRIDGE ABOUT 1100 AFTER THAT NO PROBLEMS.

(5) RAN DST WITH 7" RUBBER. PULLED ONE OFF. DST #2 OK. DST #3 – NO PROBLEMS.

TOPS:

TONKAWA	5612
MARMETON	6191
MORROW	7080
MORROW SD	7243
CHESTER	7287

FIG. 4–5. Drilling mud recap

mud costs including drainage are provided along with an itemized statement of each additive, the amount used, and the cost. Note for example the 28 units of detergent costing $1411.20 were used. This amount represents 25% of the total mud bill and hardly seems justifiable.

In addition, a high yield bentonite was used costing more than twice the regular bentonite. This practice hardly seems necessary. Also note that 31% of the mud bill is represented by an additive to control filtration. In view of the present attitude toward filtration control, this item could be reduced or eliminated. These are but a few ways in which the mud recap can aid in planning and cost reduction.

In addition to the cost data, a portion of the recap contains casing points, mud up depths, and lost circulation zones. Also summarized are the daily mud properties, hole problems, logging problems, and DST problems. All these data serve to familiarize the Drilling Engineer with the current mud practices and enables him to begin to formulate ideas for improving the system and reducing costs. Admittedly, a $3000.00 savings in mud on a $75,000.00 hole represents only 4% of the total expenditure and may seem insignificant to some observers. However, for each one hundred well program, this per well savings represents four additional wells capable of multiple earnings. It must be concluded that no savings is too small to be overlooked.

Another source of information is the scout ticket, Figure 4–6, which is available within most companies as well as commercially. Scout tickets are a particularity good source for determining the control wells productive horizon(s) and the initial potentials. Drill stem test data and pressure build up data are often included and useful in establishing pore pressures, mud weights, and casing design criteria. In addition, completion intervals and procedures are presented for those Drilling Engineers concerned with completions.

Daily drilling reports are the most valuable source for detailed time and cost analysis. Also, problems may be studied in depth. For example, fill on bottom after trips will aid in judging the effectiveness of the fluid program. Total trip time from the daily reports will allow the calculation of surge pressures. Again, the daily drilling report is the only source of information detailed enough to allow in depth study of the drilling problems and hazards which is a vital area to the success of the Drilling Engineer. Operators generally share this type of information.

Armed with all of the information provided by the presented sources, the Drilling Engineer is in a position to familiarize himself with the past drilling practices in the area of interest. All too often the Drilling Engineer stops at this point, accepts current practices, and prepares the drilling program following the established procedures to the very letter.

The responsible Drilling Engineer and manager realize that improvements are always possible, expected, and most important, virtually necessary if exploration and development are to continue in any given area. It is then the responsibility of the competent Drilling Engineer to develop and maintain an expertise in all phases of drilling operations. Local problems and drilling conditions must be analyzed with respect to this broad ever changing expertise in order to develop the best possible drilling program.

STATE: OKLA.   COUNTY: COMANCHE
OPERATOR: WESTHEIMER-NEUSTADT CORP.
WELL: 1   FARM: RYAN
SEC. 23-3N-10W   LOC.: SE SE
POOL: WILDCAT (1 Mi S OF PROD)
ELEV.: 1197 GR   1210 DF   1212 KB
CONT.: NICHOLS DRLG (RT)
API NO.: 35-031-20204
FR: 1-20-69   SPUD: 2-27-69   COMP: 5-7-69
TD: 10,366   FORM: MC LISH

COMPLETION RECORD

DRY AND ABANDONED
TD IN OLD HOLE 10,366
TD IN SIDE TRACK HOLE 9870

SIDE TRACK HOLE IS 308.18 NORTH AND
   278.26 WEST OF SURFACE HOLE LOC

CASING RECORD

SIZE: 10-3/4 , DEPTH: 1427', SX: 700

LOG TOPS: _____ (DF 1210)
DETRITAL LM _____ 6973 (-5763)
HUNTON _____ 8102 (-6892)
SYLVAN _____ 8675 (-7465)
VIOLA _____ 8882 (-7672)
FAULT TO BROMIDE DENSE ____ 9560 (-8350)

1ST BROMIDE _____ 9652 (-8442)
FAULT TO TULIP CREEK _____ 9810 (-8600)
MO LISH _____ 10,155 (-8945)
TOTAL DEPTH _____ 10,366 (-9156)

SAMPLE TOPS: _____ (DF 1210)
UNCONFORMITY HUNTON _____ 8120 (-6910)
SYLVAN _____ 8658 (-7448)
VIOLA _____ 8910 (-7700)

(03-03) SPUD 2-27-69, 10-3/4" - 1427' - 700 SAX,
   DRLG 3012'
(03-10) DST #1 (GRANITE WASH) 3914-4040,
   OPEN 40 MIN, GTS/16 MIN, TSTM,
   REC 1320' GCMSW, ISIP 1620/40 MIN,
   IFP 1099, FFP 1427, FSIP 1554/40 MIN,
   DRLG 5246'
(03-17) DRLG 7147'
(03-21) DRLG 8000'
(03-31) DST #2 (DETRITAL LM) 7870-8008,
   OPEN 40 MIN, REC 100' MUD, IHH 3709,
   ISIP 176/40 MIN, IFP 22, FFP 35,
   FSIP 136/40 MIN, FHH 3682,
   DST #3 (UNCONFORMITY HUNTON)
   8115-8210, OPEN 40 MIN, REC 3000' SWCM,
   5200' SW, IHH 3838, ISIP 3744/40 MIN,
   IFP 3294, FFP 3694, FSIP 3744/40 MIN,
   FHH 3872, DRLG 9203 LM
(04-07) DRLG 10,065 SH
(04-14) TD 10,363 ~ SCHL. 10,366
(04-21) PB 8087 - 160 SAX, DRLG CMT at 8210
(04-28) WS at 8210, DRLG 8901 LM
(05-05) at 9870 CIRC TO LOG
(05-07) TD WS HOLE 9870, SHCL. NO LOGS
   CALLED IN WS HOLE, WS HOLE IS
   308.18' NORTH AND 278.26' WEST OF
   TOP HOLE LOC

   COMPLETED 5-7-69
   DRY AND ABANDONED

FIG. 4-6. *Scout ticket*

Successful planning requires the adoption of attitudes and practices in addition to those already mentioned. These are the "Do's and Don'ts of well planning and include the following:

1. Be a skeptic. It is our responsibility as Drilling Engineers to question drilling practices which are inconsistent with sound judgment or other experience in similar areas. For example, it is common to say that increasing the bit weight will not increase penetration rate; practice says it will. Another example: It is inconsistent to believe that low water loss is necessary to drill the Springer Shale in Southern Oklahoma when troublesome shales are being drilled each day without water loss all over the world.

2. Develop expertise in every phase of drilling. Learn all you can. Learn every day about drilling practices and drilling technology from the drilling rig to the research laboratory—from the bottom of the bit to the top of the crown. If you are a young engineer, spend as much time as possible on the rig. If you're an experienced engineer, don't close your mind to new technology or the experience of others.

3. Establish realistic objectives and avoid conjecture. Stick to facts, data, and statistics. Don't allow anyone to explore the improbable "ifs." If the data support the conclusion to run 1000 feet of surface pipe, there is no justification for running 1100 feet, just to be safe.

4. Don't do anything simply because it's the established routine. Mud viscosity doesn't necessarily have to increase with depth. Recently a contractor, in using a gel-water system, was observed adding gel at the suction and Quebacho at the flow line. This practice is inconsistent with good drilling practices.

5. Time is the most important factor. All efforts should be directed at reducing time.

6. Attack general practices in view of new technology. In one area, costs were reduced from $225,000.00 per well to $190,000.00 per well by merely reducing hole sizes and simplifying completions.

7. For real savings attack the hazards. Attack the abnormal pressure problems, the deviation, the lost circulation, or the pipe sticking problems.

8. Support conclusions and recommendations with data, analysis, and calculations. Idle conversation is worthless, meaningless, and dangerous. Our responsibilities go beyond *telling* management or *telling* operations the solutions to drilling problems. We must *show* and *support* with the best data available.

9. Follow-up and honestly evaluate your efforts. Report success and generate pride in your company efforts. A raging forest fire, begins with a tiny spark.

The material that follows is an example of a drilling program. It is submitted as a guide to illustrate the principles discussed and not as a panacea. This is the minimum to be considered in preparing a well plan.

A DRILLING PERFORMANCE
PROGNOSIS
WESTHEIMER-NEUSTADT
CORPORATION'S
COMANCHE COUNTY,
OKLAHOMA, PROSPECT

CONTENTS

A DRILLING PERFORMANCE PROGNOSIS
WESTHEIMER-HEUSTADT CORPORATION'S
COMANCHE COUNTY, OKLAHOMA, PROSPECT

Introduction

The object of this investigation is to determine previous performance in the area of the prospect, the anticipated drilling environment, and the potential of improved drilling technology.

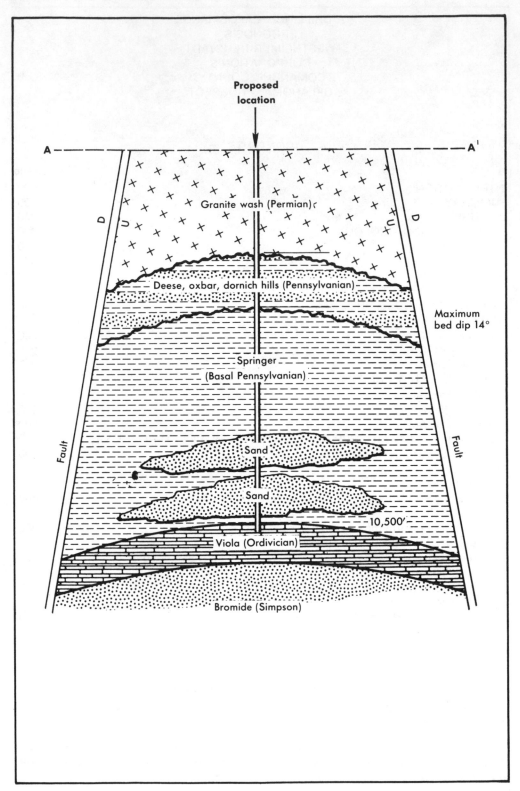

Summary Figure A

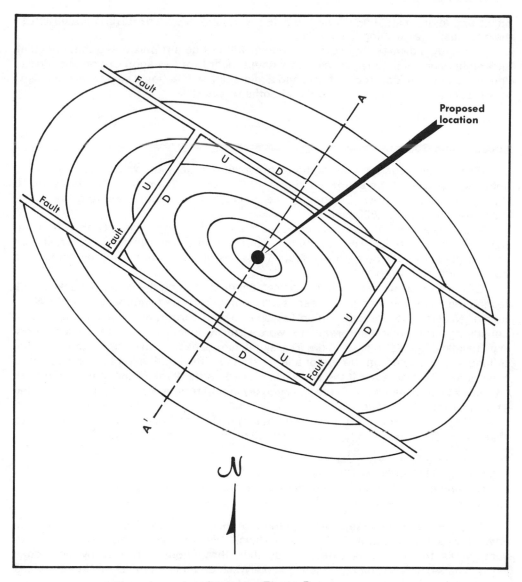

Summary Figure B

Summary

Westheimer-Neustadt Corporation's prospect is a geologic wildcat and is located in the southeast quarter of the southeast quarter of Section 23, Township 3 North, Range 10 West, Comanche County, Oklahoma. The objective is Springer Sand and the anticipated total depth is 10,500 feet (see accompanying Summary Figures A and B).

Control is very poor since the control wells were drilled in the early 1950's; however, extensive study indicates and Figure 5 in the Appendix illustrates that 97 days would be required to drill to a total depth of 10,500 feet if popular techniques of Southern Oklahoma were employed. It is predicted that a low solids, low weight, non-dispersed drilling fluid combined with carbide insert bits would reduce the drilling time to 70 days (Figure 5) or a reduction of 27.8%. It is also illustrated in Figure 5 that a successful air operation to 6000 feet would reduce the drilling time to 62 days. A discussion of the potential of air drilling is presented.

The study indicates that severe deviation will not be a problem and that optimum bit weights and rotary speeds selected without regard for deviation or potential deviation combined with deviation control and a minimum of directional surveys are justified. This practice has reduced drilling days in other areas 50%.

Discussion

Location and Geology

Westheimer-Neustadt Corporation's prospect is a geologic wildcat located in the southeast quarter of the southeast quarter of Section 23, Township 3 North, Range 10 West, Comanche County, Oklahoma. The objective is Springer Sand and the anticipated total depth is 10,500 feet.

The prospect is located (see enclosed map) in a large regional Graben trending northwest to southeast. The structure is bounded on the northwest by a thrust fault and on the southeast by a normal fault. The sides of the Graben form the boundaries on the north and south.

Extensive seismic exploration has been conducted using the latest in stacked pack techniques. The records have been studied and evaluated independently by Westheimer-Neustadt Corporation, Union Texas Petroleum Corporation, and an independent consulting firm. Their interpretations were virtually identical and are available from Westheimer-Neustadt Corporation. The significant aspect of their findings is that the drilling site is located on the top of a large structure resulting in flat bedding planes to total depth. Further, even if the site were located on the flank of the structure, dips of 14° or less would be encountered. It is equally important that no faulting was indicated at or near the drilling site.

0–6,000' — Control is good to 6,000 feet and Granite Wash should be present to that depth. This interval is typified by the following wells (logs enclosed):

Sunray Oil Company — Bentley No. 1
Sunray Oil Company — Cline No. 1
Continental Oil Company — Mansel No. 1 & No. 2
Atkinson — Brooks No. 1

This interval is generally shaley in the top portion and develops into sands in the lower portions. The Bentley, Cline, and Mansel Wells illustrate this development. In some areas, the interval is a dirty, sandy shale throughout as typified by the Brooks No. 1. The sand development in any Granite Wash area is an unknown; therefore, it is not possible to predict the extent of sand development in the prospect.

6,000'–8,700' — It is anticipated that a major unconformity will be encountered at approximately 6,000 feet between the Granite Wash section and the Hoxbar, Deese, and Dornick Hills Interval. The significance of the unconformity is that it is impossible to determine the extent of erosion; therefore, it is impossible to predict the zones to be penetrated. To further complicate the issue, this interval is similar to the first in that sand development is erratic and difficult to predict or correlate; therefore, it is only possible to typify the interval anticipated. Operator's personnel recommended the Coline Oil Company, Sessums No. 1 (log enclosed) as being most typical of the Hoxbar, Deese, and Dornick Hills section expected.

8,700'–10,500' — Another major unconformity is anticipated at 8,700 feet between the Hoxbar, Deese and Dornick Hills and the Springer Interval. Characteristically, the degree of erosion is unknown and unpredictable. As the other intervals, the Springer Interval is shale with erratically developed sands; therefore, the anticipated interval may be only typfied, and operator's personnel recommend the Texaco, Incorporated, Carr

No. 1 (log enclosed.). As the enclosed cross-section indicates, total depth will be in the Springer.

Since the degree of erosion is unknown and unpredictable, the geophysicists agreed that the entire Springer section may be eroded. If so, the seismic interpretations indicate that Viola will be encountered at 8,700 feet (again refer to cross-section); therefore, the Viola and Bromide sections would be penetrated. This prognosis is based on the premise that Springer will be encountered at 8,700 feet.

Previous Drilling Performance

Bit records, electric logs, and completion cards on all control wells previously referred to are enclosed. The previous performances of the control wells are illustrated in Figures 1, 2, 3, and 4. A composite based on the control wells is presented as Figure 5.

As illustrated in Figure 5, 97 days should be required using techniques common to Southern Oklahoma; however, it is significant that high mud weights with high viscosities were used in drilling these wells. Further carbide insert bits had not reached their present stage of development, and low bit weights were run in fear of severe deviation problems.

Casing and Hole Program

Operator's outline specifies 100 feet of $13\frac{3}{8}$ inch conductor casing. It is a common practice to drill a small hole to surface casing point and ream to the desired size; however, in view of the extensive seismic information and the fact that none of the control wells experienced deviation problems, it is recommended that a $12\frac{1}{4}$ inch hole be drilled to 1,400 feet. An $8\frac{3}{4}$ inch hole will be drilled to total depth.

Hazards

Abnormal Pressures — There is no information available from the control wells that would indicate the presence of abnormal pressures; however, blowout preventers and choke manifolds should be used from under conductor to total depth. Pressure recordings on the six pen recorder along with pit volume fluctuations should provide ample time to control abnormal zones if any are encountered. Crews should be cautioned to keep the hole full at all times during trips. Maximum pressure on the blowout preventers is 100 psig on the surface hole and 1,400 psig on the production hole. Excess pressures should be relieved through the choke manifold. Barite should be stored at the location, pending its possible use; however, it should be used only as necessary.

Hole Problems — No hole problems are anticipated until the Springer Shale is encountered. Hole enlargement is common in the Springer shale.

Deviation — As illustrated on the accompanying logs and bit records of the control wells, none experienced severe deviation. Further, extensive seismic information indicates faulting and high angle beds do not exist. The best information available indicates that no deviation problems should be encountered. Therefore, it is firmly recommended that this well be drilled as if it were not in crooked hole country until information to the contrary is obtained.

Drilling Fluid Program

Mud Program — It is recommended that a low solids, low viscosity, low weight, non-dispersed drilling mud be used to total depth. Mud properties must be controlled and dictated by hole conditions.

It is very important that the mud weight be maintained at an absolute minimum (8.3–9.0 ppg). A desander and desilter should be used as mechanical aids while water should be the primary control. Gel and Ben-Ex should be pre-mixed and added to the system to further control solids and weight. Viscosity should be controlled with the pre-mixed Bentonite which will result in maximum control with the minimum increase in solids. Pre-mixed Bentonite should be used to build viscosity and control hole problems as long as practical. Chemical dispersants should be avoided if possible since dispersion causes a system to lose its lifting capacity requiring increased solids for sufficient lift and resulting in increased mud weight which decreases in penetration rate. The estimated potential of better drilling fluids is illustrated in Figure 5.

Air Drilling — The feasibility of drilling with air through the Granite Wash Interval was evaluated. As previously described in the geologic section, this interval is very erratic and sand development is unpredictable. Five drill stem tests were run in the Granite Wash section of the control wells and only one recovered fluid. Log analysis in view of these drill stem test results indicate that, although the Granite Wash Sands might be present, they will probably be impermeable. It is my opinion that a successful air operation has approximately a fifty-fifty chance of success. The potential of a dry air operation is illustrated in Figure 5. If water is encountered and mist drilling is required (which is probable), the curve for dry air will approach that for low solids mud.

The use of air in this operation is not recommended. In the general case air operations are successful on a development scale, and random use is generally marginal to unsuccessful. Further, the tendency is to go to extremes in mud properties when mudding up which can be detrimental to the success of the remaining portion of the hole. Also, nippling up and mudding up require more time in primary efforts. Refinement of air drilling practices in a development area result in the economic advantages classically associated with air drilling. This does not mean to imply that air drilling cannot be successful in primary efforts; however, it does mean that an outstanding engineering and operating success can be economically marginal. Therefore, since the development potential is small, air operations are not recommended.

Bit Program

Since the control is poor and intervals can only be typified, it is not feasible to make specific recommendations concerning bit types; however, general observations can be made. Bit selection should be made without regard to deviation or potential deviation.

Analysis of the control wells (see logs on control wells) indicates that regular rock bits would be the proper choice in the shaley portion of the Granite Wash. If the sands develop in the lower portion, insert bits might prove economical. If developed as anticipated, insert bits appear feasible in the entire Hoxbar, Deese, and Dornick Hills section. However, control wells indicate that long shale sections may be encountered in which case rock bits would be more economical. A long shale section in the Springer should precede any sand development. Rock bits should be economical in the shale while insert bits appear profitable in the sands. If the erosion has been severe and Viola Limestone as opposed to Springer Shale is encountered, insert bits are recommended.

In air operations, regular rock bits should be used on the surface hole. Insert bits have proven to be more economical on the holes smaller than 9 inches.

Bit Weight, Rotary Speed, Deviation Control, Directional Surveys, and Hydraulics

The greatest potential savings are in optimizing bit weight and rotary speed without regard to potential deviation (50,000–80,000 pounds on $12\frac{1}{4}''$, 40,000–70,000 pounds on $8\frac{3}{4}''$ rock bits and according to manufacturer's specifications on carbide insert bits). Although deviation is not considered a hazard, a square drill collar is recommended to prevent severe dog-legs and directional surveys are recommended at 500 foot intervals. The same hydraulics program used on the Samedan Wilson (2000–2300 psig pump pressure) is applicable.

The potential of this approach in areas normally associated with severe deviation is difficult to evaluate due to the poor records which do not adequately describe the effect of the potential deviation; however, reductions in total drilling days of 50% and more have been realized in other areas where drilling practices are similar to those in Southern Oklahoma. If the normal mode of operations were followed on the control wells, the potential is as described and illustrated in Figure 5.

APPENDIX

(Note: Appendix consists of the five figures, Figure 1 thru Figure 5.)

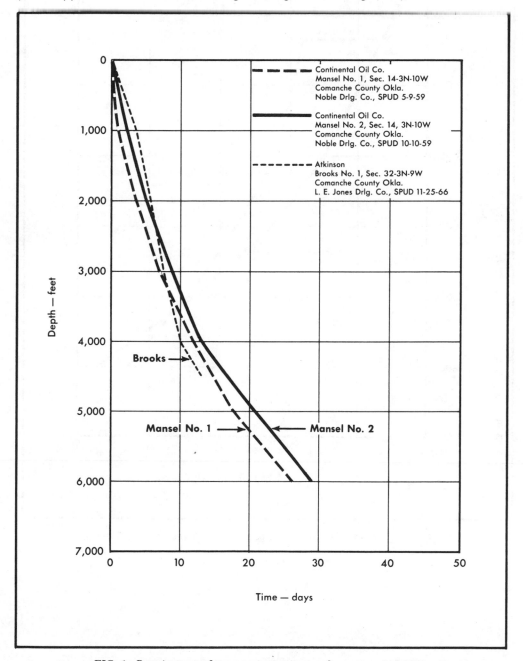

FIG. 1. *Previous performance granite wash section 0–6,000'*

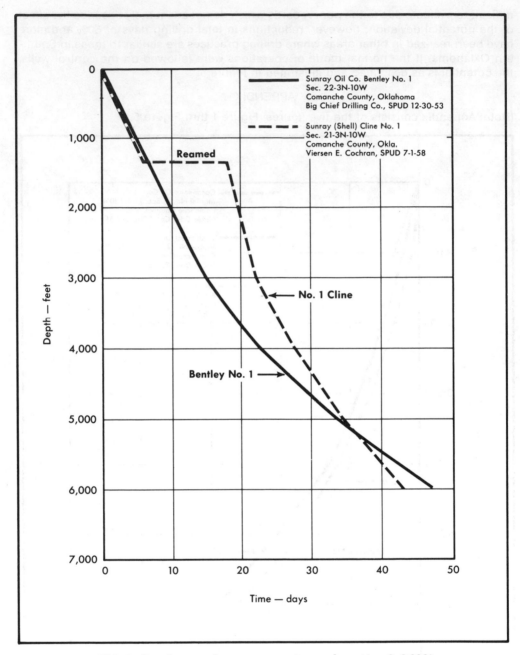

FIG. 2. *Previous performance granite wash section 0–6,000'*

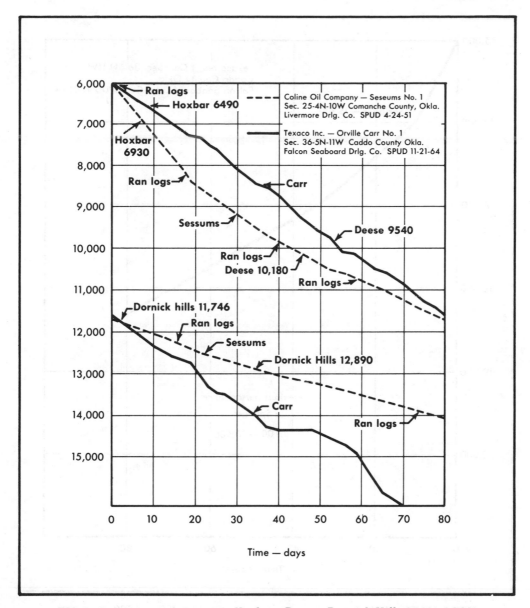

FIG. 3. *Previous performance — Hoxbar — Deese — Dornick Hills 6,000'–8,700'*

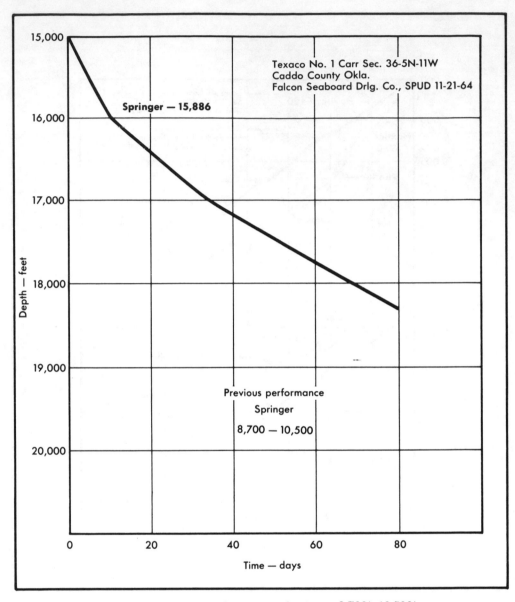

FIG. 4. *Previous performance – Springer 8,700'–10,500'*

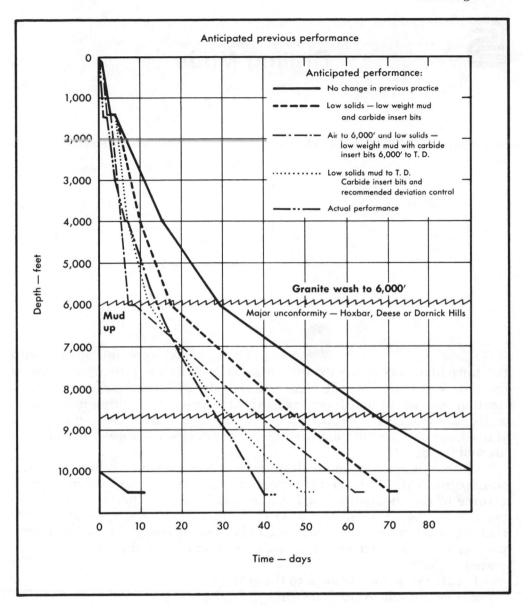

FIG. 5. *Anticipated previous performance*

5 Drilling Muds

$\mathbb{D}$RILLING muds were introduced with the introduction of rotary drilling in 1900. Initially the primary purpose of the mud was to remove cuttings continuously. With progress, more sophistication was added and more was expected from the drilling mud. Many additives for almost any conceivable purpose were introduced and what had started out as a simple fluid became a complicated mixture of liquids, solids and chemicals.

In this discussion the drilling mud will be considered first from the standpoint of the functions it serves in rotary drilling and second a brief resume of the chemistry of the system will be given. Then the basic composition of the mud will be considered, followed by the specific controls that are used in field applications and last the trends in special mud compositions will be reviewed. The primary functions of the mud are enumerated as follows:

1. Lift formation cuttings to the surface.
2. Control subsurface pressures.
3. Drill string lubrication.
4. Bottom-hole cleaning.
5. Provide an aid to formation evaluation.
6. Provide protection to formation productivity.

LIFT FORMATION CUTTINGS

Cleaning the hole is an essential function of the mud. This function is also the most abused and misinterpreted. The drill solids generally have a specific gravity of 2.3 to 3.0 and an average of 2.5 will be assumed. When these solids are heavier than the mud being used to drill the hole, they will slip downward through the mud. Specific considerations of factors affecting the slip velocity of the cuttings will be considered in the chapter on lifting capacity.

The slip velocity of the cuttings, while the fluid is in viscous or laminar flow, will be affected directly by the thickness or shear characteristics of the mud. Thus, many times when the annular mud velocity is limited by pump volume or enlarged hole sections, it is necessary to thicken the mud to reduce the slip velocity of the formation cuttings in order to keep the hole clean.

A water base mud may be thickened by adding bentonite, by adding large quantities of drill solids, by flocculation of the solids or by the use of special additives. Thus the operator is left with several alternatives and the alternative he selects will depend on the other objectives he is trying to accomplish. Bentonite is a cheap alternative; however, if the operator is also faced with water loss control problems he may be forced to add a thinner to prevent flocculation.

The net result may be a compromise that leaves very little increase in lifting capacity and the hole problem may remain. Adding large quantities of solids will also increase mud weight and the latter alternative, if not needed to control subsurface pressures, is undesirable. Planned flocculation of the mud is a cheap and easy alternative, but again this is controlled by limitations that are placed on filtration control. Special additives are acceptable alternatives but they increase the mud cost.

It can be seen that what appears to be a simple decision on thickening the mud to increase lifting capacity is complicated by the resultant effects the thickening method may have on other objectives. The specific procedures have to be determined by considering the actual problems at the time they occur.

CONTROL SUBSURFACE PRESSURES

A complete discussion on pressure control is given in the chapter on pressure control. Several problems are introduced in the selection of mud weight to control subsurface pressures. It is desirable to utilize minimum mud weights. As mud weight is decreased, drilling rates are increased and lost circulation problems are minimized. Actual mud weight requirements in the normal range vary in different sections of the country. Fresh water frequently provides adequate weight to control subsurface pressures. In Coastal areas a mud weight of 9.2 ppg may be required.

Abnormally high formation pressures require careful measurements of pore pressures to determine mud weight requirements. Even a careful measurement of pore pressure may result in wide variations in mud weights used by companies drilling wells in the same areas. This simply means their interpretations were different and some emphasis should be dedicated to a determination of why the differences exist. A discussion of mud weight control will be given in a latter section of this chapter.

DRILL STRING LUBRICATION

Lubrication and cooling have become important functions of the mud. The life of expensive equipment can be prolonged by adequate cooling and lubrication and hole problems such as torque, drag and differential pressure sticking are related directly to lubrication. Lubricants include, bentonite, oil, detergents, graphite, asphalts, special surfactants and walnut hulls. Bentonite acts as a lubricant only because it is slick when wet and reduces friction between the wall cake and the drill string.

Oil has been used as a lubricant in drilling muds for many years. Recently new limitations have been placed on the use of oil because of the associated disposal problems. Detergents have gained some attention as lubricants. Reported successful applications are regional and field evaluations are always necessary. Graphite is a good lubricant but generally must be added with oil to be effective. It is used most often in areas where high rotating torques are a problem.

Asphalt has been a mud additive for a large number of reasons. In oil base muds it is often used as a low gravity solids phase. It has also been used to help stabilize shales by a claimed process of plugging off micro-fractures in the shale, and less frequently it has been used as a lubricant. Special surfactants claimed to act as lubricants are being introduced by many companies. Success claims are regional. Most surfactants are expensive and their benefits should be carefully analyzed. Walnut hulls and similar products help reduce torque or drag; however, their use can become very expensive because in general they must be added in quantity every day.

BOTTOM-HOLE CLEANING

In general, bottom-hole cleaning is improved by having thin fluids at high shear rates through the bit. This means that viscous fluids can be potentially good fluids for bottom-hole cleaning if they have good shear thinning characteristics. A more detailed discussion of this phenomena was included in the chapter on cost control. In general, a fluid with a low solids content is the best fluid for bottom-hole cleaning.

AID TO FORMATION EVALUATION

Drilling fluids have been affected substantially by formation evaluation requirements. Viscosity has been increased to obtain better cuttings, filtration rate has been reduced to minimize fluid invasion, special fluids have been selected to improve logging characteristics, and muds have been changed to improve formation testing. Most of those precautions are taken more from habit than necessity. Oil base muds make it difficult to evaluate potential producing horizons. Salt water fluids make it difficult to use a self potential log to recognize permeable zones.

Thick filter cakes may make it difficult to obtain information from sidewall coring. Water or oil invasion certainly affects resistivity. In many cases the methods of evaluation are changed, in others the methods of measurement are not indicative of down hole conditions. Thus the emphasis placed on fluid selection and treatment in this area needs to be reviewed in detail.

PROTECT FORMATION PRODUCTIVITY

Formation productivity is of major concern and very often non-commercial hydro-carbon zones are blamed on formation damage introduced through the invasion of mud or filtrate. There is little doubt that it would be desirable to keep the down hole formation in its virgin state with no fluid of any kind entering the zone. In drilling this in general cannot be done. In some areas of West Virginia and Kentucky, productive zones are drilled with air in order to keep liquid off the formation.

This practice has been effective in maintaining formation productivity. Also in many areas, the productive horizons are drilled using an oil base mud to keep water out of the zone. This practice has also been effective; however, in gas zones it may be more damaging than a salt water fluid. Salt water and high calcium content fluids have also been used to minimize formation damage. Again to some degree these fluids have also been effective.

Even today, many companies reduce the filtration rate to very low values, below 10 cc API, to minimize filtrate damage in the pay zone. In truth this practice is not generally effective in reducing filtrate invasion and the biggest effect is the cloak of protection it provides for the practicioner. Down hole filtration rates cannot be determined by static filtration tests at the surface; particularly at very low values. More detail will be included in the section on filtration rate control.

COMPOSITION OF THE MUD

Drilling mud consists of liquids and solids. Liquids may be mixed or varied and various types of solids may be intrained in the mud. In addition very often chemicals are added. Table 5–1, shows the composition of mud.

TABLE 5–1
Liquid and Solid Composition of Mud

Liquid	Solids
1. Fresh water	1. Low gravity–Specific
2. Salt Water	gravity = 2.5
3. Oil	a. Non-reactive, sand, chert,
4. Mixtures of these fluids	limestone, some shales
	b. Reactive solids, clays
	2. High gravity
	a. Barite – specific
	gravity = 4.2
	b. Iron Ore and Lead
	Sulfide – specific
	gravity = 7.0

Fresh water is the base of most muds. Fresh water is generally accessible, cheap, easy to control even when loaded with solids and provides the best liquid for formation evaluation. Salt water has become more common because of its accessibility in offshore operations, which are expanding. Table 5–2 shows the salt required for different stages of saturation beginning with fresh water.

TABLE 5–2
Salt Conditions at 68°F

Mg/L	Salt Added lb/bbl	Percent Salt	Weight of Solution lb/gal
10,050	3.53	1	8.39
20,250	7.14	2	8.45
41,100	14.59	4	8.57
62,500	22.32	6	8.69
84,500	30.44	8	8.81
107,100	38.87	10	8.93
130,300	47.72	12	9.06
254,100	56.96	14	9.19
178,600	66.65	16	9.31
203,700	76.79	18	9.45
229,600	87.47	20	9.58
256,100	98.70	22	9.71
279,500	110.49	24	9.85
311,300	122.91	26	9.99

Table 5–3 shows the effect of temperature on the saturation level of sodium chloride.

TABLE 5–3
Effect of Temperature
on Sodium Chloride

Temperature °F	Salt to Saturate lb/bbl
80	127
120	129
160	132
200	137

It is noted that the salt required for saturation increases with temperature. Thus, it is not uncommon for salt to be deposited in surface tanks when a saturated salt water mud cools. If salt beds are to be drilled, it is frequently a practice to saturate the water before entering the salt sections. This should minimize the total hole enlargement providing the mud is stirred vigorously at the surface as it cools. In some cases operators simply let the salt from the salt beds saturate the water.

There are many disadvantages to using salt. Mud costs are higher, bentonite yield is reduced, formation evaluation methods are not as effective and potential corrosion problems may be increased. Mud costs go up because of the presence of the electrolytes such as salt, in the mud which reduces the benefits derived from most additives. A reduction in the yield of bentonite often results in poor filter cakes. It has also been noted that hole problems are frequently more severe.

Formation evaluation is affected primarily because of the reduced definition given by the self potential log opposite permeable formations. The self potential log is generated primarily by an electrochemical reaction between the mud filtrate and formation water and this is shown in Equation 1:

$$\text{Self Potential} = -k \log R_{mf}/R_w \tag{1}$$

If the mud filtrate contains the same amount of salt as the formation water, R_{mf} will equal R_w, and the log of one is zero. Thus, there will be no self potential generated. In fact this concept has been used in reverse to obtain potential logs in areas where formation waters are close to fresh water. In this case, salt has been added purposely and a positive self potential log is obtained. Corrosion is more severe with salt water because the electrolytes are good conductors of electricity and in addition it is difficult to raise the pH to the 10.0 level to minimize other corrosion problems.

Where salt sections are to be drilled, the only alternative to using salt water may be oil base muds. If salt water mud is to be used an effort should be made to minimize costs. In high weight muds, above 16.0 ppg, it may be difficult and expensive to maintain a high pH. Thus it may be cheaper to use a corrosion inhibitor that will coat the pipe. In addition if hole problems are severe, the operator may add pre-hydrated bentonite to the mud to improve the filter cake.

Some good effects may be realized from using salt water muds. The swelling of clays is reduced with an increase in salt content. Thus potentially productive formations containing swelling clays are damaged less by contact with filtrate. Also many shales may not heave or slough as much in salt water fluids as they will with fresh water fluids.

In application, salt water muds should be considered special purpose muds which are basically more expensive to use than fresh water muds. Specific applications may require that salt water muds be used and the operator should consider carefully the best and cheapest method of application.

Oil, as a base fluid, has been used for many years. Oil was probably used initially to protect potentially productive formations. Clays do not hydrate or swell in oil and formation damage in oil zones is minimized. The next recognized advantage for oil was to minimize hole problems. There are many recorded successful applications of oil for this purpose dating back more than 20 years.

Two claims were made for reducing hole problems: (1) the shale zones would not enlarge due to sloughing, and (2) the better lubrication would minimize torque and drag problems. Caliper logs until recently did not confirm the first claim and the second claim was proven readily. The only problem was that when the torque, drag and wall sticking problems were reduced, many operators also assumed the sloughing had stopped.

Because caliper logs were not run, they were not aware of the fact that sloughing had continued. It was discovered that hole enlargement continued when using oil base muds, because fresh water entrained in the oil base mud was pulled out of the mud into the formation, thus wetting the clays. The resultant effect was that the clays continued to slough. Mondshine[1] in 1966 suggested the use of calcium chloride in the emulsified water of an oil base mud to prevent the wetting of formation clays. Later in 1969 Chenevert[2] presented information on the effect of using sodium and calcium salts in the water phase of an oil base mud and introduced a simple technique to determine requirements.

Mondshine showed that if the water emulsified in the oil was fresh it could be pulled out of the mud into the formation. He also showed that if the formation water was less salty than the water in the oil base mud that water could be drawn out of the formation. Using salt in the water phase of oil muds, many times calcium chloride rather than sodium chloride, because of its greater solubility in water, proved successful in many field tests. Many shale sections that had exhibited enlargement in previous drilling were kept near bit size.

This type mud has not always been successful. Shale sections such as the running Gumbo sections in many coastal areas remain a problem. It has become common field practice to saturate or near saturate with calcium chloride the water in oil base muds without measuring the required quantities. This removes the need for field measurements to determine salt requirements and no adverse affects are noted on hole stability.

Oil muds have been very effective in reducing torque and drag and pipe sticking problems. The primary application for this purpose has been in deep directional wells where the hole size is small, less than 7 inches. Oil muds have found another application in the last few years. In general, they are more temperature stable than water base muds, particularly high weight muds, above 16.0 ppg. In deep wells where temperatures may exceed 350°F., oil muds are frequently used because they are more stable, provide better lubrication and the maintenance cost is about one-third the maintenance cost of comparable weight water base muds.

A word of caution is necessary in changing from water base to oil base mud. In a deep well with several thousand feet of open-hole, the operator is taking a big gamble in changing from a water base to an oil base mud. The open formations have generally adsorbed large quantities of water and this water becomes a contaminant in the oil base mud. Several days may be required to make the change and the operator is taking a chance on losing the hole in the conversion process. If oil base mud is to be used after using a water base mud the change should be inside cased hole.

Oil muds are considered special purpose fluids. The initial cost of oil base muds is much more than a comparable weight water base mud. Annular pressure losses, if both the water and oil base muds are treated properly, is generally higher with the oil base mud. This means the dangers of lost circulation are increased. Formation evaluation, particularly in exploratory wells, is made very difficult. In addition, the outcry on pollution makes the disposal of the oil muds difficult.

From the standpoint of operations, drilling rates using oil muds are generally considered to be 30 per cent less than for comparable weight water base muds. Specific tests are available in some deep wells where drilling rates with high weight oil muds are shown to be comparable to those with the same weight water base mud. One explanation for this deviation from the general case may be that with better lubrication the actual weight on the bit is higher using oil muds even though the surface indicator shows no change from the weight using water muds.

Many of the economic disadvantages of the oil base muds are overcome by (1) rental agreements where the mud is used and returned to the supplier and (2) the lower maintenance cost which over a period of extended use may more than repay the higher initial cost.

Solids

Solids, as shown in Table 5–1, are divided into two groups, low and high gravity. The low gravity solids are further divided into non-reactive and reactive groups. As the term infers, non-reactive solids are those that do not react to a change in environment. The low gravity non-reactive solids consist of sand, chert, limestone, dolomite, some shales and mixtures of many minerals.

These solids are in general undesirable and when larger than 15 microns in size, they may create an erosive environment which is detrimental to circulating equipment. The API classification for sand is any solid greater than 74 microns in size; however, many solids smaller than sand are detrimental to equipment. Table 5–4 shows solid sizes in microns, inches and screen sizes.

TABLE 5–4
Solid Sizes

Microns	Inches	Shaker Screen Size
1540	0.0606	12 × 12
1230	0.0483	14 × 14
1020	0.0403	16 × 16
920	0.0362	18 × 18
765	0.0303	20 × 20
		Test Screens
210	0.00827	U.S.S. No. 60
147	0.00579	U.S.S. No. 100
74	0.00291	U.S.S. No. 200
44	0.00173	U.S.S. No. 325

The reactive solids are the clays and the term reactive describes the action of these solids in water. Many definitions are given for clays. Some of these are enumerated as follows:

1. A solid with an equivalent diameter of less than 2 microns.
2. An electrically charged particle capable of adsorbing water.
3. A material that gives the appearance of swelling when water is adsorbed.

Two types of clays will be considered: (1) the most common is sodium montmorillonite, commonly called either bentonite or gel; (2) the other is attapalgite, commonly called salt gel.

Actually, bentonite as marketed is not pure sodium montmorillonite. It has been estimated that the best bentonite contains about 60 to 70 per cent sodium montmorillonite. The other 40 per cent might be calcium montmorillonite or other low yielding clays, such as Kaolonite. Sodium montmorillonite is a plate-like material, which is often compared with the pages of a book. The plates are thin and the total particle size may be less than 0.1 of a micron in size.

Hydration or swelling is accomplished primarily by the adsorption of water to the surface of the clay. The amount of swelling, observed by the measured increase in mud thickness or viscosity, that will occur depends on the available surface area and the total amount of water held on the clay. It has been theorized that a sodium montmorillonite plate would appear as shown in Figure 5–1 if none of the edges had been broken.

It is noted that the basic clay structure does not contain sodium, which is an adsorbed cation as shown in Figure 5–2. Many other cations may be adsorbed, such as aluminum, calcium, barium, potassium, hydrogen and magnesium. Because these cations are adsorbed they may be exchanged by changing the environment of the clays.

Clays with different adsorbed cations would look the same in appearance and the only way to differentiate is to establish standard tests. This has been done and bentonite must meet API specifications shown in Table 5–5.

Clay yield in Table 5–5 is defined as the number of barrels of 15 centipoise mud that can be obtained using a ton (2,000 pounds of dry clay). In this case the 15 centipoise viscosity is determined by dividing the 600 reading by 2 as taken from a standard size rotating viscometer. It is noted that the bentonite specifications require a clay yield of 91.8 barrels per ton. The normal laboratory test simulates field units by using 22.5 grams of bentonite in 350 cc of distilled water, which is comparable to 22.5 pounds of clay per barrel of water.

The plastic viscosity test is a fineness specification; yield point is a measure of clay activity, created primarily by the charged surfaces. The specification most over-looked is the one on water loss. Bentonite must have good filtration control characteristics, and surprisingly, bentonite is the best material used in the mud for the control of filtration rate.

These specification tests should be run in distilled water. As noted, the bentonite is reactive and its hydration qualities will be affected by a change in water composition. The yield of bentonite is reduced as the salt content, sodium chloride (NaCl), is increased. For example, there is no noticeable yield from bentonite in saturated salt water and the yield of bentonite is reduced by more than 50 per cent in sea water containing 50,000 ppm salt.

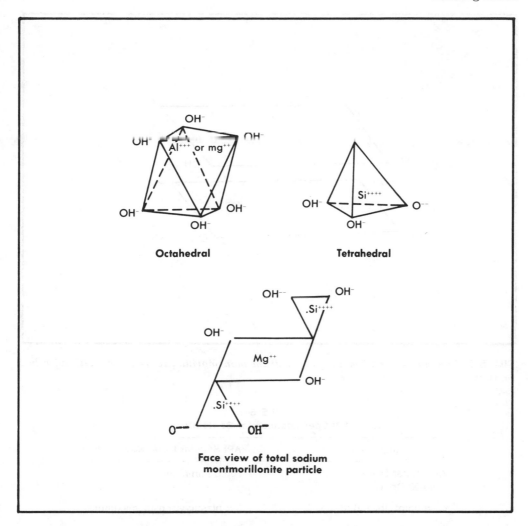

Octahedral   Tetrahedral

Face view of total sodium
montmorillonite particle

FIG. 5–1. *Schematic diagrams of sodium montmorillonite clay particle*

Exact levels of clay yield are hard to obtain as a function of salt concentration because of other impurities in both the clay and water. The effect of salt on clay yield is known, the exact causes are more difficult to explain.

Figure 5–3 shows a hypothetical sketch of the clay plates in a saturated salt water environment. All unsatisfied charges are satisfied with sodium cations. The bentonite particle becomes essentially inert. Failure of the bentonite particles to hydrate will result in very little increase in mud thickness and a minimal reduction in filtration rate as the concentration of bentonite is increased. To be effective, the bentonite needs to be pre-hydrated in fresh water before it is added to saturated salt water. Substantial benefits are also obtained by pre-hydrating bentonite before it is added to sea water. Two methods of pre-hydration are generally used:

    1. The bentonite is mixed in a tank of fresh water. The concentration will vary, but about 40 ppb is all the bentonite that can be added and

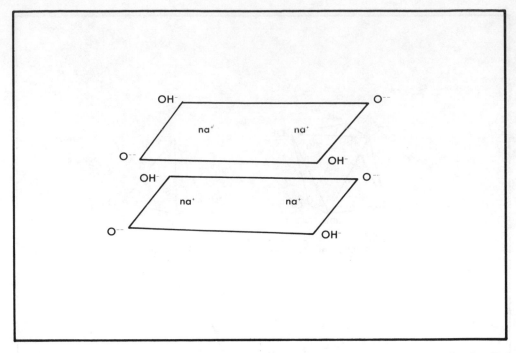

FIG. 5–2. *Schematic diagram showing sodium montmorillonite as plates with adsorbed sodium*

TABLE 5–5
API Specifications for Bentonite

Requirement	(API Standard 13A, March 1965)
Viscometer Dial Reading at 600 Rpm	30 cp minimum
Yield point, $lb_f/100 ft^2$	3 × plastic viscosity maximum
Filtrate	13.5 ml maximum
Wet screen analysis Residue on US sieve No. 200	2.5% maximum
Moisture	10% maximum as shipped from point of manufacture
Yield	91.8 bbls. of 15 cp mud per ton of dry bentonite

still obtain a pumpable slurry. It has been found that adding about one to two ppb of lignosulfonate after hydrating will prolong the effects of the pre-hydration. The time required for complete hydration may vary from 6 to 24 hours. Usually 6 hours provides an adequate time if there is some reason to hurry.

2. The bentonite is mixed with water by a pre-hydrator, which operates like a screw type pump. About 60–70 ppb can be mixed in this

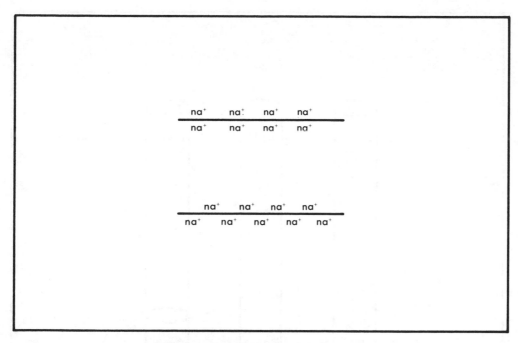

FIG. 5–3. *Bentonite clay in salt water*

fashion and advantages include faster mixing and less water dilution in weighted muds.

Disadvantages include: (1) the lack of knowledge on the actual degree of bentonite hydration, and (2) the difficulty in mixing this thick slurry into the active mud system.

The yield of bentonite is also reduced when other cations such as those of calcium, magnesium and potassium are present in the water. Calcium is often used in muds to prevent the swelling of formation clays. Muds of this type have been classified as lime, gyp and calcium chloride systems. Actually the lime treated muds had only small quantities of calcium in solution.

In the normal high pH lime mud, only about 150 ppm calcium was in solution. In the gyp muds, about 800 ppm calcium was allowed in solution. In the calcium chloride muds, as much as 3,500 ppm calcium was in solution. The effects of calcium on the swelling of bentonite is shown in Figure 5–4.

One reason for the effect shown in Figure 5–4 is hypothesized to be that the divalent characteristics of the calcium cation (Ca^{++}), prevent the separation of the clay plates, thus reducing the available surface area for water adsorption. Actually, the difference in soluble calcium in Figure 5–4 was primarily due to the pH level of the mud. Calcium solubility is affected by pH as shown in Figure 5–5. Lime muds have a pH of about 12.0 and the amount of soluble calcium was a direct result of this pH level. Almost any mud may be converted to lime mud by raising the pH and adding calcium if it is not already present.

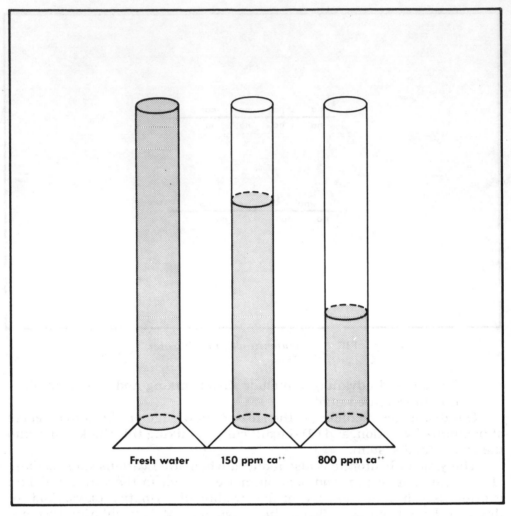

FIG. 5–4. *Effect of calcium on clay swelling*

All chemical reactions are affected by their environment. In many cases muds have all types of different materials present and this affects substantially chemical reactions. One of the most important measurable characteristics of muds is pH. For this reason, pH will be defined and examined in some detail.

pH

pH is defined as the negative logarithm of the hydrogen ion, H^+, content. This is shown as follows:

$$pH = -\log H^+$$

Further significance can be attached to this definition when it is real-

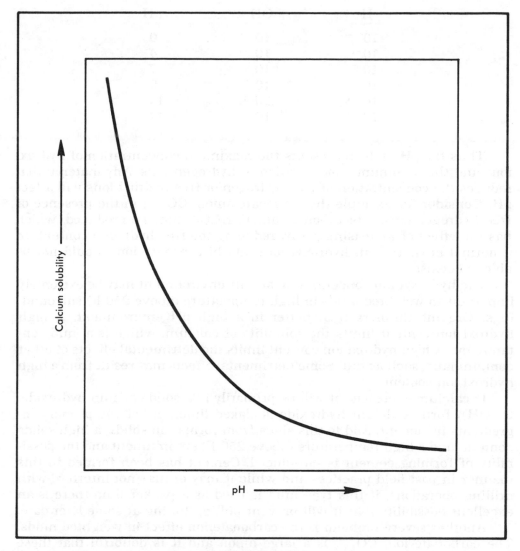

FIG. 5–5. *Calcium solubility vs. pH*

ized that the product of the hydrogen ion concentration and the hydroxl ion concentration is a constant. This is shown as follows:

$$H^+ \times OH^- = 1 \times 10^{-14}$$

The hydrogen ion represents the acidic component, the hydroxl ion represents the basic or alkaline component. Thus anything that reduces the concentration of the hydrogen ion would result in an increase in pH. A neutral solution, typical of pure or distilled water, would have the same concentration of hydrogen and hydroxl ions. This is shown as follows:

$$H^+ = OH^- = 1 \times 10^{-7}$$

For this type solution the pH equals 7. Other combinations are shown below:

H^+	OH^-	pH
10^{0}	10^{-14}	0
10^{-4}	10^{-10}	4
10^{-7}	10^{-7}	7
10^{-9}	10^{-5}	9
10^{-11}	10^{-3}	11
10^{-14}	10^{0}	14

Thus the pH of 14 represents the maximum concentration of hydroxl ions and the minimum concentration of hydrogen ions. Any material that reduces the concentration of free hydrogen or free hydroxl ions will affect pH. Consider for example the carbonate anion, CO_3^{--}; in the presence of free hydrogen ions: a bicarbonate anion, HCO_3^{-}, may be produced, which has the effect of increasing pH by reducing the free hydrogen content. In a normal environment, hydroxyl ions and bicarbonate ions would not be able to coexist.

The hydroxyl ion concentration and its environment may be extremely important in weighted muds in high temperature (above 250 F) surroundings. Organic thinners treat better in a high pH environment. A high hydroxl ion content limits the solubility of calcium, which is a mud contaminant. A high hydroxl ion content limits the detrimental effects of other contaminants such as salt. Some detrimental effects may result from a high hydroxl ion content.

If calcium is present, it will be primarily in a solid calcium hydroxide, $Ca(OH)_2$ form. Calcium hydroxide is slaked lime, one of the primary ingredients in cement. Add to this silica from formation solids, a high solids content, and a high temperature (above 250 F) environment and the possibility of forming cement is introduced. Cement has been formed in this manner in past field practices and while it may or may not interfere with drilling operations, if this type mud is used as a packer fluid there is an excellent possibility that it will prevent pulling tubing at some later date.

Another severe problem is the carbonate ion effect in weighted muds. The carbonate ion, CO_3^{--}, is a large anion and it is doubtful that these anions ever affect the clays directly. It is more probable that the carbonate anion gives a false impression of pH or alkalinity. This might have the effect of indicating the presence of hydroxl ions when their concentration was actually low. The result could be very high gel strengths and viscosity caused by poor thinning from organic thinners.

It is not uncommon for a fresh water mud to thicken when salt, calcium or some similar contaminant is added to the mud, either purposely or from formations being drilled. The mud thickens because of the flocculation of clays as shown in Figure 5–6. As noted, the unbalance of charges on the edges of the clays result in an edge to edge attraction. When this happens the mud thickens, because in addition to the adsorbed water, water is trapped between the flocs of clay as shown in Figure 5–6. The tendency for fresh water clays to flocculate is reduced appreciably as the pH is increased.

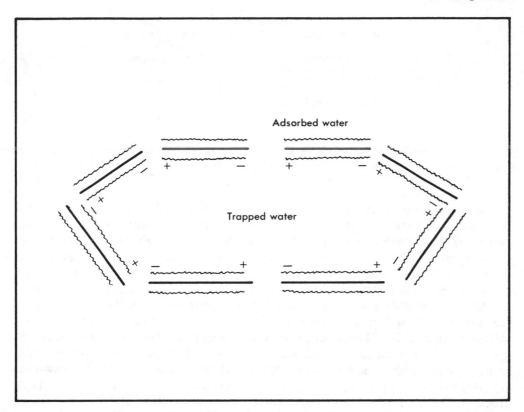

FIG. 5–6. *Clays with adsorbed water in flocculated state*

Salt Gel

Salt gel, an attapulgite clay, has often been called a needle-like material. It does not occur in plates like bentonite and yield or viscosity characteristics are obtained by vigorous agitation and cross-linking of particles. Actually, a salt gel mud is very similar in nature to a normal flocculated mud. The particles are cross-linked and water is trapped. There is essentially no water-loss control. Frequent additions of the salt gel are generally necessary to maintain a consistent mud thickness.

One question that arises is at what salt concentration should salt gel be used to replace bentonite. Bentonite is the preferred material always even in those cases where it must be prehydrated before being added to the salt water fluid. Salt gel is more expensive, does not have the lubricating qualities of bentonite, provides a poor filter cake and offers no aid to the control of filtration. Thus it is suggested that it should be used only on a special application basis.

High Gravity Solids

Barite is the high gravity non-reactive solid used to increase mud weight. Barite is primarily barium sulfate ($BaSO_4$). API specifications for barite are shown in Table 5–6.

TABLE 5–6
Barite Physical and Chemical Requirements

Requirement	Numerical Value
Specific Gravity	4.20, minimum
Soluble Alkaline Earth Metals as Calcium	250 ppm, maximum
Wet Screen Analysis	
Residue on U.S. Sieve No. 200	3.0 per cent, maximum
Residue on U.S. Sieve No. 325	5.0 per cent, minimum

It is noted that the specific gravity minimum is 4.2. This is important. Barites with a specific gravity as low as 3.9 have been detected in the field. Barite of this grade would make it almost impossible to run a 16.0 ppg mud or heavier without experiencing excessive viscosity problems. Not only does a low grade barite increase chemical thinner requirements, it may result in excessive circulating pressures in deep high temperature wells.

Other weighting materials have been used such as lead sulfide and various grades of iron ore. Some of these materials have specific gravities in the range of 7.0. These high gravity materials have been used only in special cases where high mud weights, above 22.0 ppg, were required to bring a well under control. They have not been used routinely because: (1) they are more expensive, and (2) the higher gravity would require a very fine grind or a more viscous base mud for collodial suspension.

In the past other iron ores of the same approximate specific gravity as barite have been used on a limited test basis. In many cases the cost was about the same; however, barite is preferred because: (1) the barite is essentially non-erosive, and all the iron ores are in general were erosive; and (2) the iron ores often contained toxic materials, whereas the barite is safe to handle.

CONTROL OF DRILLING MUD

Mud control can be divided into three basic categories: the control of mud weight, the control of viscosity and gel strength, and the control of water-loss.

Control of Mud Weight

Mud weight control is almost synonymous with the control of solids. In recent years, the emphasis on low solids mud was primarily an emphasis on low weight muds. The mud weight requirements should be based on that required to control formation pressures.

In coastal areas, normal formation pressures are generally in the range of 0.465 psi/ft. This equals a 9.0 ppg mud and thus to contain formation fluids during drilling operations at least a 9.2 ppg mud is required. In inland areas it is not uncommon to find normal formation pressures as low as 0.40 psi/ft which can be controlled adequately with fresh water. In some cases it may be necessary to use very low density fluids to maintain circulation. This has often been true in the Rocky Mountain areas of Utah where formations are fractured and formation fluid pressures are less than an equivalent 5.0 ppg level.

In general, the operator is encouraged to keep the mud weight just high enough to control formation pore pressures. As mud weight increases, drilling rates decrease, mud costs increase and hole problems become more common.

In fresh water, a 9.2 ppg mud has about 7 per cent low gravity solids by volume and a 10.0 ppg mud has about 13.5 per cent low gravity solids by volume. Figure 5–7 was calculated by using Equation 2.

$$M.W. = W.W. (1 - X) + 20.8 \, X \tag{2}$$

Where:

M.W. = mud weight, ppg
W.W. = water weight, ppg
X = solids fraction
20.8 = weight in ppg of low gravity solids

The use of Equation 2 is shown in Example 1.

■ EXAMPLE 1:

Assume: solids content = 0.07
all solids are low gravity
liquid base is fresh water
M.W. = (8.33)(I − 0.07) + 20.8(0.07) = 9.2 ppg

It should also be noted that the mud weight can be used to determine the solids content in an unweighted mud. Methods for controlling the solids content are given in the chapter on mechanical equipment for mud control.

While the equipment and techniques are available to control the mud weight going into the hole, the annular mud weight may become excessive. The increase in annular mud weight versus the rate of penetration is shown in Figure 5–8, where it is assumed the input mud weighs 9.0 ppg.

This annular mud weight in Figure 5–8 is based on the indicated drilling and circulation rates in a $12\frac{1}{4}$ inch hole and was calculated as shown in Example 2.

■ EXAMPLE 2:

$12\frac{1}{4}$ inch hole volume = 0.15 bpf
at 20 fpm drilling rate, 3.0 bpm of cuttings are generated
Weight of cuttings = 875 ppb × 3 bpm = 2625 lb/min

Volume of cuttings $= \dfrac{2625}{20.8} = 126.3$ gpm

Volume of input mud = 600 gpm
Weight of input mud = 600 gpm × 9 ppg = 5400 lb/min

Annular mud weight $= \dfrac{(5400 + 2625)}{(600 + 126)} = 11.0$ ppg

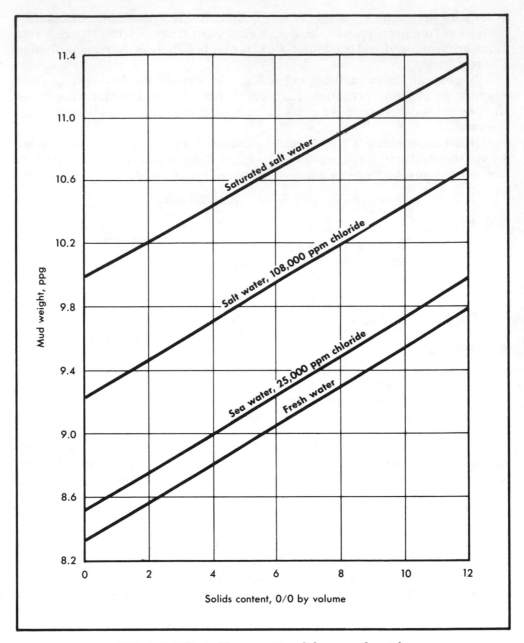

FIG. 5–7. *Effect of low gravity solids on mud weight*

It is noted that at the same drilling rate a reduction in circulation rate to 300 gpm would increase the mud weight to 12.4 ppg. Because these conditions are in a large surface hole, the friction loss change due to changing circulation rate would be small. Thus, to minimize the hydrostatic head in the annulus the operator has the choices of reducing the drilling rate or increasing the circulation rate.

It has not been uncommon to lose circulation in large surface holes

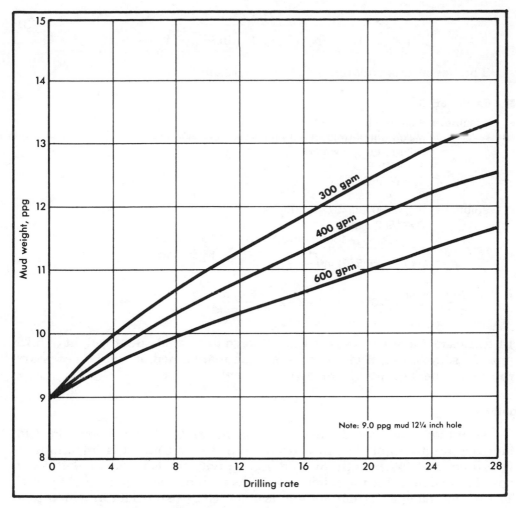

FIG. 5–8. *Mud weight increase vs. drilling rate*

while drilling at high rates. Because the mud weights are predictable and the fracture gradients can be determined, the operator should not exceed permissible limits.

Mud Weight Less Than Water Weight

Mud weight can be reduced to levels below a fresh water gradient by using oil, oil and fresh water, or by aerating the fluid column. An average weight for oil is about 7.0 ppg, thus a 50-50 mixture of only oil and water would weigh about 7.7 ppg. This assumes there are no solids. Where even lighter mud weights are required, the fluid column can be aerated. Equation 3 can be used to calculate air requirements for a given reduction in mud weight.

$$\frac{N}{100 - N} = \frac{(h\rho_m - P)Z_sT_s}{P_sZ_{av}T_{av}\left(\ln\frac{P + P_s}{P_s}\right)} \tag{3}$$

The use of this equation is shown in Example 3.

■ EXAMPLE 3:

Assume: Well depth = 5000 feet
            Maximum permissible fluid pressure = 1500 psi
            Drilling fluid = water
            T_b = 140 F Z_b = 1.1
            T_s = 60°F Z_s = 1.0
            P_s = 14.7 psia

Solution: $h\rho_m$ = (5000)(.437) = 2185 psia
            P = 1500 psia

$$\frac{N}{100 - N} = \frac{(2185 - 1500)(1)(520)}{(14.7)(1.05)(570) \ln\left(\dfrac{1500 + 14.7}{14.7}\right)} = 8.75\ \frac{\text{ft}^3\ \text{of air}}{\text{ft}^3\ \text{of mud}}$$

$$\frac{8.75\ \text{ft}^3\ \text{of air}}{\text{ft}^3\ \text{of mud}} \times 5.615\ \text{ft}^3/\text{bbl} = 49.2\ \frac{\text{SCF of air}}{\text{bbl of mud}}$$

In this case, the water weight was reduced to an equivalent weight of 5.78 ppg. In some areas, if circulation is to be maintained, it will be necessary to reduce the fluid weight to levels this low or lower.

Weighted Muds

Weighted muds are defined as those where barite has been added to increase mud weight. The measurement of solids content is important in weighted muds. With both low and high gravity solids in the mud, it is not possible to determine the solids content as a function of mud weight alone. The measurement of solids content is obtained by using a retort which generally contains no more than 15 cc of mud. A very small error may make a substantial difference in the measurements of solids content.

Where conditions are critical, more precise laboratory methods for the determination of solids content should be used as a basis for checking field results. Figure 5-9 shows the minimum solids content versus mud weight for a fresh water mud. The minimum solids curve is calculated assuming only fresh water and barite. In field practices the solids toleration level in a mud is controlled by impurities in the water, the pH level of the mud, and the type of low gravity solids. The effect of solids type on the viscous characteristics of mud is shown in Figure 5-10.

A solids concentration of bentonite of less than 4 per cent by volume will result in excessive thickening without any additional solids. Consider the solids problem in a 16 ppg mud. For a solids concentration of 32 per cent by volume, about 25 per cent by volume is barite and about 7 per cent by volume is low gravity solids. Only about two per cent of the 7 per cent low gravity solids can be expanded bentonite without excessive viscosity problems. On the other hand, if all the low gravity solids were low yielding

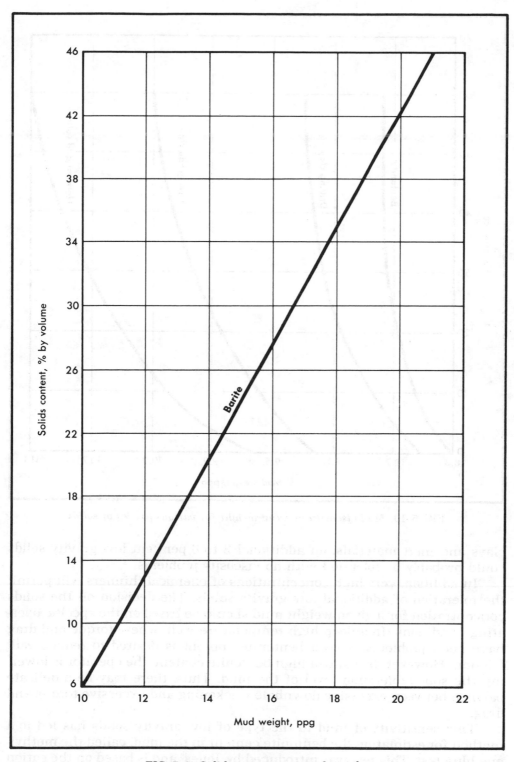

FIG. 5-9. *Solids content vs. mud weight*

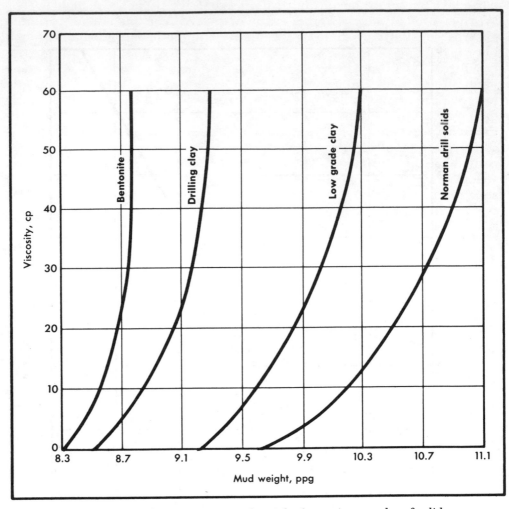

FIG. 5–10. *Mud viscosity vs. mud weight for various grades of solids*

clays and inert materials, an additional 3 to 6 per cent low gravity solids could probably be tolerated with no viscosity problems.

In addition, very high concentrations of chemical thinners will permit the toleration of additional low gravity solids. The decision on the solids concentration for a given weight mud should be based on the specific operating conditions. In a deep high temperature well, where torque and drag have been problems, a good bentonite content is desired to reduce wall friction. However, by using a high bentonite content, the operator is lowering the solids toleration level of the mud. Thus, there may be a delicate balance between excessive downhole thickening and excessive torque and drag.

This sensitivity of mud to the type of low gravity solids has led to a method for estimating the bentonite content in the mud, called the methylene blue test. This test was introduced by Jones[3] and is based on the cation exchange capacity, surface area and state of dispersion of suspended clay particles. Bentonite has a high cation exchange capacity, a high surface area per unit of weight and is easily dispersed. For this reason bentonite absorbs substantially more of the methylene blue solution than ordinary low yielding clays.

Table 5–7 from Jones[3] shows the adsorption capacities of the various bentonite samples and also of other low yielding clays. It is noted that the low yielding clays have an adsorption capacity of less than one-sixth of the bentonite. On this basis it would require at least 6 volume per cent of low gravity, low yielding solids to equal one volume per cent of bentonite. Thus, the measurement of the equivalent bentonite content must also be accompanied by an estimate of the low gravity solids content before it has any specific meaning. For example, 12 to 15 per cent by volume low gravity solids might be equivalent to 18 to 20 ppb of bentonite. Thus with essentially no bentonite it would be possible to obtain a reasonable bentonite level.

TABLE 5–7
Adsorption Capacities of Commercial Bentonites
for Methylene Blue

Sample	Adsorption Capacity g dye/g clay	Calculation Factor*
1	0.271	2.91
2	0.269	2.93
3	0.253	3.12
4	0.267	2.95
5	0.287	2.75
6	0.306	2.58
7	0.284	2.78
8	0.246	3.20
9	0.257	3.07
10	0.280	2.81
11	0.276	2.86
12	0.249	3.17
13	0.292	2.70
14	0.365	2.16
15	0.265	2.97
16	0.271	2.91
17	0.252	3.13
18	0.294	2.68
19	0.270	2.92
Kaolinite	0.05	—
Illite	0.05	—
Attapulgite	0.12	6.5
Pennsylvanian Shale	0.05	—

*For finding pounds per barrel bentonite in mud. Multiply factor by milliliters 0.45 per cent methylene blue solution needed to titrate 2 ml of mud.

In field practice, where weighted muds are often checked for equivalent bentonite content, it has been observed that the level of indicated bentonite may increase. Generally, this indicates an increasing low gravity solids content that is bentonitic in nature. In high weight muds it has been shown that there is upper bentonite toleration level for the mud.

This simply means that excessive thickening may occur at bottom-hole pressure and temperature if this level is exceeded. This is shown in Figure 5–10. The upper limit on low gravity solids at specific bottom-hole conditions will vary with mud weight, chemical treatment and the type of low gravity solids.

This discussion leads to the need for determining the quantity of low and high gravity solids in a given weight mud. The first requirement is an accurate measurement of solids content. After measuring the total solids content, the determination of the concentration of high and low gravity solids is simply a mathematical procedure. Figure 5–11 is a chart which can be used to estimate the amount of barite and low gravity solids after the measurement of solids content. This chart was prepared from a simple mass balance as shown in Example 4.

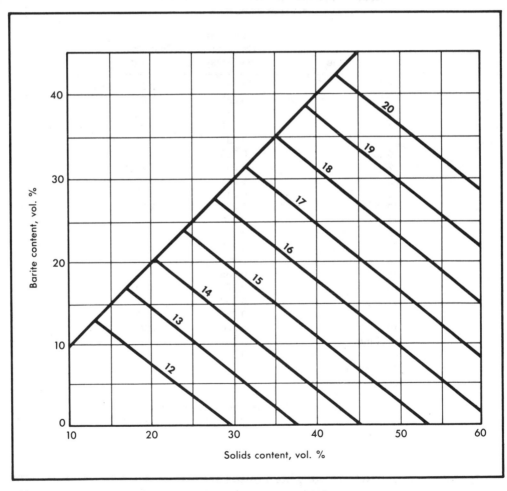

FIG. 5–11. *Barite content vs. solids content*

■ EXAMPLE 4:

Assume: Mud weight = 16 ppg
           Solids content = 32 per cent by volume
           Specific gravity of low gravity solids = 2.5 $(8.33) = 20.8 ppg$
           Specific gravity of high gravity solids = 4.3 $(8.33) = 35.8 ppg$

Solution: Let X = volume fraction of low gravity solids.
           Let .32 − X = the volume fraction of high gravity solids.
Then: $(X)20.8 + 35.8(.32 − X) + (8.33)(.68) = 16$
           X = .074 (low gravity solids) $1 − .32 = water \%$
           .32 − X = .246 (barite)

If other materials are in the mud, such as oil, it can be added directly to the same mass balance. It should be noted also that some low gravity solids weigh more and some less than 20.8 ppg and barite will still meet API specifications if it weighs 35.0 ppg. These mass balances can be used to develop general equations for the barite required to obtain a given mud weight increase. Equations 4 and 5 have been developed for this purpose.

$$X = \frac{W_2 - W_1}{35.8 - W_1} \tag{4}$$

$$X' = \frac{W_2 - W_1}{35.8 - W_2} \tag{5}$$

In these two equations, X is the barrels of barite per total barrel of final mud. After the increase in mud weight, X′ is the barrels of barite per total barrel of mud before increasing mud weight. Equation 4 has more utility because it recognizes the limits on final mud volume. Example 5 illustrates the use of Equation 4.

■ EXAMPLE 5:

Assume: Initial mud weight, W_1 = 10 ppg
           Final mud weight, W_2 = 16 ppg
           Mud volume = 1000 barrels
           Desired mud volume after increasing mud weight = 1000 barrels

Solution: (using Equation 4)
           $X = \dfrac{16 - 10}{35.8 - 10} = \dfrac{6}{25.8} = .233 \dfrac{\text{Bbls of barite}}{\text{Bbls of final mud}}$
           Required barite = (1000)(.233) = 233 barrels

This means that 233 barrels of mud must be discarded before increasing mud weight. This 233 barrels of mud will then be replaced with 233 barrels of barite. Example 6 illustrates the use of Equation 5.

■ EXAMPLE 6:

Assumptions: Same as Example 4.

Solution: $X = \dfrac{16 - 10}{35.8 - 16} = \dfrac{6}{19.8} = .303 \dfrac{\text{Bbls of barite}}{\text{Bbls of initial mud}}$
              Required barite = (1000)(.303) = 303 barrels
              Total final volume = 1000 + 303 = 1303 barrels

Equation 5 does not indicate how mud should be discarded for a final volume of 1000 barrels. In fact, starting with 767 barrels of mud, Equation 5 shows 233 barrels of barite (767 × .303 = 233), is needed to increase the mud weight to 16 ppg. Thus, only Equation 4 should be used, even though many charts are available using Equation 5. However, neither equation takes into consideration the solids content of the mud.

Special problems which may be encountered when making decisions to increase mud weight are as follows:

1. At what mud weight should barite be used to obtain further increases in mud weight?
2. Should oil be used in weighted muds for lubrication?
3. How high should the solids content be allowed in a specific weight mud before some means of reducing the solids content is used?
4. How much bentonite should be added to weighted muds?

The first problem is encountered the most often in the pressure transition zone of abnormal pressure wells. To illustrate, a set of conditions have been assumed and the results examined in Example 7.

■ EXAMPLE 7:

Assume: 1000 barrels of mud
mud weight = 10.0 ppg (unweighted)
Solids content (Figure 7) = 13.5 per cent by volume
In five drilling days the mud weight will be increased to 16 ppg. Maximum desired solids content for 16 ppg mud = 32 per cent by volume.

Solution: From Example 4, the 16 ppg mud with 32 per cent solids by volume contains 7.4 per cent low gravity solids. In 1000 barrels this means 74 barrels can be low gravity solids. In Example 5, where the mud weight was increased to 16 ppg, 233 barrels of mud was discarded. Starting with 767 barrels of 10 ppg mud means the operator would also be starting with (767 × .135) = 103 barrels of low gravity solids. The net result would be that some of the weighted mud would have to be discarded before reaching the 16 ppg level.

A quick way to determine how much mud should be discarded can be obtained by a determination of the quantity of low gravity solids that will be tolerated in a given final mud weight and divide this number by the solids content of the unweighted mud. In this example, 74 barrels of low gravity solids can be tolerated in 1000 barrels of 16 ppg mud with a 32% by volume solids content. Dividing 74 by .135, the solids content of the 10 ppg unweighted mud gives 548 barrels of the 10 ppg mud that can be retained. If the calculation shown in Example 5 was followed, some of the weighted mud would have to be discarded before reaching the 16 ppg level.

These assumed conditions give a basis for determining the maximum mud weight before adding barite. Specifically, the following additions would be required:

548 barrels of original mud
246 barrels of barite
206 barrels of additional water

Adding the water to the retained mud shows the maximum permissable weight of the unweighted mud before adding barite. This is shown as follows:

Pounds		Gallons
548 × 10	= 5480 (42)	548 (42)
206 × 8.33	= 1715 (42)	206
Total	= 7195 (42)	754

Maximum permissible mud weight before adding barite $= \dfrac{7195}{754} = 9.55$ ppg

This example has shown one of the problems the operator faces in controlling costs. His course of action will have to be determined by the actual and anticipated mud weight requirements. If his maximum mud weight will be 16.0 ppg and no unusual problems are anticipated, it may be unrealistic to emphasize a maximum solids content of 32 per cent by volume. If, on the other hand, the mud weight may eventually exceed 17.0 ppg and several weeks of abnormal pressure drilling are anticipated at temperatures above 300 F, he should have a minimum solids content before he begins to add barite.

On this basis it would not be unreasonable to specify a maximum solids content of 31 per cent by volume for the 16.0 ppg mud. With a maximum total solids of 31 per cent by volume, the mud would contain the following:

412 barrels of original mud
254 barrels of barite
334 barrels of additional water

On this basis, the mud weight before adding barite could not exceed 9.25 ppg, considering the fact that the solids in the mud are primarily undesirable solids. The best approach might be to discard all of the old mud and begin drilling at the desired weight with fresh water bentonite and barite. Many other situations could be considered; however, this would serve no useful purpose. In actual operations, the cheapest and best method of mud weight control can be determined by a few simple calculations such as those shown in these examples. Many times the guess-work applied is unnecessary and expensive.

The next problem is the use of oil for lubrication. In some cases the problem has been solved because of disposal problems. Oil emulsified in weighted muds takes up space, in other words it reduces the free water content. This has the effect of increasing thickness and gel strengths. Considering the viscosity versus solids concentration curve shown in Figure 5–10, it would simply move the equivalent solids to the right on any one of these curves.

The problem may become critical in high weight muds where the solids content is high by necessity. In general, oil is not recommended where the mud weights exceed 17.0 ppg in deep high temperature wells. Exceptions will exist when lubrication, by oil or from materials which must be added with oil, is necessary to minimize torque and drag.

The question of solids control and the use of bentonite in weighted muds is a problem that continues to plague operations. In actual fact, definite limits cannot be set for either the total solids or bentonite content. Bentonite increases the thickness of mud and in general when using weighted muds one of the primary problems is keeping the mud thin.

On this basis, bentonite would be added only when barite begins to settle. However, bentonite helps improve the filter cake and this helps reduce torque, drag and differential pressure sticking. On this basis, bentonite would be added when improvements in lubrication are needed. Unfortunately, it is not uncommon to have conflicting problems where thinning and better lubrication are needed at the same time.

Under such conditions, the operator must compromise. The compromise may be based on economic considerations or it may be based on which problem is the most pressing at the time. There are other special materials available to improve lubrication and muds may be thinned by raising pH, reducing the total solids content and increasing chemical treatment.

Viscosity and Gel Strength Control

Viscosity is a term used to describe the thickness of muds in motion and gel strength is used to describe the thickness of muds that have been left quiescent for a period of time. In scientific terms, viscosity is a proportionality constant between shear stress and shear rate for Newtonian fluids in laminar flow. Thus as a constant, shear rate would have no effect on viscosity. This is correct for true or Newtonian fluids such as water; it is not true for drilling muds. Most drilling fluids shear thin. This means the proportionality between shear stress and shear rate is reduced as shear rate is increased. As a result, the original scientific meaning of viscosity has been altered and for drilling muds it is used in a different context than for true fluids.

In drilling muds, viscosity has been adopted as the common expression for describing thickness. Unfortunately, certain thickness levels have been prescribed for drilling muds based on common usage rather than current requirements. Thus, this discussion will be dedicated first to defining objectives, second to the methods of measurement and third to the methods of control.

Objectives

Mud thickness is controlled for the following direct reasons:
1. to control circulating pressure losses in the annulus.
2. to provide adequate lifting capacity for the removal of formation solids.
3. to help control surge and swab pressures.

Indirect considerations include the following:
1. Drilling rates are higher with low solids, thin fluids.
2. Muds may be thickened to minimize erosion in some unconsolidated shale formations, because turbulent flow patterns with thin fluids may create hole erosion and excessive hole enlargement.

Methods of Measurement

The methods of measuring mud thickness are as follows:

1. The marsh funnel: This was the first method used to determine mud thickness. The measurement is made by comparing the time required for one quart of mud to run out of the funnel to the time required for one quart of water. The funnel is a calibrated instrument that is filled with 1500 cc of fluid and the fluid discharge is through a sized nozzel. One quart of water is supposed to be discharged in 26.5 seconds. The relative time for the discharge of one quart of mud is an indication of mud thickness.

   There is no quantitative basis for using this number. For example, a funnel viscosity of 200 seconds is not even proof that the mud is thicker than one with 100 seconds when both fluids are in motion. Thus, the only benefit to be obtained from the use of the funnel viscosity is to detect changes in mud properties which may be indicative of potential downhole problems.

2. The rotating viscometer: There are several models of rotating viscometers available. The field models are usually normal temperature types, which run at only two speeds, 300 and 600 rpm. The 300 rpm speed represents an approximate shear rate of 511 sec^{-1} and the 600 rpm speed represents an approximate shear rate of 1022 sec^{-1}. There are also six speed models of the rotating viscometers which in general are designed for speeds of 3, 6, 100, 200, 300, and 600 rpm.

   There are also some models of rotating viscometers that have been built to run at variable speeds and some built to run at speeds other than those listed. In addition to the normal temperature rotating viscometers, there are several high temperature rotating models being used in laboratories. The high temperature models permit the evaluation of muds while they are in motion at high temperature. Much of the work by Annis[4] was done using a high temperature rotating viscometer. These high temperature models provide a quick method of evaluating mud properties under environments similar to those encountered in the field.

   Maximum temperatures of 500 F and maximum pressures of 3000 psi are the general rule for these viscometers. The pressure limitation is not believed critical for water base muds where pressure is considered only as a factor to keep water from boiling. Pressure has a large affect on oil base muds and to obtain accurate information an instrument is required that will measure at pressures equivalent to those encountered in the field which are approaching 20,000 psi.

3. The pipe viscometer: The pipe viscometer is primarily a laboratory tool, although some field models have been used. This method of measurement meets the necessary criteria for accuracy; however, it is not easy to use in the field and fails the test for simplicity. Pressure drop is measured along a given length of pipe for mud at selected flow rates. By using heat exchangers, these pressure measurements can be obtained at any temperature and pressure permitted by the equipment in use.

The rotating viscometer is the common field instrument used to determine the effective mud thickness. However, the data obtained from the rotating viscometer must be converted to numbers that can be used on a quantitative basis. This requires either the use of equations that describe the flow pattern of the fluid or a plot of shear stress versus shear rate.

Equations 6, 7, 8, and 9 have all been used to describe the flow behavior of drilling mud.

Newtonian

$$T = -\mu\gamma \tag{6}$$

Bingham Plastic

$$T = Y + (PV)(-\gamma) \tag{7}$$

Power-Law

$$T = K(-\gamma)^n \tag{8}$$

Power-Law with Yield Stress

$$T = Y' + K'(-\gamma)^{n'} \tag{9}$$

Note: no attempt has been made to make the units consistent in equations 6, 7, 8, 9, 11, 12 and 13 and the equations are shown for purposes of illustration only.

From Equation 6, the viscosity, μ, is simply the proportionality constant between shear rate and shear stress and remains constant for all rates of shear. From Equation 7, it can be seen that there is no single term that describes thickness. Two terms are used for this purpose: the yield point, Y, and the plastic viscosity, PV.

From Equation 8, thickness is described by K and n and in Equation 9 the description of fluid behavior has been further complicated by using three terms to describe thickness: Y', K', and n'.

The primary criterion for any method of describing mud thickness is that it be reasonably accurate, have utility and be easy to use.

Figure 5–12 shows a shear stress, T, versus shear rate, γ, diagram for a Newtonian fluid. The concept is easy to understand, μ is the slope of the straight line, which goes through the origin. Figure 5–13 shows a shear stress versus shear rate diagram for a typical drilling mud.

The two speed rotating viscometer measures the shear stress in $lb_f/100\ ft^2$ at shear rates of 511 sec^{-1}, the 300 RPM speed and 1022 sec^{-1}, the 600 PRM speed. The slope of the assumed straight line between the two shear rates is called the plastic viscosity. The intersection of this straight line with the ordinate shown as the dashed line is called the yield point. The use of the plastic viscosity and yield point in this manner is known as Bingham Plastic flow behavior.

The actual behavior of most muds in the shear rate range below 511 sec^{-1} is shown by the solid curving line. Thus, the assumption of Bingham Plastic behavior is valid only if the mud is in laminar flow at a shear rate above 511 sec^{-1}, which is an unlikely condition in field operations. It will be noted that some shear force is imposed before the fluid moves. The force

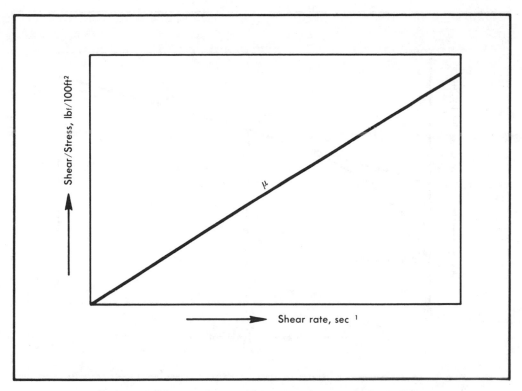

FIG. 5–12. *Shear stress vs. shear rate for a newtonian fluid*

that initiates movement of the fluid is called the gel strength. It is also noted that a tangent to any point on the curving line below 511 sec⁻¹ would show the fluid is thinning as the shear rate is increased.

The primary use of the plastic viscosity, which is measured in the units of centipoise, is to show the effect of solids content on mud thickness. It is obtained by simply subtracting the 300 dial reading from the 600 dial reading on the viscometer. The instrument has been calibrated to give the answer in centipoise. In actual fact, the difference should be multiplied by 0.937 to convert the units to centipoise. In field operations the conversion constant is generally ignored.

The magnitude of the plastic viscosity may be affected by the solids content, the size of the solids and the temperature. It is always difficult to say that a given weight mud should have a certain plastic viscosity because the solid size is also a factor. Figure 5–14, an abbreviated form of the viscosity versus solids concentration curve, shows how solid size may affect the mud thickness.

It has been noted many times that an increase in solids content at a given temperature may have little effect on total mud thickness until some critical range is reached, such as point C, on curve 1 in Figure 5–14. The same increase in solids on curve 2, also shown as point C, had little effect on mud thickness. From point A to point B the increase in solids had little effect on either curve 1 or 2. It is suspected that the performance shown for

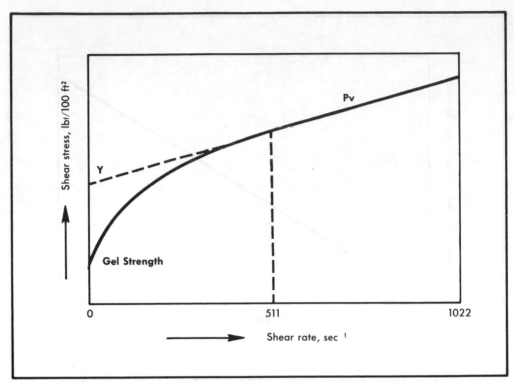

FIG. 5–13. *Shear stress vs. shear rate non-newtonian fluid*

curve 2 might be transferred to curve 1, with no change in the total solids content other than a reduction in the size of solids.

It is suggested that the level of plastic viscosity for specific muds be left an arbitrary term. In unweighted muds the solids content can be calculated, thus there is no need to use plastic viscosity as a criteria for treatment. In weighted muds the plastic viscosity for given muds will generally depend on the solids volume, the solids size and temperature. Considering temperature constant for the moment, this means that the plastic viscosity for a given weight mud may vary and still remain acceptable under certain drilling conditions.

This introduces the best way to use plastic viscosity when treating weighted muds. Measure the solids content as accurately as possible and measure the plastic viscosity. If the solids content is in an acceptable range, so is the plastic viscosity regardless of its magnitude. In critical wells measure the mud thickness frequently, watch for a change in plastic viscosity.

A sudden increase will generally indicate a solids content increase. If not detected, this increase may result in a downhole disaster. Plastic viscosity is chosen as the parameter that should be measured often, rather than solids content, because it is easier and quicker to measure and may be obtained with less danger of an error in measurement.

The yield point is a psuedo number as shown on Figure 5–13. It is obtained by extrapolating the assumed straight line between the 300 and

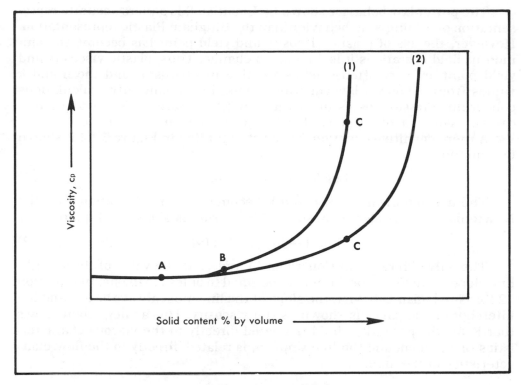

FIG. 5-14. *Viscosity vs. solids content*

600 dial readings on the viscometer. Thus, the number does not exist and its utilization in quantitative calculations immediately introduces a known error. Yield point is determined quantitatively by subtracting the plastic viscosity from the 300 reading on the viscometer.

For field purposes, yield point is used as an indicator of the attractive forces between solids or if there are no attractive forces as an indicator of the deviation of the mud from Newtonian behavior. In general field practice, the yield point is used more as an indicator of mud thickness than the plastic viscosity. In unweighted muds, the yield point is maintained at the level required for adequate hole cleaning. In weighted muds, the desired level of the yield point is generally that required to support barite.

It might be argued that the gel strength, a measured value, would be more indicative of the solids support capacity of the mud and this is probably true. In fact, the only argument in support of the yield point is that its magnitude is related qualitatively to the magnitude of the gel strength. The units of the yield point and gel strength are $lb_f/100\ ft^2$. Both are point readings taken at the ordinate of the shear stress-shear rate diagram shown in Figure 5-12. If either the yield point or gel strength is considered excessive they may be reduced by reducing solids content or by the use of chemical thinners. The equivalent mud thickness, using the Bingham flow pattern, may be determined using Equation 10.

$$\mu = PV + \frac{267Y(D_h - D_p)}{\bar{v}} \tag{10}$$

The power-law behavior shown by Equation 8 is a more accurate representation of drilling mud behavior than the Bingham Plastic representation. However, the use of plastic viscosity and yield point has become so common in field operations it is difficult to change. Thus, plastic viscosity and yield point are very often used as treating mechanisms and the n and k values from the power law equation are used for quantitative calculations concerning pressure losses and lifting capacity. Figure 5–15 shows a graphical representation of the power-law behavior shown in Equation 8.

A more definitive equation for the straight line in Figure 5–14 is shown by Equation 11.

$$\text{Log } T = \log K + n \log (\gamma) \tag{11}$$

The use and meaning of n and K become clearer if Equation 4 for the Newtonian fluid is also expressed in log form, as shown in Equation 12.

$$\text{Log } T = \log \mu + \log (\gamma) \tag{12}$$

The only difference in Equations 11 and 12 is the value of the straight line slope, n. In Equation 11, n may be equal to or less than one. In Equation 12 for Newtonian behavior, the slope of the line must always be one and the intercept on log paper is shown as the viscosity. This analogy then shows that K for the power-law fluid is related directly to the viscous characteristics of the mud, and the line slope, n, is related directly to the flow characteristics of the mud.

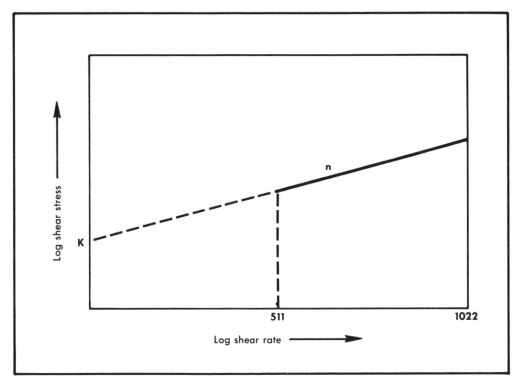

FIG. 5–15. *Shear stress vs. shear rate for a power law fluid*

If n is one the fluid is behaving as a Newtonian fluid. When n is less than one the fluid behavior is non-Newtonian. Thus, both n and K are necessary to describe the thickness of the drilling mud. Equivalent mud thickness for annular flow, using the power-law equation, is shown in Equation 13.

$$\mu = \left[\left(\frac{2.4\ \bar{v}}{D_h - D_p}\right)\left(\frac{2n + 1}{3n}\right)\right]^n \frac{200K(D_h - D_p)}{\bar{v}} \tag{13}$$

The value of n can be determined using Equation 14.

$$n = 3.32 \log \frac{\theta 600}{\theta 300} \tag{14}$$

The value of K can be determined using Equation 15.

$$K = \frac{\theta 300}{(511)^n} \tag{15}$$

The equivalent thickness for the power-law fluid behavior will normally be substantially less than that for the Bingham Plastic fluid behavior. The power-law relationship is considered the most accurate.

The power-law with yield stress shown in Equation 9 can be expressed in log form as shown in Equation 14.

$$\text{Log}\ (T - Y') = \log K' + n' \log (\gamma) \tag{16}$$

It can be seen that the only difference in Equations 11 and 16 is the shifting of the straight line on log paper by the quantity of the yield stress, Y'. The three RPM reading on the viscometer is generally taken as the first estimate of yield stress. However, the mathematical analysis using this approach becomes substantially more complicated and rather than proceed in this direction it would be more simple to plot a curve of shear stress versus shear rate to determine equivalent mud thickness. As more multi-speed viscometers become available in the field, a plot of shear stress versus shear rate will become a common method to determine equivalent mud thickness. This would be done as shown in Figure 5–16.

After plotting the viscometer readings, the shear rate of interest would have to be calculated.

An approximation of the average annular shear rate may be determined using Equation 17.

$$\text{Annular shear rate,}\ \bar{\gamma} = \left(\frac{2.4\ \bar{v}}{D_h - D_p}\right)\left(\frac{2n + 1}{3n}\right) \tag{17}$$

The calculated RPM speed may be determined by dividing the value obtained in Equation 17 by 1.7. After the determination of the average shear rate the shear stress may be taken directly from Figure 5–16, which is a plot of the viscometer data in Example 8. The equivalent mud thickness in centipoise may be determined using Equation 18.

$$\mu = \frac{479\theta}{\gamma} = \frac{(300)(\text{visc } R)}{S.R.}\left(\frac{2n+1}{3n}\right) \tag{18}$$

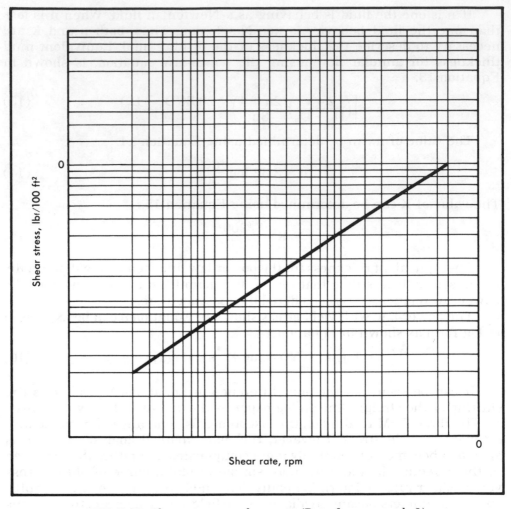

FIG. 5-16. *Shear stress vs. shear rate (Data from example 8)*

Example 8 shows how the equivalent mud thickness can be determined using the various procedures available and also shows the magnitude of differences for this particular case.

■ EXAMPLE 8:

Hole size = $8^1/_2$ inches
Drill pipe size = $4^1/_2$ inches
Mud weight = 10 ppg
Annular velocity = 150 fpm
Viscometer readings:
$\theta600 = 100$
$\theta300 = 65$
$\theta200 = 50$
$\theta100 = 32$
$\theta6 = 4.8$
$\theta3 = 3.0$

Bingham (Equation 10)

$$PV = 100 - 65 = 35 \ c_p$$
$$Y = 65 - 35 = 30 \ lb_f/100 \ ft^2$$
$$\mu = 35 + \frac{(267)(30)(4)}{150} = 35 + 214 = 249 \ c_p$$

Power-law, two speed viscometer
(Equation 14)

$$n = 3.32 \log \frac{100}{65} = (3.32)(.1874) = 0.621$$

(Equation 15)

$$K = \frac{65}{(511)^{.621}} = \frac{65}{48} = 1.35$$

Equivalent thickness (Equation 13)

$$\mu = \left[\frac{(2.4)(150)}{4}(1.205)\right]^{.621} \frac{(200)(1.35)(4)}{150}$$
$$\mu = (18.4)(7.2) - 132.5 \ c_p$$

Thickness at a specific shear rate
Viscometer data plotted on Figure 15
Average annulus shear rate (Equation 17)

$$\bar{\gamma} = (90)(1.205) = 108.5 \ sec^{-1}$$
$$\text{shear rate (rpm)} = \frac{108.5}{1.7} = 63.8 \ rpm$$

Viscometer reading or shear stress at 63.8 rpm = 23.5 lb_f/100 ft^2 *from graph with data p. 108*

Equivalent thickness (Equation 18)

$$\mu = \frac{(479)(23.5)}{108.5} = 103.5 \ c_p$$

[handwritten annotations: 300 above 479; 63 below 108.5; 131]

Example 8 shows the different procedures may give a substantial difference in equivalent mud thickness. The most reliable procedure would be the one using Equation 18, which represents the thickness at the average annular shear rate. Equation 18 requires the use of a multi-speed viscometer. It will be noted that the Bingham relationship shows an equivalent thickness of 249 centipoise, which is more than two times the thickness of 103.5 centipoise determined, using the more accurate procedure.

This example shows that plastic viscosity and yield point can be used for treating purposes only and with familiarity the power-law relationships would be preferred for this purpose. For example, a low value for n indicates a probable need for chemical treatment and a high value for K indicates a probable need to reduce the solids content. These are not precise conclusions but can be used as general guidelines.

The disadvantage of using a method such as that shown by Equation 18 to determine equivalent mud thickness is the limited information provided relative to the required mud treatment. In general, if the equivalent thickness is high some type of chemical treatment is probably needed.

Pipe Viscometer

The pipe viscometer provides a good method for determining the fluid properties at normal and elevated temperatures and pressures. It is not simple to use and unless handled carefully may give erroneous results. For this reason, it is considered at the present time a laboratory instrument. Exceptions may be the trailer mounted data units used as aids in some field operations. The pipe viscometer provides a direct method of measuring the effect of fluid properties. The pressure drop for a given tube length is measured. This pressure drop may be converted to shear stress by Equation 19.

$$T = \frac{300\ PD}{L} \tag{19}$$

After calculating shear stress, the average shear rate can be determined using Equation 17 and the equivalent mud thickness can be determined using Equation 18. A big advantage of the pipe viscometer is the determination of fluid property effects at very high shear rates above 1022 sec^{-1}, which is not possible with the normal rotating viscometer.

Temperature Effects

Temperature effects on mud thickness may be substantial. In general, the mud gets thinner with increases in temperature. However, the specific effect may be decided by the type and total solids in the mud. For example, consider curve 1 in Figure 5–14. If the solids concentration was at point B for this mud, any further dispersion of clays by high temperature might also result in flocculation and severe thickening. Whereas if the mud was comparable to that shown on curve 2, an increase in temperature would probably result in a thinning of the mud.

Several investigators have conducted laboratory tests that describe the behavior of muds at elevated temperatures. Work performed by Bartlett[5] is shown in Table 5–8. Note the plastic viscosity is reduced substantially with increases in temperature. The same table shows an erratic behavior for yield point as a function of temperature increases.

TABLE 5–8
Simulated Fann Data

Temp. F	Reading at 1022 sec.$^{-1}$	Reading at 511 sec.$^{-1}$	Plastic Viscosity, cps	Yield Point $16_f/100\ ft^2$
68	136	70	66	4
72	111	62	49	13
120	83	47	36	11
160	62	34	27	7
220	40	25	15	10
320	32	22	10	12

The reduction of plastic viscosity with increasing temperature is believed due to a thinning of the liquid phase of the mud. This belief is confirmed by Figure 5–17 taken from work by Annis[4] which shows the normalized viscosity of water as compared with the plastic viscosity of mud versus temperature. It is noted from Figure 5–17 that the thickness of the water and mud followed the same reducing trend with increases in temperature until a temperature of about 220 F was reached.

At this point the plastic viscosity of the mud did not decrease with further increases in temperature. One possible explanation was the additional dispersion of solids with temperature tended to increase the frictional effects of the solids at a rate comparable to the thinning of the liquid. The specific point of no further reduction in plastic viscosity will depend on the type mud. Note in Table 5–8 the plastic viscosity of the mud was 10 centipoise at 320 F and 15 centipoise at 220 F.

It has been mentioned that muds may thin or thicken with increases in temperature. Evidence that proves this conclusion is shown in Figures 5–18, 5–19, and 5–20 from Bartlett's[5] work. Note that in Figure 5–18, where 21 ppb of bentonite is mixed in fresh water, the mud thickens with each succeeding increase in temperature beginning with 76 F. At 300 F and a shear rate of 100 sec^{-1} the mud is more than four times as thick as it was at 76 F.

To confirm that the reaction with temperature depends on the type mud, Figure 5–19 shows a mud with 25 ppb of bentonite, 350 ppb of barite

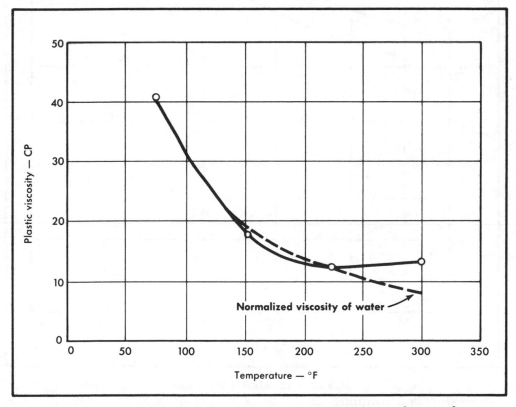

FIG. 5–17. *Effect of temperature on plastic viscosity of water-base muds*

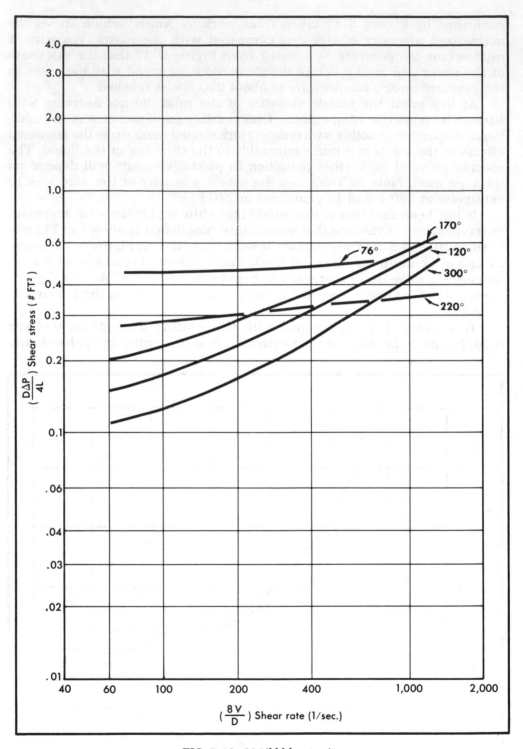

FIG. 5–18. *21#/bbl bentonite*

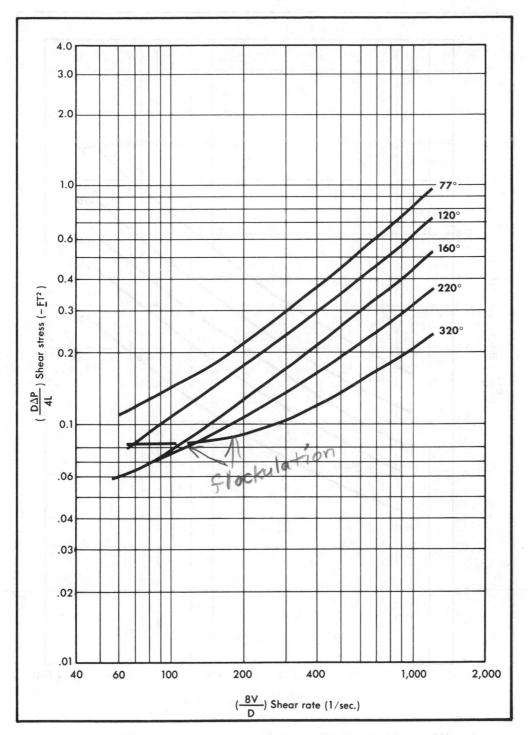

FIG. 5–19. *9#/bbl lignosulfonate, 25#/bbl bentonite, 350#/bbl barite, PH—9.5*

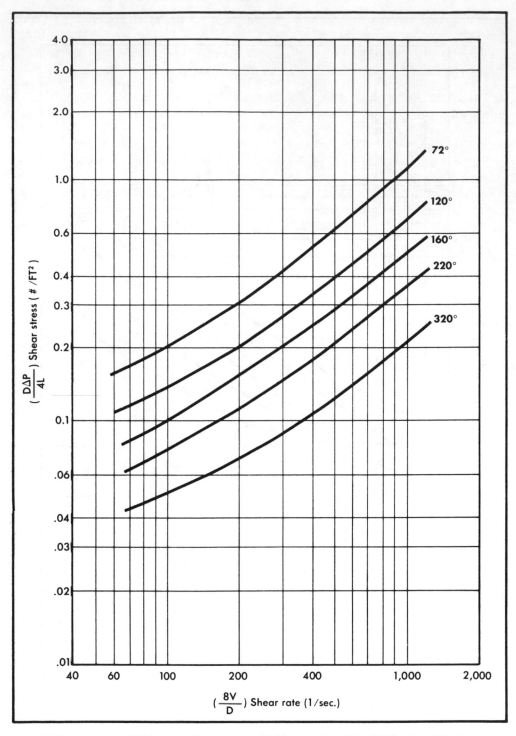

FIG. 5–20. *15#/bbl lignosulfonate, 25#/bbl bentonite, 350#/bbl barite, PH − 9.5*

and 9 ppb of lignosulfonate where thinning occurs with increases in temperature until a temperature of 220 F is reached. In Figure 5–19 it is noted that the mud becomes substantially thicker at the temperature of 320 F when the shear rate drops below 100 sec^{-1}.

The same mud composition, but with 15 ppb of lignosulfonate is shown in Figure 5–20, and the additional chemical treatment prevented the flocculation noted in Figure 5–19. The mud thinned substantially with increases in temperature to 320 F. The pH of the mud in Figures 5–19 and 5–20 was maintained at 9.5.

The effect of pH and the equivalent bentonite content is shown in Figure 5–21 from work by Annis. Figure 5–21 shows the gel strength to be uncontrollable at an equivalent bentonite content of about 27.5 ppb and a temperature of 300 F for this particular mud. Of further significance in Figure 20 is the reduction in gel strength noted when the pH was increased from 7.6 to 9.4.

These tests show that temperature has a substantial effect on the flow characteristics of drilling mud. Methods used in the field do not in general permit the measurement of fluid properties at temperatures above about 120 F. Thus the operator should always measure the mud properties at the same temperature. Table 5–8 can be used to estimate the effect of temperature on mud thickness if there is no flocculation of the mud occurring. The

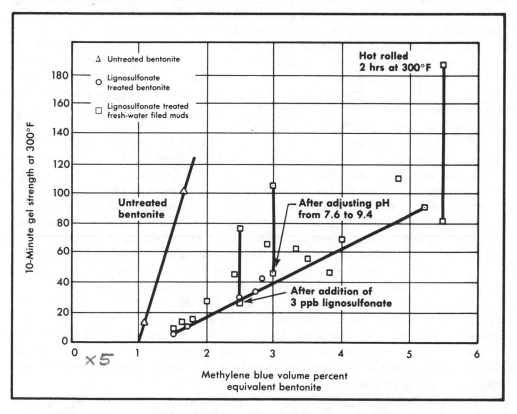

FIG. 5–21. *Effect of solids on 300F gel strengths of field muds*

data in Table 5–8 are based completely on the effect of temperature on the thinning of water as shown in Figure 5–17. Table 5–8 would have no utilization with oil base muds.

Methods Used to Control Viscosity

The control of viscosity or mud thickness will of course depend on the operator's objectives. Viscosity may be maintained at high levels to ensure adequate hole cleaning. Additives, such as some of the polymers, may be added to increase the mud thickness for adequate hole cleaning while keeping the mud weight and total solids content at low levels. With weighted muds some operators have used polymers for increasing thickness in order to keep clay solids, that provide support for the barite, to a minimum.

In general, if there are no specific hole requirements, the operator prefers to use the thinnest possible drilling fluid. Thinning is also emphasized when using weighted muds close to the fracture gradient of exposed formations.

When more viscosity is desired it may be obtained by (1) adding bentonite, (2) flocculating the clay solids, and (3) adding polymers designed for the purpose. Bentonite is the primary viscosity builder in the mud system. If fresh water is being used bentonite will increase viscosity rapidly with only small increases in mud weight. Flocculation of the clays in a fresh water mud is also a quick and cheap method of increasing viscosity. The flocculation can usually be done by adding lime or cement.

Another method for thickening mud is the use of specific polymers. There are several different polymers available. Basically, they can be divided into three groups; (1) the polymers that thicken the mud by acting on the clay solids, (2) polymers that are colloidal in nature and essentially provide a substitute for bentonite, and (3) the polymers that thicken the liquid phase and do not react with bentonite.

No attempt will be made to introduce all of the polymers that might fit any one of the above three categories. Typical examples of the specific types will be discussed. However, in many cases the extent of a given polymers utility is still being field tested and for any material to be successful in application it must not only work but also be the best material economically. Specific applications might prove any one of the polymers successful and the key is to always keep accurate well costs. It should be remembered that all results are good or bad based on comparisons with other alternatives.

A typical polymer that aggregates clay solids to increase mud thickness is formulated using a type of polyacrylates. This material aggregates bentonite, trapping water and increasing mud thickness, it also has the claimed advantage of aggregating undesirable clay solids which are then easier to remove from the drilling mud. Lummus[6] has written several papers describing the use of a polyacrylate type polymer called by the trade name, Benex. This material has been used in both low solids unweighted and low solids weighted muds.

In low solids unweighted muds the polymer, by increasing the viscosity producing characteristics of bentonite, provides the necessary lifting capacity at a very low solids content which is the same as saying at a very low mud weight. In low solids weighted muds, the low solids term, simply applies to the low gravity solids. The concept with a system of this type is to use a minimum of bentonite and increase the bentonite yield with the polymer in order to support barite. No chemical thinners should be used in either the low weight or weighted muds unless the operator intends to abandon the use of the polymer.

Polymers of the colloidal type that coat particles and provide additional lifting capacity are typified by sodium carboxymethylcellulose, abbreviated CMC. In addition, other polymers of this type are starches and gums. These materials help increase lifting capacity and also provide an aid to improving filter cake characteristics. It is not uncommon for combinations of these polymers to be used and it is difficult to evaluate all the combinations. They do not aggregate solids, as do the polyacrylates; however, they provide improved filter cakes. Many of these materials have been used in the drilling business since the 1920's. Improvements and different use guide-lines have changed their utility.

The XC-polymer, so-called because it is produced by the bacterial action of genus xanthemonas campestris on sugar, increases viscosity independent of the solids content of the mud. The material goes into solution and is capable of thickening water regardless of the electrolyte content of the water. One of the claimed advantages of the material is the thickening of the treated mud at low shear rates and the extreme thinning at high shear rates. Deily, Limblom, Patton and Holman[7] presented precise information on this polymer and related information on its use, advantages and limitations.

Polymers of all types have gained substantially more attention in recent years. The materials all tend to shear thin at high shear rates, an advantage for drilling rate, and thicken at low shear rates, an advantage for lifting capacity. Many other advantages are claimed and these can be obtained by using the indicated references. Most of the polymers have a temperature limitation, generally in the range of 300 F. Some are not compatible with certain mud additives and some require certain pH levels. Cost is always a factor. The reference to cost means total well costs. An operator should never just select a material based on the cost of that material alone or the mud cost alone. Some materials may increase mud costs and reduce well costs.

It should be further emphasized that these polymers accomplish different objectives, they cost different amounts and the success or failure for any product is always influenced substantially by associated personnel. It is apparent from the widespread use of the polymers that operators believe they have gained benefits. The claimed benefits in the case of low solids–low weight muds have in many cases been substantial. The use of polymers in weighted muds is limited; however, results in such areas as the Delaware basin of the United States should be watched carefully.

Viscosity control is generally considered in terms of reducing mud thickness. Methods used for control include mechanical aids, water dilution and chemical thinners.

Mechanical aids are discussed in detail in the chapter on mechanical aids to mud conditioning. The mechanical aids are in general solids and viscosity control devices. They help control solids in the mud and also help control the type solids, both of which are important in mud control.

There has always been a standard oil field saying, "if in doubt add water." Many mud engineers have found this to be good advice; however, at times doubt exists when it should not. Methods are available to measure solids content, determine the type solids, measure the pH and determine the degree of contamination. Tests should be run and evaluated before the miscellaneous use of water. In weighted muds, the cost of the materials is high and each barrel of water has to be weighted and treated.

Chemical thinners reduce the attraction solids have for each other. Table 5–9 includes some of the more common mud thinners.

TABLE 5–9
Drilling Mud Thinners

Chemical	pH of 1% Solution	Limitations
1. Sodium acid pyrophosphate (sapp)	4.3	Decomposes and forms flocculating agent above 175 F. Not effective in the presence of large quantities of calcium.
2. Sodium tetraphosphate	8.0	Same as 1.0.
3. Chrome Lignosulfonate	~~7.0~~ *5.2*	Material starts to decompose at temperatures above 300 F. pH needs to be at least 9.0.
4. Lignite	3.2	Material starts to decompose at temperatures above 350 F. pH needs to be at least 9.0.
5. Tannin	5.0	Not very effective if pH of mud is less than 11.0.
6. Surfactants	–	Many types, temperature stability above 300 F may be a problem. Most are more expensive than other materials.

Sodium Chromate

The phosphates are inorganic materials and they in general will treat at any pH. Phosphates are also cheap, a normal mud treatment would be in the range of 0.2 ppb. The noted limitations of 175 F and large quantities of calcium have limited the application of phosphates. Another factor that has limited the use of phosphates has been the trend towards non-thinned low solids muds in the more shallow sections of the hole.

The chrome lignosulfonates are the most common type thinners. Plural is used because there are many types available and they are sold under many trade names. These thinners are organic in nature and the base material comes from the lumber industry. Chrome is an added material and the method of manufacture may be the key to the effectiveness of these materials. For this reason, operators should run specification tests on chrome lignosulfonates if they are not sure of the source or if it is a new source of supply.

Some of the chrome lignosulfonates are similar in their degree of effectiveness; however, other materials called chrome lignosulfonate may not thin the mud. It is noted in Table 5–9 that the chrome lignosulfonates begin to decompose at 300 F. This does not mean they cannot be used above this temperature. It does mean that because of decomposition above 300 F the operator will have to increase treatment to maintain a given concentration level. The higher the pH the better these materials treat. At a pH less than 9.0, treatment with the chrome lignosulfonate will not be very effective.

Lignites are old mud additives. The chrome lignosulfonates were introduced in the middle of the decade from 1950–1960. Lignites were common mud additives at least 10 years earlier. Lignite is a hydrocarbon product. While lignites are used as thinners they are also used as water-loss control agents. They will begin to decompose at about 350 F; however, it is not uncommon to use these materials as the primary mud thinner to temperatures substantially above 400 F.

The tannins, typified by the trade name Quebracho, were the most common thinners until the introduction of the chrome lignosulfonates. To be effective the tannins have to be used in a high pH environment. This was no problem because prior to the middle 1950's high pH (12.0) lime muds were common and the tannins were very effective thinners. The reduction of mud pH to the level of 9.5 to 10.5 reduced the relative effectiveness of tannins to the chrome lignosulfonates and almost eliminated the use of tannins. If the mud pH is above 11.5 the tannins may be competitive or even better than some of the lignosulfonates.

Surfactants are surface tension reducing materials. They will thin muds, help reduce water-loss and act as emulsifiers. The most common use of surfactants is their utilization as chemical emulsifiers. They have not been competitive economically with other thinners for routine mud treatment.

Some common rules of mud treatment with chemical thinners will be reviewed. It should be emphasized that specific situations require specific techniques thus this is a general approach to mud treatment.

Mud treating decisions for weighted muds should be related to the specific downhole objectives first and to the actual cost of treating second. Mud properties, which affect decisions are the total solids content, the type solids, the pH, and the known presence of contaminants. In general no chemical treatment is recommended for unweighted muds.

Consider first the solids content. It is known that field retort measurements are approximations at best; however, this is the only method available in the field and the chances for errors in measurement simply emphasize the necessity to be accurate. While measuring the solids content accurately, determine the flow properties of the mud, using a part of the same mud sample. Then measure these mud properties every 15 minutes for a complete cycle of the mud.

Muds are generally not homogeneous in nature and this procedure provides the operator with a specific look at the entire mud system. At the same time these tests are run, the same procedures should be used to determine the pH, the type contamination and the bentonite content of the mud. Some knowledge is needed on the critical range of these mud properties.

This simply means that if everything that is measured at the surface is in an acceptable range, the properties will probably remain acceptable at downhole conditions where the temperature is much higher. Some indication of the effect of the higher downhole temperature can be determined by comparing the flow line properties to the suction pit properties. If the critical nature of the well justifies the effort, high temperature pipe viscometer or high temperature rotating viscometer tests should be run.

The flow properties of the mud should be checked frequently. If an increase in plastic viscosity is noted with no change in mud weight, this is an indication of an increase in solids content. If the increase in plastic viscosity is accompanied by an increase in yield point, the initial preferred treatment would be a reduction in solids content either by mechanical means or by water dilution.

If, however, the yield point increased without an increase in plastic viscosity, then chemical treatment would probably be preferred. These procedures are based on the assumption that no pH change occurred. The n and K values could be used in the same manner. If n is low, chemical treatment may be required. If K is high, water may be needed.

When adding the organic chemical thinners, particularly the chrome lignosulfonates and lignites, the preferred procedure is to mix these materials with caustic and water in a mixing tank or chemical barrel. These organic materials are not readily soluble, thus when mixed in the presence of caustic and water, they are solubilized more readily than when added directly to the mud. It has been mentioned that pH is important in mud treatment and one of the effects is to increase the effectiveness of these chemical thinners.

The gel strength and yield point of the mud are related to pH. In a general case, an increase in pH will reduce gel strength and yield point and conversely a low pH would result in increases in the properties. Thus in many cases where gel strength is a problem there may be a tendency to continue increasing pH. Depending on the specific circumstances, a high pH, say above 11.0, may result in large concentrations of excess lime, $Ca(OH)_2$ and the formation of a cemetaceous material if temperatures are in excess of 250 F. Needed, of course, would be a source of calcium which may be obtained by drilling cement, from calcium in make-up water, from calcium bearing formations or from calcium added to reduce carbonate contamination.

It has been generally assumed that the adverse effects of contamination were caused by cations such as calcium and sodium. The carbonate anion may be equally troublesome if not more so than the cations. This has resulted in the development of new techniques to determine the carbonate ion concentration. All of these methods are complicated by other mud additives and contaminants. If carbonate contamination is believed to be a problem, calcium may be added or the pH may be increased using caustic. In cases of severe problems, both procedures may be used at the same time.

Filtration Control

The filtration rate is generally controlled for the following two reasons: (1) to control the thickness and characteristics of the filter cake which is deposited on permeable formations and (2) to limit the total filtrate that enters underground formations.

There are basically two methods of measuring the rate of filtration: (1) static filtration rate tests, which are A.P.I. approved and (2) dynamic filtration rate tests. The static tests include the standard A.P.I. test of normal temperature (77 F) at 100 psi and the high temperature tests generally run at 300 F and pressures of 100 and 500 psi. Static tests are meant to be indicative of the loss of fluid and the build-up of filter cake when the fluid is not moving. Dynamic tests represent the loss of fluid and filter cake build-up while the drilling mud is being circulated.

Filter cake thickness and the friction between the filter cake and the drill string relate to the problems of (1) differential pressure sticking, (2) torque and drag while drilling and tripping the drill string, (3) running of wireline tools and casing strings and (4) sidewall coring. An increase in filter cake thickness increases the area of contact between the drill string and the filter cake and for a given friction coefficient increases the danger of sticking the drill string and also increases the torque and drag on the drill string.

The running of wireline tools and casing is restricted by the reduction in hole size as the filter cake gets thicker. A thick filter cake may prevent any formation recovery on a wireline test. For these problems the operator is not concerned with the actual loss of filtrate. The filtrate loss is used to indicate the potential thickness of the filter cake.

The problems of differential pressure sticking and torque and drag are related primarily to high weight drilling muds in deep high temperature environments. This was the primary reason the high temperature static filtration rate test was introduced. Efforts were being made to simulate downhole conditions and at the same time standardize the data and results to permit industry wide usage. Basic correlations were developed quickly and standard treating techniques emerged.

Two of these were: (1) the rule-of-the-thumb maximum of a high temperature water-loss of 10 cc and (2) the concept of the compressible filter cake, determined by comparing the water loss at 300 F and 100 psi with the water loss at 300 F and 500 psi. Results showed the merit of these two techniques in some field operations; however, costs were often increased and problems were increased in others by simply depending on these rules.

To obtain a 10 cc static 300 F filter loss meant in general that the mud had either a high percentage of bentonite in the low gravity solids, was highly treated with chemicals or had a very high total solids content for the specific mud weight. Thus, other methods of measurement were needed to confirm what might actually occur down the hole. The concept of the compressible filter cake was one such follow-up method. If the water-loss shown by the 300 F and 500 psi test was the same or only slightly higher than the water-loss obtained by the 300 F and 100 psi test, it was assumed that the filter cake contained a high percentage of compressible solids such as bentonite.

Another method of checking quality is to run tests for longer periods of time than the normal 30 minutes. Field and laboratory tests have shown that longer filtrate testing periods of $1\frac{1}{2}$, 3 and 4 hours do not necessarily confirm the indicated quality of the filter cake that is obtained from 30 minute tests. For example, in a series of field tests it was discovered that filter cakes from three different muds which showed a comparable thickness of $\frac{4}{32}$nd of an inch at the end of 30 minutes at 300 F, showed thicknesses of $\frac{9}{16}$-inch, 1-inch and $1\frac{1}{2}$ inch at the end of 4 hours at 300 F.

Downhole problems would be anticipated if the mud cake was $1\frac{1}{2}$ inches thick, yet the operator would have had no clue with the 30 minute test. Another method to check potential filter cake characteristics is to determine the bentonite percentage in the mud by the methylene blue test.

Equation 20 has been written to provide a better understanding of the static filtration rate.

$$\Delta V = \frac{CP^{1-b}\,\Delta t}{\mu_L rwV} = \frac{C\,\Delta P}{\mu_L\,r\,wv\,\Delta P^b} \qquad b=1 \qquad (20)$$

If a high percentage of the low gravity solids in the mud are compressible, b will approach 1.0 and pressure will not increase the volume of filtrate for a specific time interval. The viscosity of the filtrate is reduced substantially with temperature and Figure 5–16 from Annis shows that water is only $\frac{1}{4}$ as thick at 225 F as it was at 75 F. This simply means the thinning of water would provide an increase of over four to one in the filtration rate at 300 F compared with 75 F. Further dispersion of clays at 300 F may reduce the ratio in some cases to about three to one. However, if any high temperature flocculation occurs, the 300 F water-loss may be eight times the 75 F water-loss.

This can be one check on whether the operator is close to the critical solids range, where a slight increase in solids can cause severe thickening. The small r in Equation 20 refers to the type solids. The solids content is indicated by w. As shown, if the solids content increases and everything else stayed the same, the water-loss would be reduced; however, this would be at the expense of having a thicker filter cake.

None of these methods of determining down-hole static filter loss or filter cake thickness define the problems of determining requirements. Potential requirements will have to be determined by the thickness of permeable formations open to the well-bore and the pressure differentials across these formations. If less than 100 feet of permeable formations are open to the well bore, the operator may experience very few problems.

If the pressure differential across the permeable formations is very low, less than say 300 psi, again the operator may experience very few problems. The control of high temperature water-loss in the cases described above may be a waste of money and the effect on the circulating mud properties may be detrimental. An operator should never proceed blindly. He should keep an accurate record of the formations penetrated, the formation pore pressures and histories of past experience and treat the mud based on the specific requirements of the well being drilled.

Another problem of thick filter cakes is the running of wire-line tools and this problem causes trouble in shallow and deep wells. The low solids muds, in unweighted drilling fluids, have minimized the problem of running wire-line tools. In weighted muds, particularly in coastal areas, some operators will make what is commonly called a short trip before running wireline tools. If this short trip is made just before pulling pipe to run the logs, it is probably a waste of time, money and effort. After a number of hours out of the hole, trips may need to be made with the bit and drill string to wipe-off filter cake. While circulating fluid, the filter cake reaches a maximum thickness and the short trip just after circulating has stopped serves no purpose.

Operators are concerned with the total filtrate entering underground formations for the following reasons: (1) some believe that shale sloughing is caused by filtrate invasion into the shales; (2) they believe results of formation evaluation methods may be affected by filtrate invasion; and (3) they believe formation productivity is often affected by filtrate invasion.

The concept of wetting shales which results in excessive sloughing was first introduced as an engineering concept in the decade from 1920–1930. The fact that shales did slough as a result of getting wet meant that the concept was readily accepted. In actual fact, most shales are impermeable and there would be no possibility of wetting these shales by a normal filtration process. Some arguments are advanced that the shales are full of microfractures and the filtrate enters these fractures.

Many shales are in fact fractured; however, the fractures are frequently large enough for the mud to enter. The first scientific evidence that shale sloughing was not related to filtration rate was presented when caliper logs showed that hole enlargement could not be related to filtration rate.

The next big blow to the advocates of filtration control to control shale sloughing came when data were presented to show that there was no quantitative or qualitative correlation between dynamic and static filtration rate. Dynamic filtration rates are measured by circulating the mud through or over the face of a core at a given temperature and pressure differential. Outman's[8] proposed Equation 21 for the determination of the dynamic filtration rate after the filter cake reaches an equilibrium thickness.

$$\frac{\Delta V}{\Delta t} = \frac{K}{\mu_L} \left[\frac{(\tau/f)^{(-v+1)}}{d(-v+1)} \right] \tag{21}$$

It is noted in Equation 21 that the dynamic filtration increases with an increase in shear stress, thus the quantitative magnitude of dynamic water loss would depend on the flow rate, the viscous properties of the mud, the size of the annulus and the characteristics of the cake. The small f represents the friction between solids, which is an unmeasurable quantity; however, a material such as oil would decrease the value of f and result in an increase in dynamic filtration rate. The small d is the equilibrium thickness of the filter cake. Thus the thinner the cake under circulating conditions the higher the dynamic filtration rate for a given mud.

The $(-v + 1)$ is a measure of filter cake compressibility and Outman indicates the range of values to be 0.1 to 0.15. From this analysis it can be seen that many factors affect the dynamic filtration rate which are not present when measuring the static filtration rate. This accounts for the lack of quantitative or even qualitative agreement between the static and dynamic filtration rates.

Laboratory tests have confirmed the difference between dynamic and static filtration rates. Figures 5–22 and 5–23 show data obtained by Kreuger.[9] Figure 5–22 shows tests using CMC, starch and a metal lignosulfonate. Note using CMC the minimum dynamic filter loss is recorded when the static filtration rate at 500 psi and 170 F is about 16 cc. For starch, the minimum dynamic filtration rate occurred at a static water-loss of about 11 cc. The static water-loss with the metal lignosulfonate was above 20 cc at the point where the dynamic water-loss was a minimum.

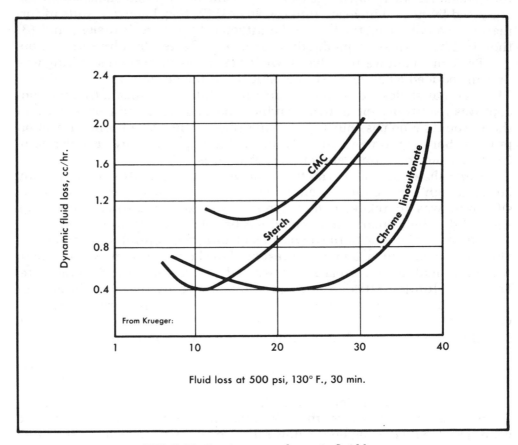

FIG. 5–22. *Static versus dynamic fluid loss*

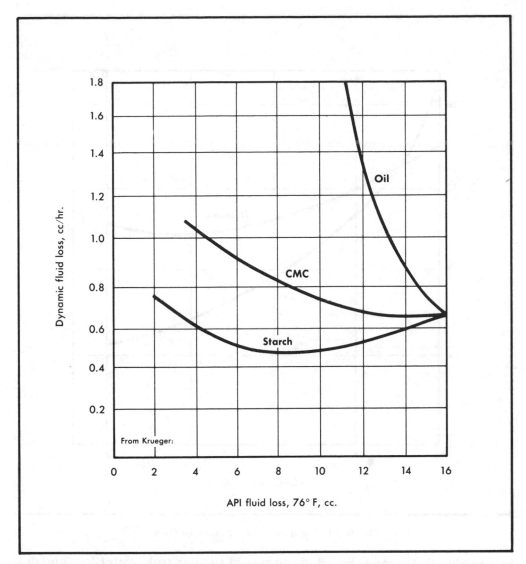

FIG. 5–23. *API static fluid loss is dynamic fluid loss*

Figure 5–23 shows additional data, this time using starch, CMC and oil. These comparisons were made between the dynamic and the standard low temperature API static water-loss. Further confirmation of these tests is shown in Figures 5–24, 5–25, and 5–26, which were obtained by Gray.[10] Figure 5–24 shows a comparison between the static and dynamic water-loss using CMC. In this test the minimum dynamic water-loss occurred when the static water-loss was just above 6 cc. Figure 5–25 shows a comparison between the static and dynamic water-loss using a metal lignosulfonate. It is noted in this test that the minimum dynamic water-loss was recorded when the static water-loss was about 13.0 cc.

Figure 5–26 shows a comparison between static and dynamic water-

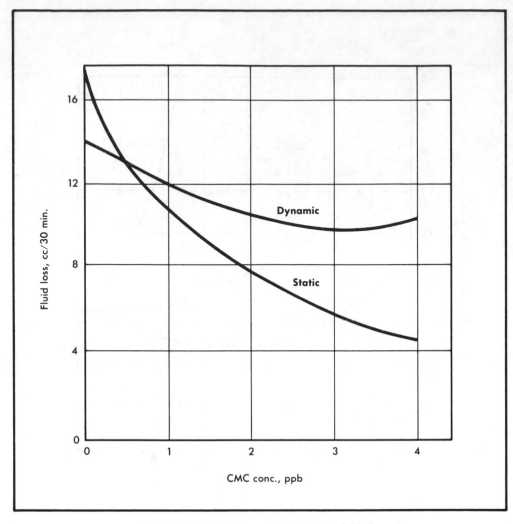

FIG. 5–24. *Fluid loss versus conc. of CMC*

loss using oil. The addition of oil increased the dynamic water-loss and decreased the static water-loss. There are no claims that this series of tests represent quantitative comparisons that can be applied in general field application. The tests show clearly that static filtration rate tests which are run in field operations provide very little information relative to the total fluid that may enter underground formations.

Why does this happen? In the case of CMC, the material increases the shear characteristics of the mud. This characteristic of CMC means the shear stress is increased, thus at some level of CMC concentration the increase in shear stress results in a thinning of the dynamic filter cake which more than offsets the reduction in filter cake permeability offered by CMC. The same argument might be proposed for the comparisons using starch, although as noted the starch used in these tests did not thicken the mud as much as the CMC.

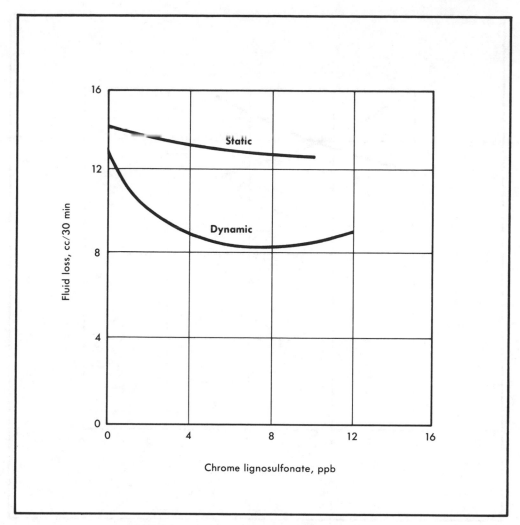

FIG. 5-25. *Fluid loss vs. conc. of lignosulfonate*

It should be emphasized that there are different grades of CMC and starch and thus the quantitative reactions with other grades of these materials would be different from those shown. The tests using oil are somewhat startling. In these tests the dynamic filtration rate was immediately increased by the addition of oil in emulsified form. Referring back to Equation 21, oil increases the shear stress and reduces the friction between particles in the filter cake. Thus, oil thins the filter cake which would be beneficial from the standpoint of problems associated with filter cake deposition but detrimental relative to the total fluid entering a permeable formation.

Almost ironically, oil has been used for many years to reduce the static water-loss to minimize the total fluid entering underground formations, while in fact if filtrate was entering such zones the addition of oil would have increased invasion. Typical examples of this practice include the use

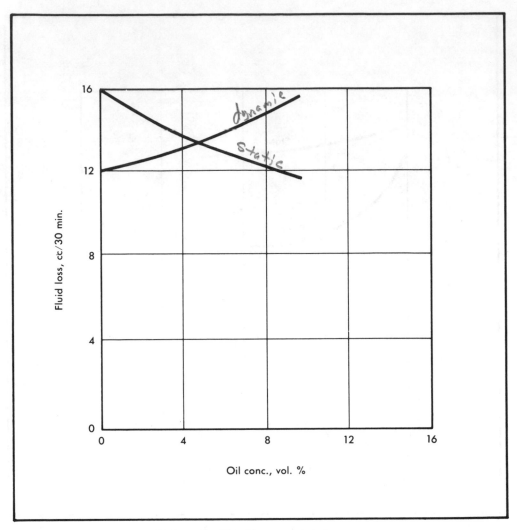

FIG. 5–26. *Fluid loss versus oil conc.*

of oil in the mid-continent region of the United States to reduce water-loss many times to values less than 4 cc's, to prevent shale sloughing. If the primary purpose is to prevent the wetting of shales by filtrate invasion, the practice is ridiculous and every bit of scientific information available would indicate the potential of more filtrate invasion, if in fact any invasion of filtrate occurs.

The most recent information on the wetting of shales indicates that the wetting takes place by an imbibition or osmotic pressure process, which is not related to filtration rate. The causes of shale sloughing are discussed in the chapter on hole problems

It can also be seen that the static filtration measurement is not a reliable tool relative to the amount of potential filtrate invasion in permeable zones. In fact, when demands are made for static filtration rates, as measured under normal temperature API conditions, to be below 10 cc's, there is an excellent possibility that filtrate invasion will actually be increased. If there is a strong doubt, samples of the mud can be tested in the laboratory under dynamic conditions.

Additives Used to Reduce Filtration Rate

Additives in common usage to control filtration rate are shown in Table 5-10.

TABLE 5-10
Additives for the Control of Filtration Rate

1. Bentonite—	Provides a better basic filter cake.
2. Metal lignosulfonates—	Reduces filtration by deflocculating the mud. Also increases the viscosity of the filtrate, . Starts breaking down at 300 F. Not very effective above 350 F.
3. Lignite—	Reduces filtration rate by deflocculating and plugging open spaces in filter cake. Starts breaking down at 300 F. Effective to reduced degrees above 400 F.
4. Sodium Carboxymethylcellulose, CMC—	Reduces filtration rate by coating solids and sometimes by minimizing flocculation. Starts breaking down at 300 F.
5. Starch—	Reduces filtration rate by coating solids. Will spoil in a fresh water environment where pH is less than 11.5. Also a salt concentration of 250,000 ppm will prevent spoiling in any pH water. Breaks down rapidly when temperatures exceed 275 F.

The most reliable method of reducing filtration rate is generally to maintain a good percentage of bentonite in a completely deflocculated mud. Often some other problem may prevent the addition of bentonite, particularly to weighted muds and treating compromises have to be accepted. For example, in a deep high temperature well where the total pressure exerted by the mud column is approaching the fracture gradient of an underground formation, the operator would be reluctant to add bentonite which would probably thicken the mud and increase chances of losing circulation.

The most common agents used to control filtration rate in weighted muds are bentonite, metal lignosulfonates and lignites. Bentonite as stated is an excellent material for this purpose. The metal lignosulfonates are primarily deflocculants and are used commonly. When temperatures exceed

350 F, it is not uncommon to use primarily lignite because of its greater temperature stability. In unweighted muds, it is generally not necessary to control filtration rate, if it is controlled then bentonite, starch and CMC are common additives.

Trends in Special Muds

Special muds are generally introduced to combat specific hole problems, to reduce the cost of drilling and to help in formation evaluation. It is not uncommon for special muds to be re-introduced after being discarded in prior years. It would be impossible to name all the special mud mixtures available, thus this discussion will be limited to general trends. In recent years the industry has moved towards the use of polymers, the re-application of inhibited muds used in past years and new mixtures of inhibited muds.

Polymers

Polymer type muds were discussed in the section on viscosity control. Polymers first became popular because of the trend towards low solids muds. Their use was extended when the concept of non-dispersed muds was introduced. They became even more popular when they were used as a part of inhibited mud systems. In low-solids, low weight muds polymers were used to replace bentonite and other low-gravity solids. In the inhibited muds the polymers have been used to encapsulate the clay solids to minimize hydration.

The polymers have helped in many cases and increased costs without helping much in others. Because changes are still being made and because material costs fluctuate, no recommendation can be made on the general use of polymers.

Inhibited Muds

Inhibited muds used many years ago include lime muds, Gyp muds and calcium chloride muds. The lime muds were very thin even at high mud weights and were appealing to the rig crews. Their use was discontinued because of high temperature solidification, slow drilling rates and poor control of high temperature water-loss. Lime muds have been re-introduced because they are very thin, a desirable feature in deep high temperature wells with high pore pressures. Also, the lime muds have a very high toleration level for solids, which is a help in drilling through gumbo shale sections.

The lime muds currently being used include the following:
1. a high pH, 11.5 to 12.5
2. 4 to 6 barrels of excess lime
3. 6–9 ppb of chrome lignosulfonate for thinning
4. Special filtration control additives
5. Close control of total solids.

The chrome lignosulfonate is used instead of tannins used in previous years, probably an assist to prevent high temperature solidification. Special filtration control materials have been an assist to prevent excessive filter cakes and close control of solids has minimized previous objections on slow drilling rates. Operators still must be cautious, because it will be easy to fall into previous traps offered by lime muds. It should be remembered that looks may be deceiving and something pretty is not always the best material. Field results currently show some advantages to the use of lime muds and they should be watched carefully.

Gyp and calcium chloride muds are being used on a limited basis and in all probability their future use will be limited. The use of potassium chloride muds is being expanded. Potassium chloride was used many years ago in completion fluids to prevent formation damage. Most of the potassium chloride muds in current use are being mixed with some type of polymer. Their use is extensive in Western Canada and they have been used on a limited basis in coastal areas. The potassium cation is used to inhibit clay swelling and the polymers are used to help inhibit and provide lifting capacity.

There are many other inhibited muds, such as the complex aluminum systems, magnesium systems and saturated salt systems. All are in use and have been helpful in many applications.

The final test for any mud system is whether total well cost is reduced and potential drilling risks are minimized. The answers are obtainable only in actual field tests. Operators should always make honest evaluations based on facts rather than following the general trends in an area. It should be remembered that in general the most simple mud system is the cheapest and the best.

NOMENCLATURE

K = Constant dependent on temperature
R_{mf} = Resistivity of mud filtrate
R_w = Resistivity of formation water
N = Volume of air at the surface
h = Well depth in feet
ρ_m = Mud gradient in psi/ft
P = Pressure exerted by air or gas cut mud, psi
P_s = Pressure at surface, psi
T_s = Temperature at surface, degrees Rankine
Z_{av} = Average air or gas compressibility factor
T_{av} = Average temperature, degrees Rankine
T = Shear stress, $lb_f/100 \ ft^2$
μ = Viscosity, centipoise
γ = Shear rate, sec^{-1}
θ = Viscometer readings, $lb_f/100 \ ft^2$
PV = Plastic viscosity, centipoise
Y = Yield point, $lb_f/100 \ ft^2$
n = Slope of shear stress versus shear rate straight line on log paper
K = Intercept of shear stress versus shear rate straight line on log paper
μ_L = Viscosity of liquid filtrate in centipoise
V = Filtrate volume in cubic centimeters
Γ = Solids type

W = Weight of solids per unit volume of filtrate
Δt = Change in time, seconds
C = Proportionality constant

REFERENCES

1. Mondshine, T. C. and Kercheville, J. D., "Successful Gumbo Shale Drilling," The Oil and Gas Journal, March 28, 1966.
2. Chenevert, M. E., "Shale Control with Balances Activity Oil Continuous Muds," paper SPE 2559, Fall Meeting, Sept. 29, 1969.
3. Jones, Frank O., Jr., "New Test Measures Bentonite in Drilling Mud," The Oil and Gas Journal, June 1, 1964.
4. Annis, Max R., Jr., "High-Temperature Flow Properties of Water-Base Drilling Fluids," Paper SPE 1968, University of Texas, Jan. 25, 1967.
5. Bartlett, L. E., "Effect of Temperature on the Flow Properties of Drilling Fluids," Paper SPE 1861, Fall meeting of SPE Oct. 1, 1967.
6. Lummus, J. L. and Field, L. J., Non-Dispersed Polymer Mud: Petroleum Engineer, March 1968.
7. Deily, Lindblom, Patton and Holman, "New Low-Solids Polymer mud Designed to Cut Well Costs," World Oil, July 1967.
8. Outmans, H. D., "Mechanics of Static and Dynamic Filtration in the Borehole," SPE Transactions, 1963.
9. Krueger, R. F. "Evaluation of Drilling Fluid Filter Loss Additives under Dynamic Conditions," SPE Transactions, 1963.
10. Gray, J., "A Dynamic Filtration Study of Drilling Fluids," Master's Thesis 1965 at the University of Oklahoma.

Drilling Fluid
Solids Removal

GEORGE S. ORMSBY
Pioneer Centrifuging Co.
Houston

FOR many years drill solids were removed with seldom anything more than settling pits and vibrating shaker screens, few of which had openings less than one-sixteenth of an inch across. The mechanical processing of rotary drilling fluids entered a new phase with field-practical decanting centrifuges[1] in 1952, efficient 6″ hydrocyclones[2] in 1954, much more efficient 4″ hydrocyclones[3] in 1962, and very fine shale shaker screens[4] in 1966. Spin-offs and special equipment are being developed constantly.

The purpose of this chapter will be to reduce the confusion and to lay out available approaches to solids removal for better mud control, and at reduced mud cost in many instances.

SYSTEM MATERIAL BALANCE RELATED TO MUD COST

Anyone familiar with drill site operations is aware that many active mud systems tend to increase in surface volume. The excess must be disposed of by some means after being hauled away, or jetted out, etc. Examine Figure 6–1 with the following:

The subsurface mud system consists only of the hole. This subsurface system increases at the rate of drilling new hole, plus caving or sloughing rate. Drilling the hole does not change the level in the surface tanks, as every 100 barrels of hole drilled adds 100 barrels of cuttings and results in 100 barrels of hole to be filled with mud. The same is true of *caved* volume, and whether the formation might be bentonite or granite does not change this.

Filtration to the formation tends to reduce mud pit level at rates varying from rare cases of hundreds of gallons per minute drilling sands and silts with "clear water," to small losses using filtration control, or to no filtration loss in impermeable zones. Solids removal tends to reduce system level. The reduction of system level can be more than the drilling rate, even though the net solids removal rate is always less than the drilling rate. This is because solids are not removed nearly as dry as they exist in the formation, so the volume of solids removed plus the liquid necessarily removed with them may be much more than the hole drilling rate.

Drill joint addition slightly increases the level in the tanks, each 31 foot drill joint amounting only to approximately one cubit foot, or 7.5 gallons or 28 liters, of steel. During the drilling of that 31 feet of hole the cuttings formed amount to much more volume than the steel added (in a $7\frac{7}{8}''$ hole it is roughly 10 times as much as the steel volume).

The drilling of a gas sand displaces some liquid from the hole by expansion of the gas rising in the annulus. Liquid level in the surface tanks will increase *temporarily*. If a gas sand were *continuously* drilled, the level in the tanks would begin to decrease as the formation gas porosity was replaced with drilling fluid in the new hole.

The gas breaking out at the surface reduces total system content. Once the gas zone is past and the gas circulated out, the surface mud system level would return to a point lower than when the gas zone was first encountered. The fact that the gas expansion in the annulus is a major temporary affect should not mask the underlying and continuing minor reducing effect on the surface volume and level.

There are other very minor effects that are sometimes confused. Solubility is a negligible factor causing a decreasing trend in surface level too small to be noticed in the field. Temperature expansion should increase a system in the early stages of shutdown, assuming the subsurface is hotter than ambient.

Reverse the statement if subsurface is cooler than ambient, but in any normal field case it would be very difficult to detect. Excluding entrained gas, already discussed, there is no liquid compressibility affecting the circulating system. The column of liquid changes only by drilling deeper and at a rate of absolute change that would preclude field observation. None of the effects in this paragraph have any significance other than theoretical, and will not be discussed further.

The natural trend of a surface mud system is to decrease in volume as the hole is drilled, formation fluid entry being the only natural factor capable of reversing this trend. In normal overpressure drilling, with mud hydrostatic head greater than the formation (pore) pressure, any excess surface system volume that we must drain off, jet out, or vacuum truck away, can be caused only by factor No. 5 in Figure 6–1, "Addition of Liquid and Commercial Solids Materials." Even water is a cost item, and though the scarcity varies more than most other materials it is a major item in some areas.

Since buying material to increase pit level is made doubly expensive by the problem of disposing of the excess, there should be powerful valid rea-

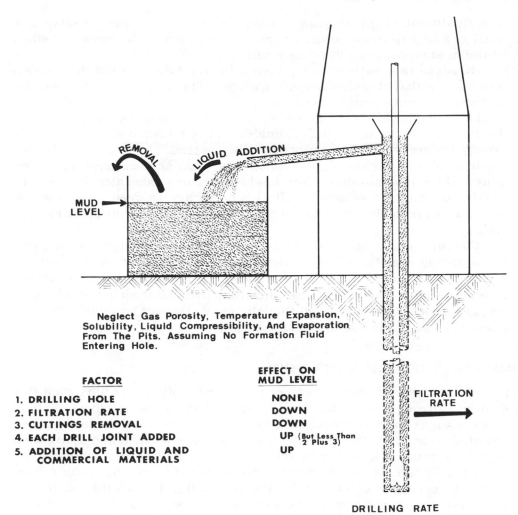

FIG. 6–1. *Mud system material balance*

sons for willfully adding so much to the problem and the cost. Before the advent of modern solids removal equipment, the valid reason for "dumping," or "watering back"; was to control mud properties. Today it is usually a holdover in custom or the result of failure to understand: (1) the problem and (2) the solids removal equipment available.

MUD TREATMENT SIMPLIFIED

There are three basic ways of treating liquid drilling fluids, all relating to solids: (1) adding solids or their equivalent, (2) removing solids, and (3) treating solids (chemically).

The unavoidable *addition* of solids is from the drill bit. For specific controls, commercial colloidal and soluble solids are added to increase Yield Point, Gel Strengths, and Plastic Viscosity, and to decrease filtration rate with minimum weight increase. Heavy and coarse commercial solids material is added to increase the density of the slurry with minimum effect on the mud properties and solids volume.

Oil added to a water base mud emulsifies and the effect of the droplets is much like that of colloidal solids particles. The same is true of the water fraction of a continuous oil-phase mud.

Removing solids is necessary before a drilling fluid can be returned to the drill bit. Recirculated drilling fluids must be returned to as near the optimum drilling properties as economically practical. The more effective the mechanical solids removal, the less dilution and chemical treatment is required. If the total annular return fluid is being discarded after one pass by the bit, of course no separation effort is necessary. In most situations this latter is not possible for either economic or ecological considerations, or both.

Treating solids, "mud chemistry" or mud treating," is the science and art of adding specific soluble materials to a drilling fluid to alter the behavior, directly or indirectly, of some specific solids in the slurry. The chemicals act on the colloidal particles, including hydratable shales—and not on the larger inert particles. Even though chemicals may be soluble, such as a salt, they do add to the true total solids.

DYNAMIC FILTRATION AND WATER DILUTION

The addition of water does not necessarily dilute a drilling mud in a drilling situation. The water leaving the mud as filtrate in a dynamic fluid loss situation can be very high in some muds. Down hole filtration has been reported[8] as high as 400 GPM in long stretches of open hold in massive sand and silt sections and with very little selective solids control. Failure to replace the filtrate loss would cause a decrease in volume and resultant increase in per cent solids, weight, viscosity—all independently of drilling rate. However, if the filtrate loss is constantly replaced, the mud is constantly restored from the effects of filtration. This water is not causing a change in mud—it is only restoring, much like drinking water restores body fluids being lost by sweat. Whether this "pass-through water" amounts to most or a minor part of the total water requirement depends upon the ratio of the dynamic down hole filtration rate to the solids removal rate.

Water added to replace solids removed from a system does change the system total solids. Water added over and above that necessary to balance filtration does dilute the total solids. Solids added by the bit do increase total solids per cent, but they do not increase the surface system level. If surface level is held constant, then total solids content is being altered only by the addition and removal of drilled and commercial solids regardless of the amount of water being added. All water being added is then replacing solids removed, liquid lost with the solids, and/or restoring filtrate lost down hole.

PLASTIC VISCOSITY AND TOTAL SOLIDS CONTENT OF WEIGHTED MUDS

Frequently articles are circulated indicating that Plastic Viscosity of a weighted drilling fluid of a given density will vary with total solids content,

THE "EFFECT" OF LOW SPECIFIC GRAVITY
SOLIDS ON THE PLASTIC VISCOSITY
OF WEIGHTED MUDS[8]

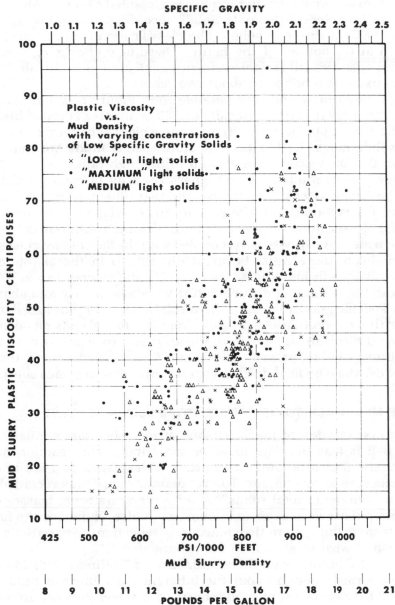

FIG. 6–2. *Total solids vs. plastic viscosity—field weighted muds*

or with drill solids, or low specific gravity solids, content. Or it may be implied that a high Plastic Viscosity is a sure indicator for the need to reduce low specific gravity solids, or drilled solids. Figure 6–2 is a plot of over 350 random, bona fide field checks of drilling fluids in the coastal area of the Gulf of Mexico, in which the muds are classified "Low," "Medium," and "High" in Low Specific Gravity Solids according to the Average Specific Gravity ($\overline{ASG}$) of the total solids. The classification is purely arbitrary and is shown at the bottom of the figure. These field checks represent many wells and operators, all service companies and are typical for all the various muds in use over a period of about two years.

No strong conclusions are possible from Figure 6–2. For example, although the highest plastic viscosity of 100 centipoise is for a "high" solids mud, the lowest plastic viscosities in any weight range are seldom for the "low" solids muds. Rather the "medium" solids muds are found consistently in the low viscosity range, and in the high range as well. Although the "high" solids muds are the highest viscosity in each weight and are never the lowest viscosity in any weight, that statement must be tempered with the observation that the variations and overlap of viscosity in and between all classifications is more striking than their differences.

The wide range of viscosity, coupled with the lack of clear classification would indicate there must be great differences in the character of the solids in drilling fluids. And so there is. The differences in size and shape cause varying effects. And the greatest variations in size and shape are in the low specific gravity solids, or light solids, whether purchased or drilled. It is for this reason that conclusions from laboratory tests on "drill solids" are often difficult to support in the field. It also explains why so many diverse things can be "proved" in laboratory experiments involving drilling fluid solids, and why field drilling conclusions are sometimes contradictory.

THE USE OF MUD STILLS TO DETERMINE TOTAL SOLIDS CONTENT

The first reaction of many persons upon seeing data similar to that of Figure 6–2 is that the data must be inaccurate. The reaction has some validity. The field mud still, or retort, was developed in 1952 to obtain a rough measure of oil content. The condensate "gap" was recorded and assumed to represent "total solids." After some years time, masses of these data were published with median lines as "guides," ignoring the fact these figures were nothing more than inaccurate reflections of variations, with no discussion of what might cause the variations.

The API Committee on Standardization of Drilling Fluids Material became concerned enough about the inherent inaccuracy of mud stills, as compared to the margins of solids control many operators are attempting to hold, to set up a "Task Group on Determination of Liquids and Solids" at the annual 1968 meeting. The later work[5] of this group would seem to confirm that dependable accuracy of field mud retorts does not justify the actions and expenditures many operators make based on mud solids tests. There is a good reason to believe that field readings of per cent solids can and do vary from down two percentage points, to up four percentage points, from the true non-evaporate solids in a mud. Perhaps more significant, from a remedial point of view, is the fact that retorting affords no clue to the size or shape of the offending solids, or indeed that they are offending.

PARTICLE SIZE REFERENCES AND TERMINOLOGY

To appreciate any discussion of small particles some sense of size is necessary. The micron is the most convenient unit to use in discussing small particles. Table 6–1 shows the micron size of common materials. The finger tip sensitivity can be used to determine the effectiveness of coarse solids removal by rubbing a test filter cake between the fingers. Since red blood corpuscles are very small, they make a good thinking reference when looking at tabulated sizes. Particle size distributions of the solids in drilling fluids will show that most single particles are invisible.

TABLE 6–1
Micron Relationships – Familiar Things

Material and/or Sense	Diameter of Size, in Microns	
Human Hair	30–200	
Pollen	10–100	
(Portland) Cement Dust	3–100	
Milled Flour	1–80	
Talcum Powder	5–50	
Red Blood Corpuscles	$7\frac{1}{2}$	
Human Eye Resolution, Normal	40	Minimum
Cosmetic Powder	35	Maximum
Face Skin Sensitivity	35	Minimum
Lipstick Solids	35	Maximum
Lip Skin Sensitivity	35	Minimum
Finger tip Sensitivity	20	Minimum
Between-teeth Sensitivity	6–8	Minimum
Human Eye Resolution, Absolute	6–8	Minimum

It is sometimes inconvenient to refer to all sizes in exact micron sizes or micron ranges, so various disciplines have adopted some standard terminology for convenience. When an industry or group lacks a standard, individual investigators must use improvised or borrowed terms. Figure 6–3 illustrates the terminology problem and compares some usages often heard in the field. Also shown is the first recommendation of the API Committee on Standardization of Drilling Fluid Materials in API Bulletin 13C[11] on terminology for strictly mud solids particle *size*. Inter-related disciplines may object to the use of "colloidal" for particles up to two microns, or of "ultra fine" for particles up to 44 microns, but the standard no doubt will be improved if usages indicates the need.

The word "clay" will be used occasionally in this chapter, and it will consistently refer to particles less than two microns, interchangeably with the new API standard term "Colloidal," as Figure 6–3. "Silt" as used here

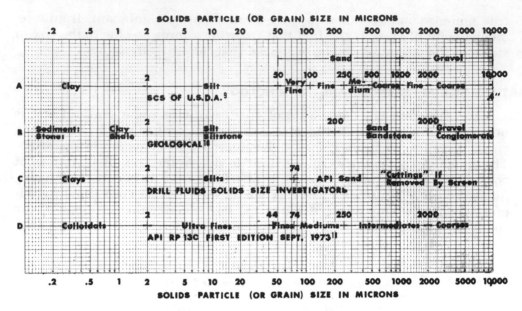

FIG. 6–3. *Some particle size terminology and usage*

will cover the 2 to 74 micron range combining the new "Ultra fine" and "Fine" classification. "Sand" and "API Sand" will be used interchangeably with each other and will refer to particles larger than 74 microns, as would be found by an API Sand Test, and will be used interchangeably in that sense with the new classifications (combined) of "Medium," "Intermediates," and "Coarses."

These size usages will not change with the nature of particles, whether drilled limestone, quartz, dolomite, basalt, chert, or commercial bentonite, barites, or steel from a milled fish.

A source of confusion in the application of solids removal equipment has been the use of the word "clay" by most mud treating authorities in reference to all commercial solids and all drilled solids in the system, i.e., to all solids except commercial weighting material. It is still found in the following mud solids equation:[12]

$$\frac{(100\%)}{\text{Sp. Gr. of Solids}} = \frac{\% \text{ clay by wt.}}{2.6} + \frac{\% \text{ wt. Mat'l by wt.}}{4.3} \tag{1}$$

This is one of the most basic equations in drilling fluid solids work, and the equations to be presented in this chapter are derived from it, but the terminology is extremely unfortunate. The relationship expressed here concerns the specific gravities of solids materials, their weight percentages in a mixture of these solids, and the average specific gravity of the mixtures. The confusion arises because the term "clay," used here in this mathematical expression as general for materials of 2.6 specific gravity, has been recognized and used by many solids investigators and by earth-related sciences as meaning a size range, whether or not the size ranges are in exact agreement. Refer again to Figure 6–3.

The damage from this unfortunate usage has come in this way: It is true to say "Decanting Centrifuges can separate liquid and clay from the larger solids particles in the drilling muds processed through the machines," using the term "clay" strictly in the size sense. It is completely erroneous to say "Decanting Centrifuges can separate low specific gravity solids (or drilled solids) from barites in the drilling muds processed through the machines." Yet the erroneous statement has been written and widely accepted because of the ambiguous use of the word "clay" in recognized mud treatment works.

THE RELATION OF SOLIDS PARTICLES SPECIFIC SURFACE AREA TO MUD PROPERTIES

The effect solids particle size alone can have on mud properties is illustrated in Figure 6–4 from Ritchey.[13] Although another "regular" barite might exhibit some variation, and "3 microns or less" barites in another test would not follow the same line as those in Figure 6–4, a similar basic difference would exist. The difference in results when particle size is varied in a mud slurry is a matter primarily of surface area, for the surface area adsorbs, or "ties-up," water. Obviously, if there is more surface area, more water is adsorbed. Table 6–2 shows how surface area increases as particle

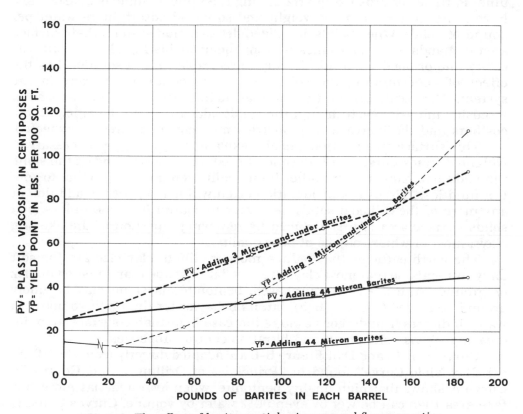

FIG. 6–4. *The effect of barite particle size on mud flow properties*

TABLE 6-2
Effect of Particle Size & Shape on Surface Area

Particle Diameter or Equivalent Diameter Microns	Type Particles	Square Feet per Pound	Square Meters per Kilogram
50.0	Glass Spheres	234.5	48.0
5.0	Glass Spheres	2,345	480.4
1.0	Glass Spheres	11,725	2,402
"	Crushed Quartz	17,160	3,515
0.5	Glass Spheres	23,450	4,804
"	Crushed Quartz	34,320	7,030
0.1	Glass Spheres	117,250	24,018
"	Crushed Quartz	171,600	35,151
?	Bentonite (hydrated)	?	?
0.003	Silica Gel	2,930,000*	600,192*

* Includes interior area of sponge type pore openings.

size decreases, even though the shape and quantity (net volume and weight) remain exactly the same for one pound of glass balls.

Table 6-2 also shows that if shape is changed, the surface area is affected. The sphere shape permits the least area possible for any given volume. Particles of crushed quartz having the same volume as quartz balls, have more area per unit of weight and so would adsorb more water per pound of solids. When solids are added dry to a mud system, their surface area demands water and takes it from the available liquid. If plastic viscosity and/or yield point are already measurable, this adsorption has the effect of concentrating the solids (including bentonite) already in the system. The result can be the same as adding bentonite (an increase in viscosity and gels) even though the solids added may not be efficient viscosifiers and the increase in properties may not be desired or expected.

This surface area relationship also explains why it may be necessary in weighted muds to remove fine solids to reduce viscosity, and at the same time add bentonite or some other high quality commercial product to control fluid loss. Viscosity consideration often will not permit the addition of any more of the colloidal solids necessary to control filtration unless total solids surface area is first reduced by removing a portion of the existing clays, most of which may be of poor quality.

The earth particles that reduce most readily to clay size are the soft flaky minerals, so that most clays are of a flatter shape than crushed quartz or ground barites, etc. As purchased, bentonite claystone may be finely ground and sized (see Fig. 6-5), but its behavior for the first two hours in water indicates it undergoes a huge increase in available surface area, increasing its viscosity effect and filtration control properties.

Curve A, B, C, and D of Figure 6-6 are adapted directly from the "Typical Clay Yield Curve" found in "Principles of Drilling Fluids Control."[12] Curve A shows that high quality bentonite, when hydrated, has more surface area than can be wet by even 90% water by volume. Curves C and D show that some materials marketed as "clay" have questionable value as viscosity builders, apparently having a low potential surface area.

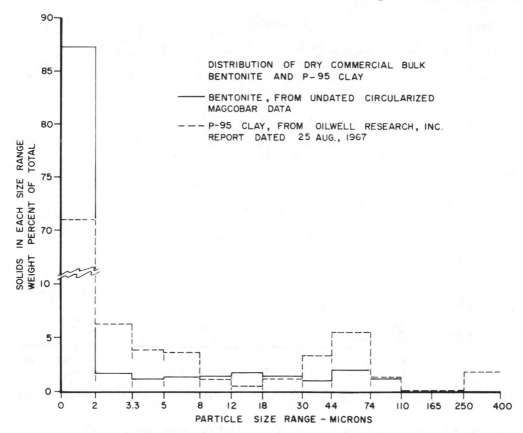

FIG. 6–5. *Distribution of dry comm. bulk bentonite and P-95 clay*

Curve E of Figure 6–6 is drawn as an approximation of the probable minimum Plastic Viscosity possible at any given per cent solids in a water-base mud having filtration rate controlled to less than 15 c.c. by API standard low-pressure, low-temperature test. This is based on the field data points shown, which are the minimum found in over one thousand field tests screened for validity and covering all mud types. The only points below the curve are of muds mixed quickly to control a "Kick," with little or no control of filtration.

If the plastic viscosity of a water base mud is significantly higher than Curve E, its plastic viscosity can be reduced by reducing the colloidal content; if the viscosity is near Curve E, the plastic viscosity probably can be reduced only by reduction of total solids. This latter may or may not be possible, according to the slurry density necessary, and the amount of low specific gravity solids ("light" solids) present. The size distribution of the light solids present will determine the economics of reducing total solids.

The particle size distribution of barites milled to meet API specifications are somewhat similar over the world, as might be expected. A maximum of 3% by weight is allowed to be of API Sand size, and at least 5% must be too large to pass a U.S. Standard No. 325 screen, or greater than 44 microns. This latter requirement is not because large particles are de-

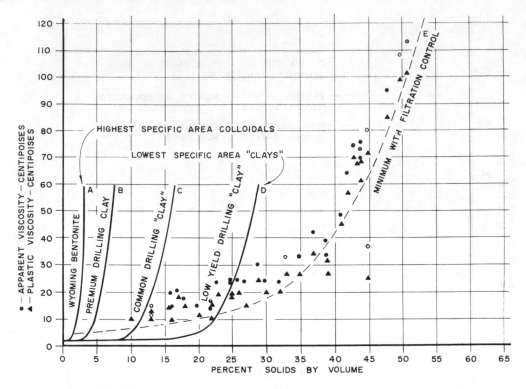

FIG. 6–6. *The effect of particle size and shape (solids area) on viscosity vs. solids content*

sirable, but to make sure the average grind is coarse enough to minimize the colloidal particles. However, the difference in the amounts of barites in the colloidal range are sometimes noticeable and also sometimes create viscosity problems. As already discussed, this is due to the specific surface area phenomenon. Typical variations in barite particle size distribution from suppliers of API grade barites are shown in Figure 6–7. Sometimes barite suppliers add phosphates, tannins, lignins, etc, at the bulk plants, or at the mill, to counteract the viscosity-increasing effect of the water adsorption by the colloidal barite fraction added. In Figure 6–8 are shown special particle size distributions ground for use to build some of the highest drilling mud weights in the world, in Iran.

STOKES LAW AND SOLIDS SEPARATION

Stoke's Law, which gives the mathematical relationship of the factors governing the settling velocity of spheres in a liquid stated in its simplest form as:

$$J_B = \frac{d_B^2(D_B - D_L)G}{18\,\mu} \tag{2}$$

where

J_B = the terminal or settling velocity of the ball

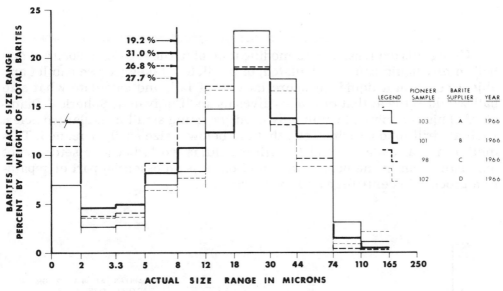

FIG. 6-7. *Particle size distribution — API barites — typical*

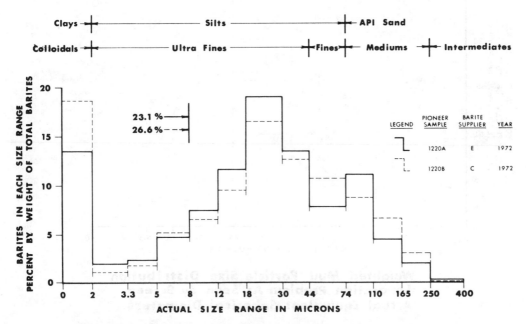

FIG. 6-8. *Barite particle size distribution — special grinds from Iran*

$$d_B = \text{diameter of the ball}$$
$$D_B = \text{density of the ball}$$
$$D_L = \text{density of the liquid}$$
$$G = \text{acceleration of gravity}$$
$$\mu = \text{viscosity of the liquid}$$

Using this expression, or a modification of it, the settling velocity of any ball in any liquid can be calculated, or predicted. Likewise, we can let any object settle in a liquid, measure its rate of fall, and calculate what size ball would settle at that rate. This gives us an "Equivalent Spherical Diameter." This is a method commonly used for sizing small irregular particles. such as drill solids, barites, etc. that are below a size easily screened. This method is particularly valid for drilling fluids work, because settling not only can occur in the hole or in a mud pit, it is an essential part of separation process in centrifuge and hydrocyclones.

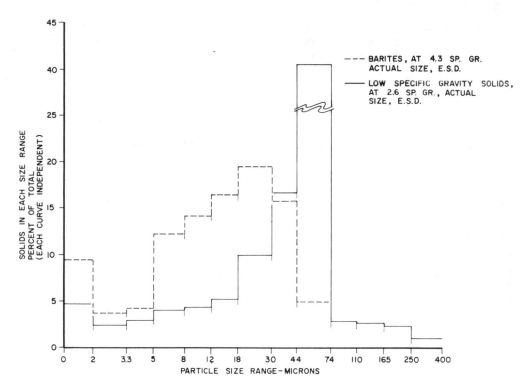

**Weighted Mud Particle Size Distribution
-Separation Problem As Seen By Screen
-Actual Equivalent Spherical Diameters**

FIG. 6–9. *Actual distribution — screen separation problem*

Two solids particles of known different specific gravities, or densities, will settle at the same rate if their size, or Equivalent Spherical Diameter, are of a certain mathematical relationship. This can be shown quite simply for two particles c and b if we assume their settling velocities are the same, or $J_c = J_b$. Then

$$\frac{d_c^2(D_c - D_L)G}{18\,\mu} = \frac{d_b^2(D_b - D_L)G}{18\,\mu} \tag{3}$$

$$d_c^2(D_c - D_L) = d_b^2(D_b - D_L)$$

$$d_c = d_b\sqrt{\frac{D_b - D_L}{D_c - D_L}}\,; \quad \text{or} \quad d_b = d_c\sqrt{\frac{D_c - D_L}{D_b - D_L}}$$

If the normal range of values for barite density, light (or drill) solids density, and the liquid phase of the mud system are substituted in this equation the result will always indicate that the light solids particle will be approximately $1\frac{1}{2}$ times the diameter of the barite solids particle when the two settle together. A 10 micron barite particle and a 15 micron light solids particle will not be separated by any settling device, including those utilizing centrifugal force.

Although Stokes Law as presented here is for Newtonian fluids, the $1\frac{1}{2}$ to 1 relationship is as valid in non-Newtonian muds as any rule-of-

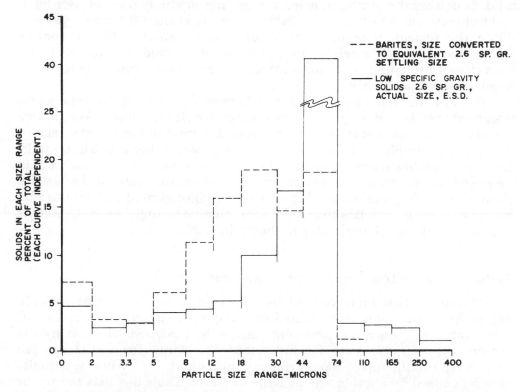

FIG. 6-10. *Strokes law equivalent distribution—settling separation problem*

thumb yet devised, and predictions using it agree well with field results.

Figure 6–9 presents unusual data in that high density solids were removed from a weighted oil mud by flotation, and then particle size distributions were run independently on each type solids.[15] The actual sizes are shown in Figure 6–9. The sample is from mud that has passed through a 12 × 12 shale shaker screen and the graph shows that a finer screen might have removed some more drill solids and perhaps without cutting deeply into the barites (assuming a fine screen could function properly in the situation).

Figure 6–10 shows the problem as it refers to any settling type separation device. The barite distribution has been re-sized (has been multiplied by approximately 1.5) to the equivalent settling size of the low specific gravity solids. This makes it obvious that any removal of light solids, at either the colloidal end or at the medium and intermediate end, by any settling device, will also take some barites. This is not necessarily bad, and the removal may be very desirable, but the sacrifice of barites must be expected and understood.

THE CONTRAST BETWEEN WEIGHTED AND UNWEIGHTED MUDS

Water as a drilling fluid does not qualify as a mud. If there are no hole (formation) problems that prevent its being the most economical drilling fluid; if neither the geologist, palentologist, nor production supervisor have valid objections; and if it is available, water is seldom if ever surpassed. When the formation requires, or a supervisor demands, filtrate control and/or viscosity and/or gels in the drilling fluid, a "mud" is built. Or if the fluid density required is too high for salt water alone, mud properties are required to suspend barites.

The first and most basic difference between an unweighted mud and a weighted mud is that money is spent on the unweighted mud to keep slurry density *down*, and money is spent on weighted mud to keep slurry density *up*. Another simple, and usually correct, criterion is that if a drilling fluid does not contain barites it is unweighted and if it does contain barites it is weighted. In the search for simplicity, it is sometimes missed that as the change is made from an unweighted to a weighted mud system the philosophy, or logic, of solids treatment also must be changed as the problems almost completely change. This is shown in Table 6–3.

The Factors Governing Total Solids Control in Weighted Muds

Although great improvement has been made in solids removal techniques in twenty years, in weighted muds there are overriding factors that cause total solids to vary, sometimes more than purposeful effort by the operator. If the reasons are not understood, the blame (for credit) may fall on the type mud, supervision, removal equipment, etc., when it actually may be caused by another one or none of those. Table 6–4 lists the major and minor governing factors.

TABLE 6–3

The Difference in Mud Solids Problems Between Weighted and Unweighted Water Base Muds

Items	Weighted Mud	Unweighted Mud
Costliest Portion	—Weight material	—Liquid, solubles & colloidals
Size of:		
Ideal Solids	—Mostly ultrafine & fine	—Mostly colloidal
Solids from Bit	—Colloidal to coarse	—Colloidal to coarse
Specific Gravity of:		
Ideal Solids	—High	—Low
Drilled Solids	—Low	—Low
Source of:		
Needed Solids	—Commercial ultrafine and fine barites —Commercial colloids	—Commercial colloid —Colloidal cuttings (variable)
Detrimental Solids	—Drilled fines and larger —Drilled colloidals —Fine and larger barites —Colloids mechanically degraded from larger particles.	—Drilled ultra fines and larger
Detrimental Effects of:		
Colloidal Cuttings	—$\overline{PV}$ increase (major)	—Density (minor)
Colloidal Barites	—$\overline{PV}$ increase (major)	—Not applicable
Size Degradation of All Solids	—$\overline{PV}$ increase (major)	—No problem with good removal system
Ultra Fine and Fine Cuttings	—$\overline{PV}$ increase (minor) —Source of degraded colloids	—Same as for fines and larger cuttings, see below:
Medium and larger Cuttings	—$\overline{PV}$ increase (minor) —Abrasion (variable) —Filter cake character —Source of degraded colloids	—Density (major) —Abrasion (major) —Filter cake character (major) —$\overline{PV}$ increase
Ultra fine and fine barites	—$\overline{PV}$ increase (minor) —Source of degraded colloids	—Not applicable
Medium and larger barites	—Slight $\overline{PV}$ increase —Abrasion —Filter cake character —Source of degraded colloids	—Not applicable

Contrary to popular belief, total solids do not always and consistently tend to run "high" in weighted muds. In drilling a medium hard formation with toothed bits, the necessity to mix new mud for the volume of the high percentage of large cuttings removed maintains a minimum solids content. When the fact is faced that the most economical and trouble-free method of drilling an extremely hard, or an extremely soft, formation with weighted mud will at best involve some higher total solids content than drilling a medium-hard formation, attention can be concentrated on the optimum

TABLE 6–4
The Factors That Control Total Solids in Weighted Muds
(Numbered in Order of Decreasing Influence)

Item	Contributes to:	
	Low Total Solids	High Total Solids

FIRST ORDER OF IMPORTANCE

Item	Low Total Solids	High Total Solids
1a. Type formation	—Medium hard	—Unconsolidated —Very hard
1b. Bit cutter type	—Long teeth	—All —Single diamond teeth
2. Mud density	—Minimum	—Above Minimum
3. Bit jet horsepower	—Adequate	—Inadequate
4. Annular lift	—Adequate	—Inadequate
5. Rig shale screen	—Constant efficient operation	—Bypassed

SECOND ORDER OF IMPORTANCE

Item	Low Total Solids	High Total Solids
6. Full flow hydrocyclone removal of mediums and larger solids	—Effectiveness varies with formation & bit	—Not applicable
7. Rig screen mesh	—Fine*	—Coarse*
8. Fine screen used to return to system part of the liquid, colloids, ultrafines, and fines from hydrocyclone underflow (6. above)	—Secondary separation can not reduce solids	—Variable increase in total solids, but more than centrifuge salvage (10. below); no viscosity reduction.
9. Removing direct from system a fraction of colloids & liquid with centrifuge.	—Variable total solids effect, but good viscosity reduction	—Primary separation cannot increase total solids.
10. Centrifuge used to return to system the liquid and colloid from the hydrocyclone underflow (6. above).	—Secondary separation can not reduce solids.	—Variable increase in total solids, but less than screen salvage (8. above); no viscosity reduction.
11. Chemical treatment to prevent shale cutting dispersion	—Variable, but may help screen removal	—Normally does not, but decreases viscosity-to-total solids ratio.
12. Chemical treatment to disperse shalestone to colloidal size	—Variable, but can help centrifuge removal of shale.	—Normally does not, but increases viscosity-to-total solids ratio.

* Whether a finer screen will help noticeably in this primary separation depends upon the comparative size relation between the cuttings reaching the surface and the screen mesh and whether or not the finer mesh can be maintained in proper operation. If a "fine" screen cannot operate properly at full flow, a coarser screen will maintain lower total solids than a finer screen that is bypassed.

solids removal with equipment available, control of critical mud properties, and on the truly controllable drilling and completion parameters.

BASIC SLURRY

The equations that follow are basic to mud solids relations. From them can be derived literally countless mathematical expressions of these relations. The many and various nomographs available are based on derivations from these equations. The symbols and subscripts used agree with the list recommended in API Bulletin 13C as far as that list extends.

$$\frac{100\%}{S} = \frac{100\%(F_c)}{2.6} + \frac{100\%(F_b)}{4.3} \tag{1}$$

$$\frac{100\%(D_m)}{8.345} = (\%V_s)(S_s) + (\%V_w)(1) + (\%V_p)(.8) \tag{4}$$

$$W_s/Bbl. = \frac{(\%V_s)}{100}(S_s)(8.345)(42) \tag{5}$$

$$W_b/Bbl. = W_s/Bbl. \times F_b \tag{6}$$

$$W_c/Bbl. = W_s/Bbl. \times F_c \tag{7}$$

If it is assumed that $S_c = 2.6$ and that $S_b = 4.3$, then:

$$W_b/Bbl. = (8.8687)(\%V_s)(S_s - 2.6) \tag{8a}$$

$$\%V_s = \frac{W_b/Bbl.}{(8.8687)(S_s - 2.6)} \tag{8b}$$

$$S_s = \frac{W_b/Bbl.}{(8.8687)(\%V_s)} + 2.6 \tag{8c}$$

The symbols used in the equations thus far are:

D = Density, Lbs. per Gallon
F = Weight fraction of total solids weight
S = Specific Gravity
%V = Per Cent by Volume, by subscript
W = Pounds (weight quantity)

The subscript usages are:

b = barites
c = Low Specific Gravity Solids or Light Solids (non-barites)
m = Slurry, of mud or mud component
p = Oil
s = Solids as may exist in the slurry in question
w = Water

The variations between graphs and nomographs based on different assumed specific gravities for the low specific gravity solids and for the barites are not important. The margin of error in solids determination far outweighs these minor differences. But beware of graphs that attempt to classify muds as "good" and "bad" on the basis of specific gravity relationships. They are an invitation to high mud bills whether or not drilling benefits result.

SHALE SHAKERS

If we were to list the primary solids removal devices[17] in their best order of series operation, it would be: shale shaker, sand trap (or settling tank, or shale tank, etc.), hydrocyclone desander, hydrocyclone desilter, and centrifuge. Regardless of drilling and mud economics improvement with the other mechanical devices, none at present can operate properly and continuously without the protection of the shale shakers. The sand trap should serve to protect the downstream equipment temporarily if something happens to a shale shaker screen.

The term "shale shaker" is used in drilling mud work[11] to cover all the devices that in another industry might be differentiated as "shaking" screen, "vibrating screens," and "oscillating screens."[7] All three of these types are in use, although most would probably fall in the "vibrating screen" classification. The old cylindrical rotating, or "squirrel cage" screen is rarely seen.

The trend in the last decade to "fine screen shakers" (using screen cloth finer than 30 mesh[11]), whether replacing the conventional mesh screens or operating in series down stream on the liquid-undersize solids discharge of the coarser mesh conventional shale shaker, has magnified all the problems of operating shale shakers. Many disappointments were suffered when solids problems often were more severe with the new fine screen shakers than with the older shaker and coarser screens. Out of the confusion and disappointments some better and long-overdue information on shale shaker operation has become available.[11,16]

The particle size a shale shaker can remove depends almost completely upon the size and the shape of the mesh openings in the screen cloth. See Figure 6–11. An oblong mesh allows a heavier gauge wire, for the same minimum dimension opening as a square mesh, without excessive sacrifices in open area. However, it is obvious that under some conditions the square mesh will make a separation finer than the "equivalent" oblong mesh.

Brandt cites an example of a square 40 mesh and an "equivalent" number 40 oblong both of which will separate approximately 420 micron dry material, while in a 45-second Funnel Viscosity drilling mud the number 40 oblong cloth separated particles down to approximately 300 microns and the square 40 cloth separated down to about 200 microns. He ascribes this difference to a coating of fine particles on the wires, as shown in Figure 6–12.

Obviously, this finer separation due to coating of the wires is a step in the direction of coating of the apertures and complete loss of mud, and there is more tendency for this problem in the square mesh.

If a shale shaker unit (one vibrating mechanism including screen cloth and all appurtenances) has multiple screens in a series arrangement, as Figures 6–13A and 6–13B, the particle size separation will be determined by the finest mesh screen, which should be the bottom screen. If a shale shaker unit had multiple screens in a "parallel" arrangement, as Figures 6–14A

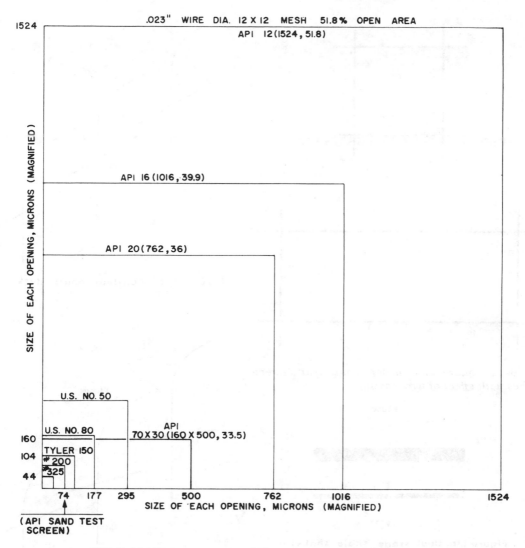

FIG. 6-11. *Common screen cloth openings – scale comparison*

and 6–14B, the size separation will be determined by the coarsest screen in the unit through which liquid passes.

The capacity of a shale shaker to handle oversize solids is determined primarily by the screen amplitude and motion. The amplitude, or one-half the stroke, of a shaker is determined by the vibrator eccentric weight, and is of primary importance in moving oversize solids. Whether the motion is eccentric or circular at various parts of the cloth also affects the transport of solids.

Less effective motion can be overcome by sloping the deck towards the solids discharge end (see Figure 6–15A), but this may be at the expense of wetter oversize solids carrying over more drilling fluid. The balanced mo-

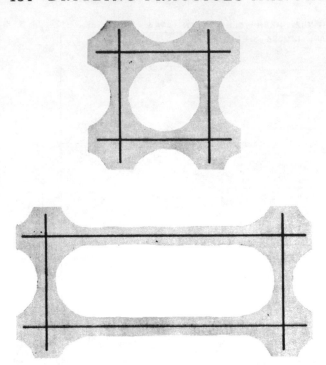

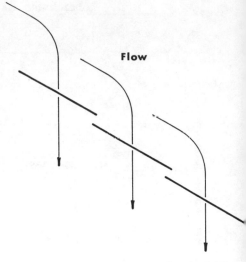

Figure 14a. Constant angle deck[16]

FIG. 6–12. *Square and oblong "equivalent" screen meshes with effect of wire coating*

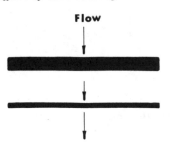

Figure 13a. Dual stage shale shaker[16]

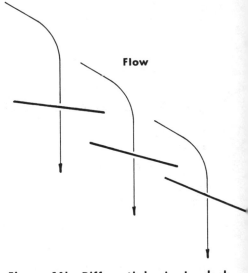

Figure 14b. Differential single deck

FIG. 6–14. *Screen cloths arranged in parallel*

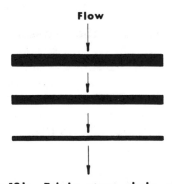

Figure 13b. Triple stage shale shaker

FIG. 6–13. *Screen cloths arranged in series*

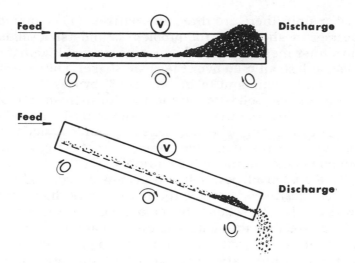

15a. Unbalanced motion creating combination circular and elliptical motion and showing the deck inclined to overcome the resulting solids pileup.

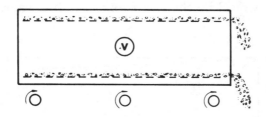

15b. Balanced motion, with uniform circular motion and resultant even solids flow irrespective of deck angle.

FIG. 6–15. *The effect of shale shaker motion*

tion, with the vibrator in the center of gravity of the vibrating mass imparts a uniform circular motion capable of moving the oversize solids even slightly uphill, the transport being relatively independent of the angle. See Figure 6–15B.

Liquid throughput capacity, if mud properties and formation solids in the feed remain equal, depends first and foremost on the transport of oversize solids from the screen cloth. If the mesh openings plug with near-size particles, or if the mesh openings coat over with sticky, undersize particles in a viscous mud, there is negligible liquid throughput capacity, and the screen cloth is "blinded." For this reason selection of screen cloth mesh can be critical to the entire solids removal installation in mud system.

If the screen installed on a shale shaker creates a blinding problem for the shale shaker unit, there are three alternatives: (1) change the screen mesh to a coarser mesh cloth if the problem is undersize solids coatings, or possibly to a finer mesh if the problem is near-size plugging, or (2) continue to blind and lose all mud over the shale shaker, endangering the job and wasting a sizeable amount of money, or (3) bypass the shale shaker, permitting all formation solids to pass to the downstream removal equipment, immediately plugging the desanders and/or desilters and endangering the hole and drill string. The only sensible course of action is number one. Unfortunately, the resultant chain of events not being visibly obvious, the most common choice in the field is number three.

Even a shale shaker of best solids transport design and fitted with a screen or proper mesh selection for the particle size distribution can be overloaded with solids. The solids to be removed always varies with the hole (volume) drilling rate, plus the sloughing rate if that is occurring. If a solids overload occurs, it is also reflected in a reduction in liquid throughput capacity. Figure 6–16 illustrates this effect both for series and for single screen arrangements. (Review Fig. 6–13). Figure 6–16 must be considered general and indicative, as screen size and other shale shaker details, viscosity and other mud specifications, hole size, etc. are not specified.

For example, if the cuttings were so fine, or if the first stages of screens in a series were of such a coarse mesh, that no cuttings were removed before the slurry reached the last (finest mesh) screen cloth, the series screen curve in this figure would be the same as for a single screen of the same

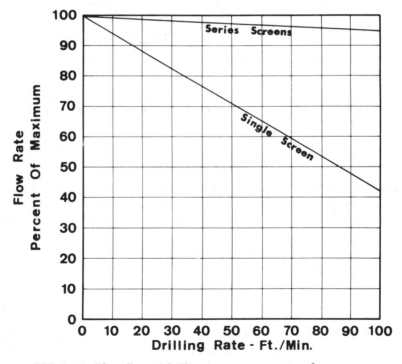

FIG. 6–16. *The effect of drilling rate on screen performance*

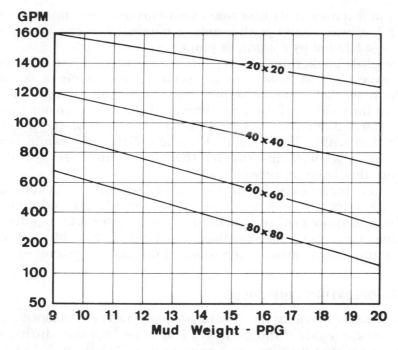

FIG. 6-17. *The effect of mud weight on screen cloth flow rate*

mesh. This does sometimes occur in the field, and then no benefit is realized from the series arrangement.

Figure 6–17, representing the effect of slurry weight on the liquid throughput capacity of a screen, is general for the finest mesh cloth of a series screen, or of a single screen, since the effect of cuttings load is omitted. The effect portrayed here is a combination of that of viscosity and of fine particles as well as the reduction in effective screening area for some typical meshes.

The frequency and severity of screen overload problems (or bypassing) is explained in great measure by the fact that the effects shown in Figures 6–16 and 6–17 are additive to each other. Screen liquid capacity decreases as drilling rate increases, as mesh size decreases, and as mud weight increases. Also as cuttings size decreases toward mesh size, and as viscosity increases, the liquid capacity decreases, both being additive to those conditions already mentioned. If the specification of shale shakers for an operation, or for a rig, is left to someone familiar only with the usual "catalog specs" without consideration for the most adverse combination of shale shaker conditions probable, the shale shakers will be inadequate.

From casual observation of conventional shale shakers operating with screen cloths of mesh coarser than 30 × 30, it has been assumed for years that shale shakers remove solids with relatively little loss of mud. Recent preliminary investigations[18] strongly indicate that mud loss ranges from one gallon of mud per gallon of net dry cuttings removed under the best conditions and coarser screens, to four gallons of mud loss per gallon of net dry solids removed with severe conditions and very fine screens. This range of loss was found with primary removal screens operating "efficiently," not blinded by coating or plugging.

Some shale shaker designs no doubt must be superior to others in removing excess liquid from oversize solids. For example, a screen cloth over which the solids move uphill should logically salvage more liquid mud (from the solids surface area) through the screen than one moving the oversize solids downhill; however, no investigation of the difference is available, and with a heavy solids loads any difference would tend to be lessened. Also, as particles smaller than "medium" size lose their free liquid film, they become very cohesive.

Economic limits exist for all methods of separation. As conditions become more severe, the shale shaker design principles and construction details become more critical, and maintenance costs climb rapidly. Anyone responsible for shaker specifications must familiarize himself with the field operating characteristics and limits of the various machines available.

SAND TRAPS—OPERATING PRINCIPLES

If the shaker screen were always adequate, never developed a tear that passed oversize solids through, never had to be bypassed during drilling, etc., the major justification for a sand trap (or "shale trap," or "settling tank") would disappear. Since that is unlikely in the near future, the sand trap is extremely important. Some information has been published on the subject,[3,19] but the following points are cardinal:

1. The sand trap usually receives the liquid slurry passing through the shale shaker, and should receive all liquid slurry bypassing the shale shaker and going to the active mud tanks.
2. Being a gravity settling compartment, it is not to be stirred and it must *not* be used as a suction compartment for *any* removal process.
3. A sand trap must have a discharge control easily and quickly opened and closed, so the settled solids can be dumped with minimum whole mud losses.
4. The sand trap should only be dumped, not "washed out." If the bottom is not sloped to the solids pile angle, the settled solids should be left to form their own sloped sides. "Cleaning the bottom," other than possibly at moving time, serves no purpose but increases loss of mud and mud cost.
5. Since Stokes Law applies in a sand trap, large quantities of barites (as well as API sand) may be settled from weighted drilling fluids. Provision for bypassing the undersize screen discharge slurry from the carrying pan direct to the next processing compartment is also advisable. As all compartments except the sand trap are stirred in a well-designed active system, this will prevent settling out of barite. The sand trap must *not* be bypassed if there is any problem with any other solids removal apparatus.
6. The mud exit from the sand trap should be over a retaining weir to a stirred compartment.

Rig operators who have engineered a proper sand trap into their removal systems have thereby increased the utility of the other removal equipment and many times have been able to continue drilling progress under unexpectedly severe solids conditions.

DEGASSERS

Degassers sometimes are essential to the solids removal process.

Shale shakers remove a good portion of the gas from a badly gas-cut mud, especially if the Yield Point is as low as 10 lbs per 100 Sq. Ft. It is generally accepted among experienced field drilling personnel that special degassing equipment is not necessary if the Yield Point of the mud is six or less, although with prevailing high figures the knowledge is academic.

Hydrocyclones are fed by centrifugal pumps. Slurry-handling centrifugal pumps for abrasive oil-field muds are not able to maintain efficiency when pumping gas-cut muds. Hydrocyclones do not function properly if pumping (feed) head is not constant or if there is gas or air in the feed. Therefore provision for degassing should be in a stage between the sand trap and the first hydrocyclones.[19]

HYDROCYCLONES

Hydrocyclones can perform the finest cut of any primary separation equipment operating on the full-flow circulating rate of an unweighted mud system. Some understanding of the most common designs is essential to proper operation. The hydrocyclone (see Figure 6–18) has a conical shaped portion in which most of the settling takes place, and a cylindrical feed chamber, sometimes called the "barrel," at the large end of the conical section. At the apex of the conical section is the underflow opening for the solids discharge. In operation, the underflow opening is usually at the bottom, for convenience, and "bottom" or "top" as used in this chapter concerning hydrocyclones will refer to that mode.

Near the top end of the feed chamber, the inlet nozzle enters tangential to the inside circumference and on a plane perpendicular, or nearly so, to the top-to-bottom central axis of the hydrocyclone. A hollow cylinder, called the "vortex finder," extends axially from the top into the barrel of the hydrocyclone past the inlet. The inside of the vortex finder forms the overflow outlet for the liquid discharge or effluent. The overflow outlet is much larger than the underflow opening. The "size" of a hydrocyclone is determined by the largest inside diameter of the conical portion. All dimensions are critical to the operation of any specific design and size.

The hydrocyclone obtains its centrifugal field from the tangential velocity of the slurry entering the feed chamber. An axial velocity component is created by the axial thrust of the feed stream leaving the blind annular space of the feed chamber. The resultant is a downward spiraling velocity represented by "S" in the velocity diagram in Figure 6–18, in which "T" represents the tangential and "A" the axial velocity component.

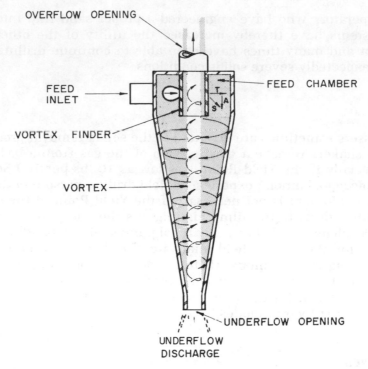

OVERFLOW OPENING

FEED INLET

FEED CHAMBER

VORTEX FINDER

VORTEX

UNDERFLOW OPENING

UNDERFLOW DISCHARGE

FIG. 6–18. *Hydrocyclone nomenclature from API bulletin 13C*

Balanced Design

In a hydrocyclone the stream spirals down along the wall of the conical section to the underflow opening, reverses axial direction, and then spirals upward to exit at the overflow opening. A balanced design hydrocyclone can be adjusted so that when a clean liquid is fed there is no underflow discharge. When separable solids particles are fed to such a hydrocyclone, solids are discharged at the underflow opening, each particle taking with it a free liquid film covering its surface area.

The solids particles in the turbulent downward-spiraling feed stream are settled outward toward the wall, following modifications of Stokes Law. When the liquid stream reverses to spiral upward, the solids particles continue spiraling downward through the underflow opening due to their greater mass and higher inertial forces. See Figure 6–19.

When a balanced design hydrocyclone is adjusted with the underflow opening too small in relation to the balance point opening, a "beach" of dry cone wall exists between the edge of the actual underflow opening and the balance point, or the point of liquid turnback. If the hydrocyclone is capable of settling any ultrafine solids (below 44 microns), this adjustment or design will cause sticking and plugging as these ultrafine particles lose their liquid film attempting to traverse the beach. This undesirable adjustment is described as "dry bottom" and the trouble it causes is known as "dry plugging."

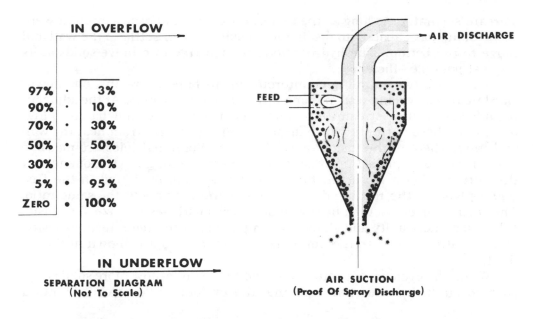

SEPARATION DIAGRAM
(Not To Scale)

AIR SUCTION
(Proof Of Spray Discharge)

SEPARATED SOLIDS FREE TO DISCHARGE AT UNDERFLOW
MUD LOSS PROPORTIONAL TO SURFACE AREA OF REMOVED SOLIDS
HIGHEST SOLIDS REMOVAL EFFICIENCY OPERATION AND ADJUSTMENT
LOWEST BLADDER AND VORTEX WEAR RATES

FIG. 6–19. *Balanced design hydrocyclone in spray discharge*

If a balanced design hydrocyclone is opened to an underflow opening greater than the balance point diameter, a cylindrical shell of spiraling liquid will discharge unless there are enough separable solids to take up the unbalanced discharge space. This adjustment is termed "wet bottom" and unless it is excessive, it does not, contrary to appearances and general belief, result in high mud losses in normal unweighted muds and at normal drilling rates. As a matter of fact, a wet bottom adjustment will run a cheaper mud, a better drilling mud, and a better hole-conditioning mud than will a dry bottom adjustment or design. The ideal balance point adjustment is not always known and is not absolutely necessary.

Figure 6–19 depicts a balanced design hydrocyclone operating in spray discharge. The air being sucked in at the underflow opening is the proof of spray discharge, and it can be felt with the end of a finger for absolute verification. This air is simply entering the low pressure zone created by removal of the air from the vortex by the friction of the high-velocity, upward-spiraling inner stream, which takes the vortex air out with it through the overflow opening. The flow rates in mud hydrocyclones are so high that total retention time, from inlet to either exit, is less than one-third of a second.

The opposing flows of two fluids through the underflow opening in spray discharge (air in, and a liquid-solids slurry out), indicate the underflow opening is not acting as a choke. It is a circular weir, or ring dam (as is the overflow opening), well known in liquid-liquid separation such as in a cream separator. As long as the underflow opening acts as a circular weir, the cyclone has capacity to discharge all solids it can settle to the wall and move to the bottom. The hydrocyclone is then free to remove solids at its highest possible efficiency.

There are two spiraling countercurrent turbulent streams, one downward near the cyclone wall and one upward near the vortex. At the adjacent boundaries of these streams eddy currents are created somewhat as in Figure 6–19. These currents sweep up some solids particles from near the wall and move them to the inside upward stream. Some particles are massive enough not to be picked up at all; some are small enough to seldom get out the bottom. Those too small to be separated at all are found in the free liquid phase of the feed, underflow, and overflow in about the same ratio. The separation curve for a hydrocyclone varies with design, size, feed head, solids specific gravity and shape, percent solids in the feed, feed viscosity, and adjustment. Illustrative (only) separation curves are shown in Figure 6–21.

When the cyclone can settle and send to the underflow more solids than pass through the underflow opening, the cyclone is "overloaded" and a

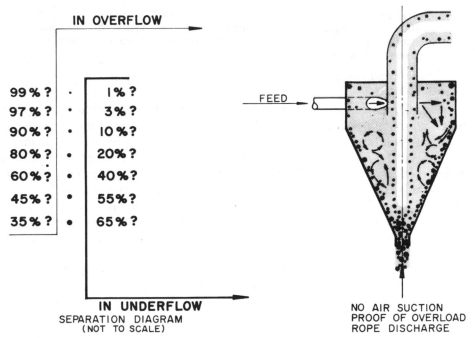

IN OVERFLOW

99 % ?	·	1 % ?
97 % ?	·	3 % ?
90 % ?	·	10 % ?
80 % ?	·	20 % ?
60 % ?	·	40 % ?
45 % ?	·	55 % ?
35 % ?	·	65 % ?

FEED

IN UNDERFLOW
SEPARATION DIAGRAM
(NOT TO SCALE)

NO AIR SUCTION
PROOF OF OVERLOAD
ROPE DISCHARGE

UNDERFLOW SOLIDS DISCHARGE RESTRICTED

MUD LOSS PROPORTIONAL TO SURFACE AREA OF REMOVED SOLIDS

LOW SOLIDS REMOVAL EFFICIENCY

HIGH BLADDER AND VORTEX WEAR RATES

UNDERFLOW PARTLY PLUGGED

UNDERFLOW WILL INEVITABLY AND SOON PLUG COMPLETELY

FIG. 6–20. *Balanced design hydrocyclone operated in rope discharge*

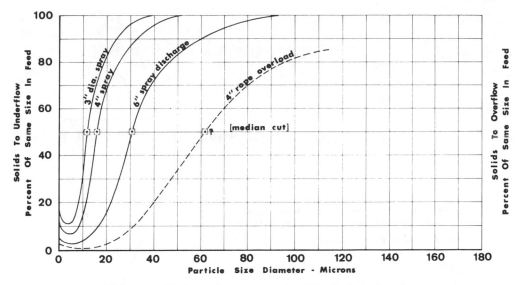

FIG. 6–21. *Hydrocyclone separation. Typical perforation*

balanced design underflow will take on the characteristics of a rope, or sausage links. From this we get the terms "rope discharge" and "sausage discharge" for the overload condition. There is no air suction at the underflow opening in rope discharge, and in a balanced design this is the proof test for overload. See Figure 6–20. The underflow opening is no longer a weir; it is now a choke.

In a rope discharge, not all particles reaching the underflow opening can exit. Due to the inertial separation situation, the smaller ones must exit at the overflow, and immediately. The difference between rope and spray underflow discharge is predictable. Smaller particles have more surface area per pound, therefore the rope underflow has *less* solids surface area per pound of solids than a spray discharge underflow. The liquid accompanying the solids in a balanced cyclone is a function of surface area, therefore the rope discharge has less liquid per pound of solids and is heavier in slurry weight than a spray discharge underflow. It is not an indication of greater efficiency. Unfortunately, the finer solids remaining in the system do increase mud weight just as effectively as coarser solids and the finer solids particles still will take their liquid when they are finally removed, unless a secondary device is used. See Figure 6–21 again for an example of typical separations in a 4″ hydrocyclone operation changing from spray to rope discharge.

Flood Bottom Design

Another design of hydrocyclones is shown in Figure 6–22. This can be described as a "flood bottom" design, as it has no balance point and both the overflow and underflow openings act as chokes at all times under all conditions. For example, if clear liquid is fed to a flood-bottom cyclone with

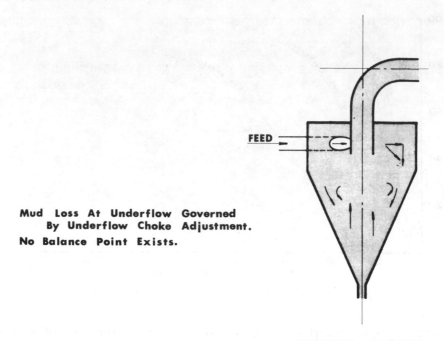

FEED

Mud Loss At Underflow Governed
By Underflow Choke Adjustment.
No Balance Point Exists.

**PRESSURED DISCHARGE
FLOOD BOTTOM**

FIG. 6–22. *Flood bottom hydrocyclone design*

a ten millimeter bottom opening, it will discharge a forced underflow stream ten millimeters in diameter. Adding separable solids to the feed will cause solids to discharge in the underflow, but has little effect on the discharge volume rate.

Liquid underflow discharge is not a function of solids discharge unless an overload condition is reached so that not all reporting solids can exit in the underflow. The best adjustment when operating a flood bottom design hydrocyclone will be a compromise between the minimum opening required to prevent underflow discharge of all mud during the low-solids feed period (making drill pipe connections), and the maximum opening required to remove cuttings during the short high-solids feed period (pumping the kelly down) in fast hole.

Hydrocyclone Capacity

As with the shale shaker and all other mechanical separation devices, the hydrocyclones have two capacities—feed slurry capacity and underflow solids discharge capacity. As a design problem, the two capacities are related, but the resultant products available commercially have an extreme variance in their relative values.

"Feed capacity" is the volume rate of feed slurry a specific hydrocyclone will accept at a specific feed head. Head may be expressed in feet or meters; it cannot be expressed as pressure unless feed slurry weight is also specified.

As an example, we might find some size and make of hydrocyclone rated as having a feed capacity of 100 GPM at 75 feet of head. If the feed were water, which has a hydrostatic head of 0.433 PSI per foot, the feed pressure would be 32.48 PSI. If the feed slurry weight were to increase to 16.64 PPG (Specific Gravity of 2.0), the hydrostatic head gradient would be doubled to 0.866 and 75 feet of feed head would amount to 64.95 PSI. The volume feed rate would be the same with a 16.14 PPG slurry at 64.95 PSI as it would with a 8.32 PPG slurry at 32.48 PSI, because the head is the same.

Centrifugal pumps operating at constant speed (electric motor or closely governed engines) and feeding the same equipment, automatically maintain a constant head as slurry weight changes. Further reading on this subject[19] is strongly advised for anyone operating or installing hydrocyclones.

Since pumping mud through a hydrocyclone is of no economic or physical benefit, and is expensive, the solids removal ability is the real key to the value of the equipment and the operation, not the feed capacity. Solids removal ability consists of two things: first, the high separation forces and internal flow pattern to settle finer particles (most difficult ones) to the underflow opening; second, a large balanced underflow opening so the cyclone can discharge the settled solids at a maximum rate before reaching solids overload. If a balanced cyclone does reach the overload condition the underflow will rope and removal ability is ruined as shown in Figure 6-21. Obviously, a balanced design can be operated economically with a much larger opening than can a flood bottom design.

A rare series of hydrocyclone performance tests were run in 1964 under the auspices and using the facilities of the National Science Foundation Mohole project, with the actual designer or head engineer responsible for field performance adjusting his own hydrocyclone in each test. Since that test series at least one of these cyclones has been discontinued in the oilfield market, at least three have been modified, and at least three new models have been introduced by these manufacturers. However, the difference and dilemmas pointed up by these tests still are found and they provide valuable insights.

Table 6-5 presents unaltered summaries[20] of performance results, in order of decreasing solids removal. Rating "A" gives top rating for removing the highest percentage of solids entering; Rating "B" penalizes for liquid in the underflow. These are contrary ratings, and the two multiplied together give a "total" performance rating, which surprisingly was in almost exactly the same order as Rating "A," or per cent removal only. This arrangement is in the approximate order of practical assistance to a field man desiring to maintain minimum solids as economically as possible in his unweighted mud system. The cyclones in the higher ratings will literally remove solids at a cheaper rate per pound than those in lower ratings given the same feed materials. Practical limits do exist in operational mechanics.

TABLE 6–5
Mohole Hydrocyclone Evaluation
(From Timed and Measured Data Taken During Tests)

Original order of removal efficiency. (Also approximate increasing order of equipment cost for equal underflow solids removal rates).

Make	Cone Diameter Inches	Feed GPM	Rating A UF Solids as Vol. Rate % of Feed Solids	Rating B UF Solids % by Wt. of Total UF	A × B Total Ratings
W	3	16.76	11.72	56.25	670
X	2	22.03	11.69	49.30	575
W	4	57.12	8.94	56.83	507
X	6	108.39	5.95	47.26	281
Y	4	62.65	4.55	47.96	218
W*	6*	88.73*	3.28	59.31	194
Z	4	44.90	1.55	65.25	101
X	12	407.00	1.80	54.46	98
Z	8	167.70	.746	66.70	49.8

* Company "W" requested, prior to any evaluation work, that their 6" cone test be deleted as it was not taking the proper feed volume due to an undetermined malfunction. (Request denied.)

In Table 6–6 the data has been rearranged in the approximate order of increasing purchase cost per volume rate of feed. In specifying new equipment, this "initial cost" often determines the choice, as the specifier is seldom field experienced and is usually pressured to hold down the initial equipment cost for totally processing (feed rate) some specific maximum rig pump capacity. It is immediately apparent that the cost per unit of feed rate capacity is almost diametrically opposed to the cost per unit of solids discharge ability.

TABLE 6–6
Mohole Hydrocyclone Evaluation
(Rearranged in Order of Increasing Cost per Volume Rate of Feed)

Make	Cone Dia. In.	Feed GPM	UF Solids as Vol. Rate % of Feed Solids	Total Ratings
X	12	407.00	1.80	98
Z	8	167.70	.746	49.8
W*	6*	88.73*	3.28	194
X	6	108.39	5.95	281
Y	4	62.65	4.55	218
W	4	57.12	8.94	507
Z	4	44.90	1.55	101
X	2	22.03	11.69	575
W	3	16.76	11.72	670

* Malfunctioning.

The inexperienced are quite often unable to accept that there are big differences in performance between hydrocyclones of the same size, as shown in Table 6–5. In Table 6–7 the cyclones are in the same order of rating as in Table 6–5, but they are columned by manufacturer, or by design philosophy. It is apparent from Table 6–7 that smaller cyclones of the same design have greater separation ability, but also that different design philosophies can cause even more difference than size. There is no substitute for observing comparative equipment performance in the field and issuing specifications based on performance.

TABLE 6–7
Mohole Hydrocyclone Evaluation
(In Original Order But Arranged in
Show Effect of Design)

					A × B
	Cone Make & Size			Feed Head Feet	Total Ratings
W	X	Y	Z		
3				76.1	670
	2			105.7	575
4				76.1	507
	6			76.8	281
		4		76.8	218
6*				76.1	194*
			4	67.1	101
	12			67.1	98
			8	80.7	49.8

* Malfunctioning.

Hydroclone Plugging Effects

The results of underflow plugging in hydrocyclones is as disastrous as blinding of shale shakers, but the effects sometimes are not as obvious. Figure 6–23 illustrates the change in balanced hydrocyclone operation from normal to plugging at the underflow. To look at it, a cyclone plugged in this manner offers no clue to the damage it is inflicting. Actually, it is undoing the cleaning work of all other cyclones on the unit by mixing unprocessed mud directly into the cleaned overflow mud from the other cyclones on the same header. At the same time, the inner parts of the cyclone and the overflow pipe are rapidly being cut to pieces by the particles that should have been discharged out the bottom. A plugged cyclone must be unplugged or removed from the manifold immediately.

Underflow plugging is usually caused either by a dry beach or by solids overload. The dry beach effect can be from improper design (dry bottom), or from too small an adjustment at the underflow. Solids overload can be overcome by adjusting to a larger underflow opening if possible, or installing more cyclones in the stage, or removing more solids in a stage ahead

NORMAL OPERATION

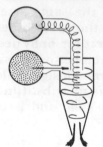

UNDERFLOW PLUGGED

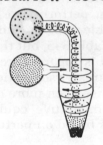

High Feed Velocity

Strong Cyclone Action

Balanced Operation

Good Solids Separation

Good Solids Removal At Underflow

Clean Mud To Overflow And Rig Pumps

Normal Hydrocyclone Wear

Lowest Cost Operation

High Feed Velocity

Strong Cyclone Action

Balance Not Possible

Some Solids Separation

No Solids Removal At Underflow

Oversize Solids To Rig Pumps

Rapid Wear On Hydrocyclone And Rig Pumps

Very High Cost

FIG. 6–23. *The effect of plugging the underflow discharge of hydrocyclones*

FEED PARTLY PLUGGED

FEED COMPLETELY PLUGGED

LOW FEED VELOCITY

NO CYCLONE ACTION

FLOOD (PRESSURED) BOTTOM

NO SEPARATION

WHOLE MUD LOSS AT
    UNDERFLOW, HIGH RATE

NO HYDROCYCLONE WEAR

VERY HIGH COST

NO FEED VELOCITY

NO CYCLONE ACTION

FLOOD (PRESSURED) BOTTOM

NO SEPARATION

WHOLE MUD (CLEANED BY
    OTHER CONES) LOSS AT
    UNDERFLOW, HIGH RATE

NO HYDROCYCLONE WEAR

VERY HIGH COST

FIG. 6–24. *The effect of plugging the feed inlet of hydrocyclones*

of the roping hydrocyclones, or by purchasing cyclones of a better design with larger balanced underflow openings.

Figure 6–24 shows the effects of feed plugging. With partial plugging, mud is feeding, but has a low velocity in the feed chamber. The cyclone body acts as a swirling funnel and the mud lost at the underflow is the same as pit mud and may be discharging at a volume rate from 10 to 200 times the volume rate of a normal balanced underflow discharge. The underflow may even contain some cleaned mud from the overflow headers. The cost of this mud loss, and the aggravation of the end-of-location cleanup problem cannot be tolerated. A cyclone flooding feed and overflow mud at the bottom must be unplugged, repaired, or removed immediately.

Complete plugging of the inlet will result occasionally in no discharge at the underflow (air suction through the apex into the overflow manifold). Most often, it results in a high rate discharge of cleaned overflow mud to the waste pit. This must be stopped immediately.

Feed plugging is caused by poor housekeeping on the mud system. Bypassed shale shakers, or torn screen cloths are the most common offenders. Special large open area suction screens installed on the end of the centrifugal pump suction inside the suction pit can reduce plugging 90 per cent, but even this is not enough to cope with a bypassed shale shaker, especially if there is not a good sand trap.

DECANTING CENTRIFUGES

The decanting centrifuge is the only liquid-solids separation device used on drilling fluids that can remove (decant) all free liquid from the separated solids particles, leaving only adsorbed liquid or "bound liquid," on the surface area. This adsorbed liquid is not prone to contain solubles, such as chlorides, nor colloidal suspended solids, such as bentonite. The dissolved and suspended solids are associated with the continuous free liquid phase from which the decanting centrifuge separates the inert solids, and are removed with that liquid. The adsorbed liquid can only be removed from the separated solids by evaporation, which has been neither desirable nor practical so far in drilling mud work.

Although centrifuges of other types, notably the nozzle solids discharge type with vertical axis, has been tested periodically in this country since the early thirties for use on weighted water base muds, the decanting centrifuge was first tested and proved adaptable as a practical field tool in 1953.[1] It moved rapidly into the rental field and is now recognized universally as a useful tool, though perhaps not as well understood as it might be.

A more complete description of present decanters would be "continuous conveyor, conical bowl, decanting centrifuges." A schematic cutaway of the actual separating chamber is shown in Figure 6–25. The size of the machines is usually given as maximum bowl diameter and length. Decanting centrifuges in drilling fluid use are the Pioneer Centrifuging Company 18″ Dia. × 28″ Long; Bird Machinery Co. 18″ × 28″ (Baroid and others); Sharples 14″ × 20″ (independent); and the Gilreath approximately 14″ × 22″ (Dresser-SWACO and others). The principle of operation is the

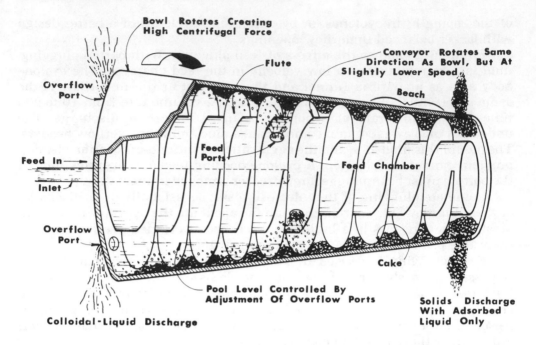

FIG. 6–25. *Decanting centrifuge; sectional view of separation*

same for all these. All else equal, the larger machines have higher load capacities at equal separation ability, and finer separation ability at the same loads. The Gilreath machine has departures from conventional design that increase its feed capacity at a heavy sacrifice of separation ability.

Operating Principles

Referring to Figure 6–25, the separation occurs inside the bowl, which is rotated at speed ranges that vary (according to the bowl size, conveyor design, application, and preventive maintenance program) from 1500 to 3500 RPM. Inside the bowl is usually a double lead conveyor connected to the bowl through an 80:1 gear box so the conveyor will lose one revolution for every 80 revolutions of the bowl. Thus there is a relative rotation of 22.5 RPM when the bowl speed is 1800 RPM. This relative rotation is the conveying speed, and it can be varied by using other gear ratios.

The mud slurry to be treated is metered to the centrifuge through a feed tube centered in the hollow axle to the feed chamber. From there the feed passes through feed ports into the leads in the separation chamber. The scrolls, or blades (one in the single lead mode), form the walls of the leads (or lead). As the slurry is slung outward into the annular ring of mud called the pool, it also accelerates to the approximate rotational speed of the conveyor and bowl. The depth of the pool is determined by the adjustment of the eight to twelve overflow openings or overflow ports.

The slurry flows through the leads toward the overflow ports under very high centrifugal force and in laminar flow in most machines and applications. Following Stokes Law, the solids settle outward to the outer walls, the most massive particles settling first; and the lesser ones further along the leads. Those not having committed to the solids cake against the wall by the time the slurry stream is increasing in velocity to exit through the overflow ports, will not be settled. They will pass out the overflow with the liquid. In the normally recommended operating ranges, retention time in the machines will vary from 10 to 80 seconds (less for the Gilreath-designed machine).

As the solids are settled outward against the wall, the conveyor continuously plows or scrapes them toward the small end of the bowl, out of the pool, and across that portion of the wall between the pool and the solids underflow discharge openings called the "beach." As the solids cake crosses the beach, all free liquid is removed by both centrifugal squeezing of the cake. The free liquid drains back to the pool, taking with it the colloidal and dissolved solids. As the separated solids discharge out the underflow, they contain only adsorbed, or "bound," liquid.

Dilution of the feed to a decanter has only one purpose—to reduce the viscosity of the feed to aid in maintaining separation ability. The most convenient rule of dilution is to add only enough dilution water to reduce the effluent to 37 Marsh funnel seconds but do not add any to lower the viscosity to less than 35 seconds. More dilution will decrease settling time out of proportion to the increase in settling rate.

Dilution is applicable only to water base muds. Continuous oil-phase muds respond better to heat, and stove-oil dilution for centrifuging is not advised. Heating above 90°F. (32° Celtius) for centrifuging usually is neither economical nor necessary, but separation does benefit at a decreasing rate up to at least 140°F. (60° Celtius).

Capacities

The solids discharge capacity of a decanting centrifuge is limited by the volume rate of solids cake that can be moved by the conveyor and discharged by the underflow openings. The liquid discharge capacity of a decanter is limited by the capacity of the overflow ports to discharge liquid at whatever pool depth adjustment they may be set. The feed rate capacity of the decanting centrifuge will be determined by the volume of separable solids in the feed or by the free liquid and colloid content of the feed, according to which discharge limit is being approached—liquid or solids. This is to say that a feed mud low in solids has the feed rate limited by liquid discharge capacity; a feed very high in separable solids (a heavily weighted mud) has the feed limit determined by the solids discharge capacity.

If a very fine separation is desired, it may be necessary to hold down feed volume rate in low solids feeds to stay within laminar flow in the settling chamber and also to maintain retention time for fine separation ability.

Separation Limits

An effective decanting centrifuge for weighted water base mud work should be capable of making a separation approximately at the two micron barite particle size. Due to the settling behavior according to Stokes Law, light solids would be separated down to three microns ($1\frac{1}{2} \times 2$ microns) in the same machine under the same conditions. In a continuous oil phase mud, the effective viscosity of the oil is greater than that of water in a water base mud, therefore the separation in the centrifuge is not as fine.

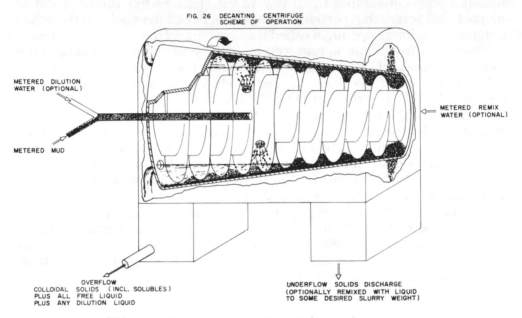

FIG. 26 DECANTING CENTRIFUGE
SCHEME OF OPERATION

METERED DILUTION
WATER (OPTIONAL)

METERED REMIX
WATER (OPTIONAL)

METERED MUD

OVERFLOW
COLLOIDAL SOLIDS (INCL. SOLUBLES)
PLUS ALL FREE LIQUID
PLUS ANY DILUTION LIQUID

UNDERFLOW SOLIDS DISCHARGE
(OPTIONALLY REMIXED WITH LIQUID
TO SOME DESIRED SLURRY WEIGHT)

FIG. 6–26. *Decanting Centrifuge, Scheme of operations*

In field operation, the decanting centrifuge is fitted with a housing over the bowl, liquid and solids "hoppers" to take the liquid and the solids discharges of each respectively, a slurry feed pump, dilution water connection, and remix water connection if desired. See Figure 6–26. The feed is metered with a calibration of one kind or another. The bowl speed usually can be varied, but variation is best left to the centrifuge technician, as it is seldom either necessary or desirable. To reduce damage to expensive gears, there are shear-pins, trips, limit switches, etc. to shut down the equipment automatically in case of a solids overload.

Dilution water connections should be equipped with a flow gauge or meter and a flow adjusting valve. Remix connections are usually provided. "Remix water" can be added in the housing to aid in softening the underflow solids for more rapid mixing into the system, or for reducing the underflow solids slurry density to the range of system density, as desired.

PERFORATED ROTOR CENTRIFUGES

The Perforated Rotor Centrifuge[21] was developed in the middle 1960's in an effort to provide a liquid solids separator with all the advantages and none of the disadvantages, real and imagined, of the decanting centrifuge. It does not decant, and so has never been widely accepted, but it should be understood as a tool.

Operation

The perforated rotor centrifuge design consists of a 6 inch outside-diameter clindrical shell rotor 42 inches long, with 474 one-half inch holes drilled in the shell and evenly spaced, spinning in a horizontal position inside an 8″ inside-diameter stationary pipe housing fitted with fluid seals at each end for the axles of the rotor to protrude. See the schematic diagram, Figure 6–27. A slurry to be treated is fed by a positive displacement metering pump and is mixed with dilution water (assuming water base mud) from a second metering feed pump. At least part of the dilution water is used also to flush the rotating seals. The diluted feed enters the annular separation chamber through an opening into one end of the annular space.

Under centrifugal force, larger solids are concentrated against the outer annular wall by Stokes Law. At the opposite downstream end of the annular

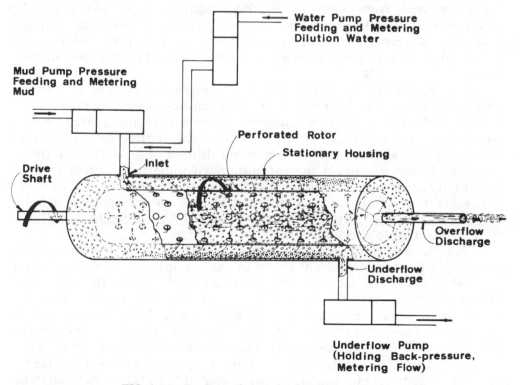

FIG. 6–27. *Perforated rotor centrifuge separator*

space, and on the periphery, is an underflow solids discharge opening. All feed material, both liquid and solids, would discharge out this opening were it not for a positive displacement pump which is arranged to hold *back-pressure* on this underflow opening and to permit the underflow solids slurry to discharge at a metered rate.

Separation Parameters

According to the rate setting of the underflow discharge metering pump as a ratio of the rate setting of the feed metering pump, a volume fraction of the feed is forced out the overflow; or equally true, that volume portion of the feed not permitted to leave at the underflow discharge can only exit at the overflow, assuming rotor seals are in good repair. When rotating at constant, near-design speed, the particle size of the median cut varies approximately by the square root of the product of the overflow volume rate times the overflow viscosity. On water base mud, normal operational variations within recommended range will result in a median cut point varying in the range of 3.5 to 5 microns, 2.6 specific gravity (equivalent to 2.4 to 3.5 microns median cut on barites).

The principle potential benefit of the perforated rotor centrifuge is portability. For this, it is necessary to discharge the underflow as a very fluid stream in order to be able to flow this solids discharge, under its own discharge head, from some spot on the drilling location easily accessible to a trailer, over to the active mud system, without excessive line plugging problems. Operating literature recommends an underflow solids slurry weight approximately the same as the undiluted feed (active system mud) weight.[22]

If the perforated rotor machine were operated at the condition of underflow weight equal to feed weight at zero dilution, there would be 100% underflow discharge of all feed and zero overflow of anything. At the recommended operating dilution of 70% of the feed volume rate, the liquid-colloidal phase of the feed mud is being diluted at an average ratio approximately one-to-one. At the recommended underflow weight equal to undiluted feed weight, the removal of undersize material (clay or colloidal) to the overflow discharge is approximately 50%. If dilution is doubled to two-to-one, the clay fraction to the overflow is 67%; at three-to-one dilution the clay fraction to overflow is 75%.

Clay fraction to the overflow can be increased at any dilution rate by decreasing the liquid-colloidal phase in the underflow (increasing underflow weight), which increases the plugging problems in the underflow. This might be alleviated by installing the machine directly over the active mud system receiving tank, but it would eliminate the only advantages of the machine over the decanter. At the one-to-one recommended dilution the perforated rotor machine must process twice the amount of mud as a decanting centrifuge to remove the same amount of clay-sized particles. Recommended capacity calculates only about 100% higher than the larger decanters, but can be pushed at some sacrifice of separation ability. The perforated rotor machine averages a slightly poorer salvage of weight material than the better decanting centrifuges, but much better than the poorest designs of decanters available for drilling muds.

EQUIPMENT APPLICATIONS—PRIMARY SEPARATION

The influx of machines and ideas concerning liquid-solids separation in the last decade requires some classifications of process for any ordered discussion. Primary separation, as used here, refers to the use of liquid-solids separators to *remove* either a liquid-undersize solids fraction or an oversize solids fraction from the active mud system. Any separator taking its feed directly from an active system and returning either the liquid (less solids than the feed stream) or solids (more solids than the feed stream) back to the active system has in effect removed a selective fraction from the system, and is performing primary separation. Figure 6–28 is a schematic diagram of typical primary separation equipment related to a system.

From prior to 1952 to the present, improvement in primary removal equipment has had considerable impact on general mud weighting practices, as shown in Figure 6–29. The lowest drilling mud weights economically possible, the cost of maintaining a given weight mud (assuming the same type chemical system), the lowest weight at which barites are used, and the highest weight at which barites are not used, have all moved to the left and downward on this diagram. At the same time, the high limit of drilling mud weight that can be maintained, as a matter of practicality,

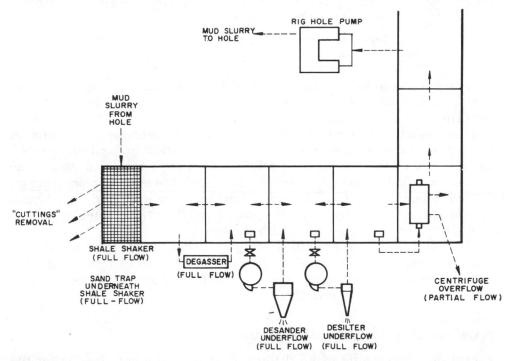

FIG. 6–28. *Primary separation equipment in proper schematic relationship to a drilling fluid system*

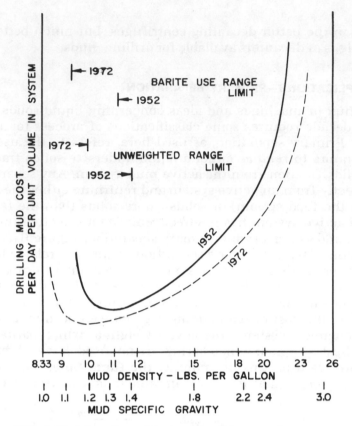

FIG. 6–29. *The effect of improved solids control on water base drilling fluids*

has moved to the right. It is not possible to insert actual cost figures because of innumerable variables, however for a specific area one could be built easily from actual records.

This reduction in cost at any given weight is because of the greater efficiency of mechanical removal devices over dilution, or "watering-back," which was once the mainstay of solids control after mud had passed the conventional shale shaker. The technique of dilution is to increase mud pit level with liquid addition, discard a fraction of whole mud, add liquid, discard mud, etc. etc. Obviously this procedure effects no change in particle size distribution—the good goes with the bad.

Mechanical devices selectively remove solids, permitting us to alter the particle size distribution. They permit the removal of more of the un- wanted solids with less loss of beneficial (for the purpose at a specific time) solids than is possible by dilution.

FULL FLOW PROCESSING–UNWEIGHTED MUD

Full flow processing of all drilling fluid from the bore hole is for the purpose of formation solids removal, and selectively those in the inert (larger than clay or colloidal size) range. In this application the best 4″ hydrocyclones, or desilters, can perform the ultimate removal to the economic limit. For the desilting to be performed economically and efficiently, balanced design hydrocyclones must be used with spray type discharge, also called "umbrella" discharge because of its appearance. Solids discharge overload, or rope discharge, must be prevented.

Whole Mud vs. Selective Solids Discharge

Figure 6–30 is a working nomograph comparing the volume losses in removing a given amount of solids from an unweighted system by discarding whole mud (feed mud to a liquid-solids separator) against removing the same amount of solids at the underflow discharge of a mechanical separator. In the dotted line example, three underflow solids discharges from 9.5 PPG, 10.0 PPG, and 11.0 PPG density underflows are compared to the equivalent discard in each case of 9.0 PPG system mud, the same amount of solids assumed to be removed in each separate case.

These ratios of 1.7, 2.35, and 3.7 in the example represent only the total increased volume in waste pit size, or vacuum truck hauling, etc. over discarding underflow discharge slurry. The saving in the value of the discard, using mechanical equipment, is even higher in each case than the ratios indicate because as the solids content (a direct function of density in unweighted muds) increases, the liquid phase becomes less. Each barrel of underflow discard contains less liquid than a barrel of system, or feed, mud.

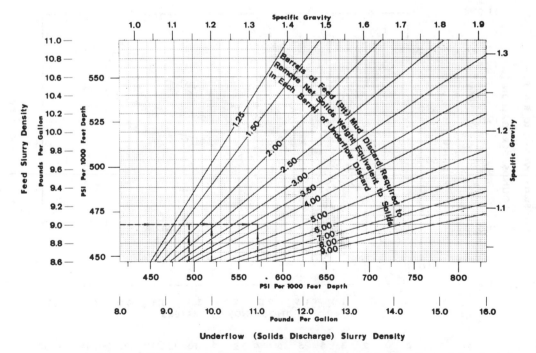

FIG. 6–30. *Reduction of total discard using solids separation equipment in weighted muds*

Finally, still referring to Figure 6–30, the underflow contains selectively the coarsest and usually the most troublesome of the inert solids in the system, and far less of any beneficial solids, than an equal volume of feed. Figure 6–30 is valid for any type solids removal device in an unweighted mud.

The knowledgeable drilling supervisor with an adequate removal system at his command will not permit whole mud discard from the active system under any normal drilling condition. In emergencies he will attempt to save as much as possible in steel storage equipped with adequate stirrers and will salvage back into the active system as quickly as practical when the emergency is over.

Oil as a Weight Reducer

If oil were added to the system at any point, it would temporarily decrease mud weight due to its low density. Within a very short time the system weight would return to original projection and then increase above it, due to: (1) the increased viscosity caused by the oil, (2) the oil particles acting in the cyclone as solids particles and interfering with separation, and (3) the increased underflow waste problem due to the oil. *For minimum drilling mud weights with water base muds, oil should not be added.*

Oversize and Overload Protection

At slow drilling rates, diamond-drilling for example, cuttings are small and their volume cannot possibly begin to overload a proper full-flow in-

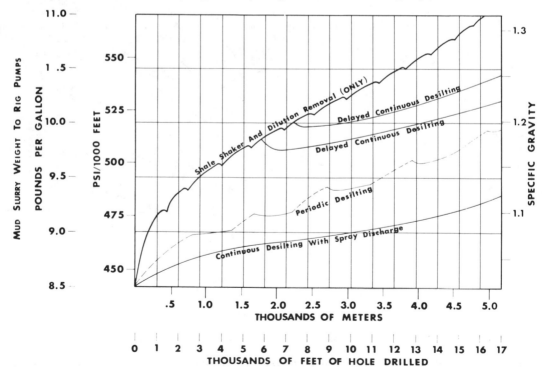

FIG. 6–31. *The effect on mud weight of various solids removal techniques*

stallation of balanced desilters. If it could be assured that no formation cavings could enter the feed, no other full flow separation device would be necessary for this situation. This cannot be assured, and since some part, if not a major part, of every hole is usually drilled at a rapid penetration rate, protection against both oversize and solids volume overload is necessary for the desilters.

The result of various full flow solids removal procedures is generalized in Figure 6–31. Optimum solids removal in unweighted muds is obtained by continuous spray discharge underflow from adequate balanced desilters handling the full circulation rate as a minimum. This presupposes a shale shaker upstream adequate to protect against oversize solids particles that could plug the desilter inlets, and whatever other solids removal equipment (such as desanders) that may be necessary to prevent solids underflow discharge overload of the desilters.

Delayed and Intermittent Desilting

The upper line in Figure 6–31 represents the shale shaker-plus-dilution technique necessary before efficient hydrocyclones were available. To "save money," operators often wait till mud weight is fairly high to begin desilting. The result of this delayed desilting is also shown. The additional solids in the system, representing the increased mud weight over a continuously desilted mud, have decreased in size due to mechanical degradation and most no longer can be removed by the desilters.

Delayed desilting has another penalty besides increased mud weight. The specific surface area of solids particles, on a net weight or volume basis, increases with decreased size particle (Review Table 6–2). The volume of liquid film associated with the surface area of all solids particles removed from a mud by hydrocyclones and by screens must vary directly with the surface area per pound of solids removed, i.e., more liquid phase will always be lost proportionately with finer solids. Delay in removing the solids will always result in greater liquid loss and more expense for those that can still be removed.

Periodic desilting is another false "money saver." It never results in minimum mud weight or best mechanical characteristics. The resulting weight fluctuations, especially weight decreases, can be extremely detrimental to the stability of an open hole. With balanced cyclones, the cost per pound of solids removal will be minimized when desilting is continuous.

Particle Sizes in Desilted Muds

Figure 6–32 shows the particle size distribution of two desilted (low silt, or low silt size) drilling fluids from very different situations. Curve (a) is the distribution of solids in a drilling fluid (at the rig pump suction) that has been processed first through a 12 × 12 mesh shake screen, and then through well designed balanced desilters at a very low underflow solids load, always in spray discharge. This is probably as "clean" as a desilted mud will be. Note the small volume of particles that are larger than eight microns. There were no desanders on the job, and had there been, it would

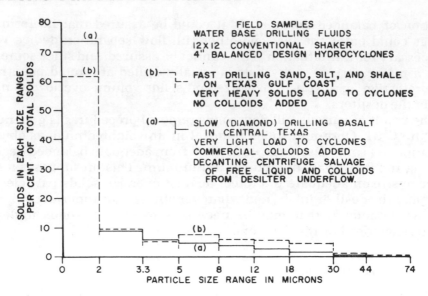

FIG. 6–32. *Comparison of solids removal – screened and desilted slurry*

not have changed the solids in the mud as the light load to the desilters permitted them to operate at maximum efficiency.

Curve (b) is the distribution of solids in a drilling fluid that has been processed through the same type of well-designed desilters but at very high underflow solids loads. Most of the drilling time – until shortly before the sample was taken – the underflow discharge was roping periodically. There were no desanders. The conventional shale shaker had 12×12 mesh cloth, and separated very little material because the formation was poorly consolidated sand, silt, and shale. The shale shaker prevented any feed plugging of the cyclones and they were operated with full opening (slightly wet bottom).

Had there been a stage of good desanders ahead of the desilters, or had there been more desilters to provide greater underflow discharge capacity, the solids distribution might have been as fine as that of curve (a). In this drilling condition (fast drilling a sticky shale with a water base mud) it is doubtful that very fine shale shaker screens could have functioned well enough to have kept the desilters in spray discharge without blinding and losing mud. Very good 6″ desanders certainly could have helped. Nevertheless, the mud was well enough desilted that low silt mud drilling benefits[3] were obtained.

Most of the 73% colloidal material in mud (a.) of Figure 6–32 was purchased to build viscosity and gels and to reduce filtration rate. There were no good clays in the formation which would add any desirable properties to the mud. In mud (b.), there are no commercial clays. The formation is that coastal area contains enough bentonite clay to provide filtration control, viscosity, and hole stabilization, but only when these clays are constantly saved in the system by good desilting practice. In any formations where beneficial clays are not present, they must be purchased and added if mud properties are needed.

Particle Sizes in Desanded Muds

Solids distributions in two desanded muds are shown in Figure 6–33. Both these muds have passed through conventional 12 × 12 mesh shale shakers and then through full flow 8″ hydrocyclones. Curve (b.) is from a mud being used to drill a very hard formation in Central Southern Oklahoma, using a flat toothed bit. Drilling was very slow, and the desanders never approached solids overload. There is a small amount of API sand size particles which these particular hydrocyclones could not remove, even at the low solids load. Note the much higher percentage of solids above eight microns compared to either of the desilted muds in Figure 6–32.

Curve (a.) of Figure 6–33 is from an unusual drilling situation found in the first few thousand feet of drilling in the delta area of Nigeria. Here the formation is coarse sand and fine gravel, with almost no formation particles less than 74 microns. Curve (a.) shows the desanders have left some of the API sand size particles. All the colloidal material in the mud is commercial bentonite, added at the rate of one 100 pound sack per foot of hole primarily to increase gels and reduce whole mud loss through the coarse formation pores.

If the Nigerian mud in curve (a.) were more thoroughly desanded, it would have the characteristics of a desilted mud, simply because there is such a small amount of formation particles between eight and 74 microns. Either finer screens ahead of the desanders to prevent hydrocyclones underflow solids overload, or better desanders with more underflow solids capacity, would have removed all the sand size solids. On the other hand, in curve (b.) the solids load had nothing to do with the problem. Adequate desilters would have produced a low silt mud with nothing else needed except a shale shaker to protect against oversize particles. The slow drilling rate should not overload desilters.

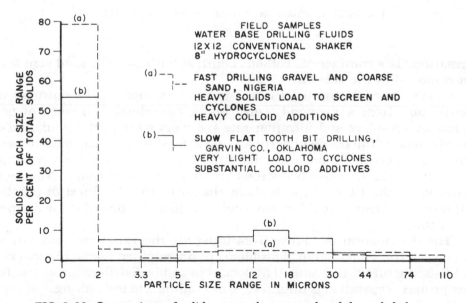

FIG. 6–33. *Comparison of solids removal—screened and desanded slurry*

Particle Size in Coarse Screened Mud

Figure 6–34 affords a study of two extreme solids distributions in unweighted muds. Curve (a.) is a mud from the Niger delta area that has been through the 12 × 12 mesh shale shaker screen only. While not from the same drilling rig, it is from the same area and is very similar to the mud of curve (a.) in Figure 6–33 *before* it had been desanded, i.e. feeding to the hydrocyclone desanders. The small amount of solids particles between two and 74 microns below the shaker screen is striking and also obviously is due to their being hardly any in the unconsolidated, clean, gravel and sand

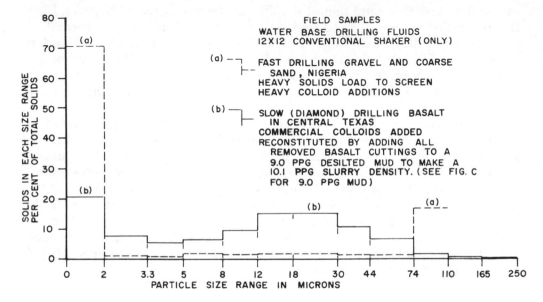

FIG. 6–34. *Comparison of solids removal—screened only*

formation. The commercial colloids contrast with the API sand size solids not removed by the screen.

Curve (b.) of Figure 6–34 is a very good example of the cuttings size distribution from a diamond (core) bit. A considerable amount of commercial viscosifiers and filtration rate reducers were in the mud, and obviously most of the colloidal particles and at least half the particles between 2 and 3.3 microns are not diamond bit cuttings. Just as obviously, almost all of the particles above 5 microns are diamond bit cuttings. None were removed by the 12 × 12 mesh shale shaker cloth. The finest of practical full-flow rig screens would remove only a minor fraction of these diamond bit cuttings.

The distribution in Figure 6–34 illustrate the variations that can and do exist in the solids removal problems in unweighted muds. Figures 6–33 and 6–32 picture the change that occurs in solids distributions as the full-flow primary separation is improved to desanding and to desilting.

WEIGHTED MUDS—FULL FLOW PROCESSING

In muds weighted with barites, the problem of solids removal is more complicated. Study again Figures 6–7A and 6–8A. Compare the API barite distributions (which are consistently similar) with the light solids distributions in Figure 6–8A. and in Figure 6–34. It is evident that removing light solids coarser than barites with fine screen cloths without removing any barites should be fairly effective in some cases and impossible in others, with varying results between the extremes.

Problem of Barites Loss

It will be recalled that the reduction in screen cloth liquid flow capacity, due to weighted mud effect, can be a problem. For this reason, a few operators have for several years used screen cloths coarse enough to take total flow without blinding problems and then have processed the weighted mud through desilting hydrocyclones to remove the coarsest drill solids with acceptable losses of coarse barites. Recalling the one and one-half to one ratio of diameters effect of Stokes Law at equal settling rates of barite and drilled solids, study Figures 6–8 and 6–9 to understand how difficult it would be not to lose some of the coarser end of the barites. At the same time, many of these operators report their losses of barite at the underflow to be negligible under most conditions.

Several facts are pertinent here: (1) The API barite standards limit API sand size barites, and a loss of barites (as other solids) in this size range no doubt is very beneficial in reduction of equipment erosion and filter cake problems. The same is true of that portion of barites above actual 44 microns, or 66 microns equivalent to light solids. (2) The desilters are not as efficient in separation ability in heavy weighted muds, so their removal is not as severe as would otherwise be expected, especially if there is a slight restrictive adjustment at the underflow. This must not be overdone. (3) If there is a moderate load of coarse drill solids in the API sand range to remove, it is very easy to adjust to lose virtually no barites below 44 microns actual size. (4) It is not difficult to test the hydrocyclones underflow to determine the cost of replacing whatever solids are being removed.

Benefits From Size Refinement

If the larger particle size fraction (some fraction of the 44 to 74 microns and all larger than 74) can be removed continuously from a weighted mud, any one or a combination of several possible benefits occur: abrasion in the rig pumps, manifolds, etc., is reduced; mud filter cake becomes smoother in character, and may be slightly thinned; differential pressure sticking is usually reduced if it has been a problem; the buildup of low grade colloidal material, both light solids and barites, is reduced in rate, easing (but not eliminating) the viscosity control problems; total solids content may be reduced in some conditions; overall drilling progress may be speeded up. Although which of the above factors receives credit may depend upon the observer's previous benefits from desilting unweighted mud,[3] it is no mere coincidence. Such benefits in unweighted muds have been the result of solids particle-size-distribution changes, not solids average-specific-gravity changes.

Barites vs. Light Solids Replacement

Improving the full-flow primary separation removal in weighted muds using hydrocyclones desilters downstream from the shale shakers does involve an added cost that has been misunderstood and misconstrued. Many persons are wary of the cost of barite losses. Most are not at all aware that every 100 pounds of light solids removed must be replaced by 80 pounds of barites and 20 pounds of water (assuming barites specific gravity of 4.3 and light solids specific gravity 2.6). It will help to realize that in this process the larger end of the particle size distribution will surely be removed, and the solids average specific gravity might be increased. Under no circumstances will the solids specific gravity be decreased. A decrease in the coarsest material certainly will improve the drilling operation. Our object then is to remove the coarser end of the solids particle size distribution in the mud. An example of the two extreme removal situations possible, with one between the two, will help put the economics in perspective. Assume in each case that 1500 pounds of coarse solids are removed in the hydrocyclone underflow each hour. Assume water cost is negligible (though this is not always true) and that barites cost $3 per 100 pounds, U.S.

Case 1. All removal is light (drilled) solids. Cost to replace the weight and volume effect of 1500 pounds of light solids:

$$1500 \times \frac{\$3}{100} \times 80\% = \$36 \text{ per hour}$$

Case 2. All removal is barites particles. Cost to replace the weight and volume effect of 1500 pounds of barites:

$$1500 \times \frac{\$3}{100} \times 100\% = \$45 \text{ per hour}$$

Case 3. Removal is one-half barites, by weight. Cost to replace the light solids:

$$\frac{1500}{2} \times \frac{\$3}{100} \times 80\% = \$18$$

Cost to replace the barites:

$$\frac{1500}{2} \times \frac{\$3}{100} \times 100\% = \$22.50$$

Total replacement cost:

$$\$18 + \$22.50 = \$40.50 \text{ per hour}$$

Case 1 is as unlikely as Case 2, with Case 3 more closely averaging real situations. The most important thing to realize is that if our object is to remove most of the particles larger than 44 microns in a weighted mud, replacement cost is not much more for barites than for light solids and the benefits are about the same. Also it must be realized that first barite loss on a "new" system will be reduced as the larger barites are removed. Further barite loss will be almost completely restricted to makeup barites. This should be performed on the full-flow rate, never on partial flow.

PRIMARY SEPARATION—PARTIAL FLOW

Partial flow processing in primary removal is, or should be, only to reduce the colloidal (or sometimes the soluble) solids content of the mud. Obviously, to remove the colloidal phase from the total circulating rate of a weighted drilling mud is courting immediate disaster whether the liquid phase is replaced without colloids or an attempt is made to replace a liquid phase with needed properties at the full rig circulating rate.

Decanting Centrifuge

The decanting centrifuge has been such a popular colloidal removal machine simply because it assures, by its very nature, a complete removal of colloidals from the small fraction of the circulating rate it was intentionally selected to process.

Although the machine is most often referred to as a "barite salvage" machine, that is a misnomer. All primary separation machines have a removal function. Barites can be "saved" merely by not pumping mud from the pit at all. The primary problem in a water base weighted mud is viscosity increase. The primary cause of this problem is the natural and constant increase in colloidal size material, usually not of high quality. Their presence and quantity prevents addition of better filtration control colloids and may severely retard drilling progress.

Dilution and discard to remove these excess low-quality colloids is effective but also expensive due to the high cost of barites to replace the total weighting material (including inert-size low-specific-gravity particles) in the discard. Unpublished data from several parts of the world indicate the minimum dilutions (or equivalent mechanical treatment) amounts to an average of 5% of the total system each day a weighted mud is in use. A rare maximum apparently exists representing the maximum rate at which a colloidal phase can be rebuilt and controlled successfully. It is in the range of 10% per circulation. The larger decanting centrifuges, operated for a cut point of two microns barite particles, have a feed capacity ranging from extremes of about 3% to 12% of the circulating rate, depending upon circulation rate and mud weight.

An example will be used to illustrate the economic reasons and the comparative effects of "dilution and discard" against "decanting" a weighted water base mud to keep the colloidal phase under control.

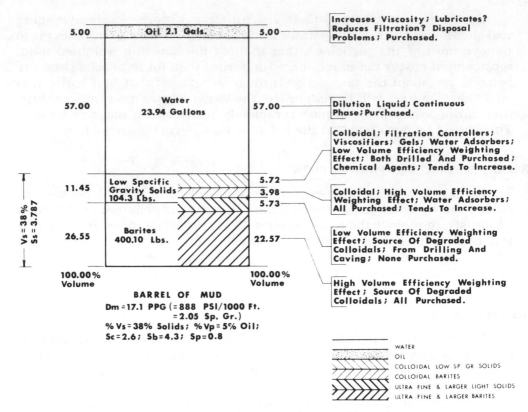

FIG. 6–35. *Weighted water base mud barrel*

Figure 6–35 is a diagram that combines the certain, non-variable, mathematical facts (on the left side) concerning a barrel of 17.1 PPG barites-weighted, water base mud containing 5% oil and 38% solids, the specific gravities being 4.3, 2.6, and 0.8 for the barites, light solids, and oil, respectively. On the right is shown some of the important functions, facts, and trends concerning the liquids and the colloidal and the inert solids fractions.

To calculate the right-hand information, two basic assumptions were necessary: 1) The light solids colloidal fraction, assumed at 50% of the total solids, and 2) The colloidal barite fraction, assumed at 15% of total barites. Once these assumptions (which are within a realistic range), were made, derivations of the solids equations presented earlier were used to calculate all fractions, whether by weight or by volume, absolute or percentage.

If a 10% dilution and discard (or more economically and realistically, "discard and make up" in that chronological order) is made, the make-up purchase is only 5 pounds (.55% by volume) of low specific gravity filtration control material, 479.9 pounds of barites, 26.3 gallons of water, and 2.1 gallons of oil per barrel. See Figure 6–36. When added to the 90% remaining from the original system, the reduction in total solids is not impressive, and the decrease in colloidal phase from 9.7 to 9.1 per cent by volume is

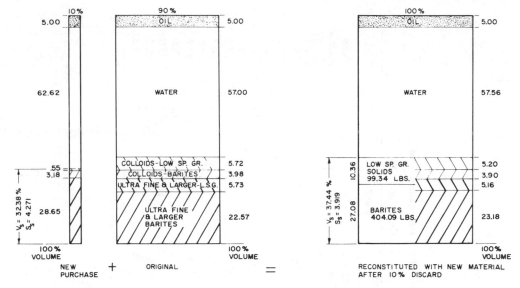

FIG. 6–36. *Discard and makeup of weight water base mud for viscosity control*

disappointing. The reason is the 10% assumed colloidal content of the newly added barite.

Study the centrifuging process in Figure 6–37. All colloidals from the 10% of the system centrifuged are removed. The same 5 pounds of fluid loss control agent per barrel, the same 2.1 gallons of oil, and much less barites are purchased compared to discard makeup in Figure 6–36. Total

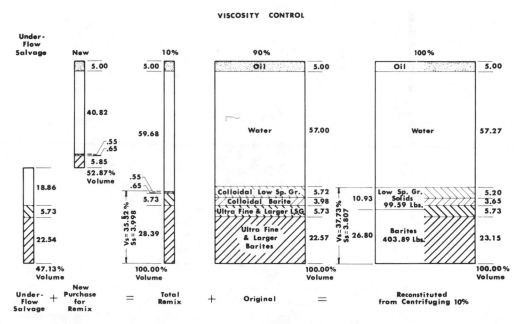

FIG. 6–37. *Centrifuging weighted water base mud*

TABLE 6-8
Typical Discard and Makeup v.s. Centrifuging 17.1 PPG Weighted Water Base Mud

Materials*	Original System Mud	Discard 10%		Centrifuge 10%			
		Makeup Purchase	New System	Underflow Salvage	Makeup Purchase	Total	New System
Oil, gallons	2.1	2.1	2.1	—	2.1	2.1	2.1
Water, gallons	23.9	26.3	24.2	7.9	17.1	25.1	24.1
Colloidal Low Sp. Gr. Solids, %	5.7	.5	5.2	—	.5	.5	5.2
Colloidal Barites, %	4.0	3.2	3.9	—	.7	.7	3.7
Total Colloidal Solids, %	9.7	3.7	9.1	—	1.2	1.2	8.9
Ultrafine-plus L.S.G. Solids, %	5.7	—	5.1	5.7	—	5.7	5.7
Ultrafine-plus Barites, %	22.6	28.7	23.2	22.5	5.9	28.4	23.1
Total Light Solids, %	11.4	.5	10.3	5.7	.5	6.2	10.9
Total Barites, %	26.6	31.9	27.1	22.5	6.6	29.1	26.8
Total Solids, %	38.0	32.4	37.4	28.3	7.1	35.3	37.7
Total Light Solids, Lbs.	104.3	5.0	94.4	52.2	5.0	57.2	99.6
Total Barites, Lbs.	400.1	479.9	408.1	340.1	98.0	438.1	403.9
Total Solids, Lbs.	504.4	484.9	502.5	392.3	103.0	495.3	503.5
Total Solids, Avg. Sp. Gr.	3.79	4.27	3.92	3.96	4.17	4.00	3.81

* All based on 42 U.S. gallons per API barrel; per cents by volume; U.S. pounds per applicable API barrel.

solids reduction is less; viscosity reduction is more. This is because total light solids removal is only 50%, but colloidal removal is complete and the small makeup requirement of barites reduces the colloidal barite addition. Table 6–8 summarizes these and other details, including the difference in barite purchased of 480 pounds compared to 98 pounds for discard makeup and centrifuge makeup respectively.

Assuming there is a viscosity problem, the colloidal barites can be as troublesome as low-grade colloidal light solids, on a net volume basis, and must be removed. The cost of replacing the colloidal fraction removed is very close to the same regardless of whether the removed fraction is barites, or light solids, or a mixture of the two as in this case. The key to economical colloidal phase removal is a fine cut and to remove all possible free liquid from the processed fraction, and with as little dilution (water and cleanup cost) as possible.

Perforated Rotor Centrifuges

The perforated rotor centrifuge in the recommended mode of operation, as already discussed, separates the colloidal fraction on a dilution basis. To remove the same volume of colloidal material it must process approximately double the amount of the same mud necessary for a decanting centrifuge. The dilution, per unit of raw mud feed, will average being double or three times that required for the best designs of decanting centrifuges.

The portion of the diluted liquid-colloidal phase reporting to underflow is almost the same total volume as the liquid-colloidal phase in the undiluted feed. Although the overflow volume needing to be disposed of, or to be used elsewhere, is only two to three times that of a decanter, per unit of undiluted feed, it may easily be four or five times that of a decanter to accomplish the same colloidal removal, or viscosity reduction. Refer to Table 6–9.

This is not to say the perforated rotor centrifuges have no application. If dilution water compatible with the mud is available in quantity and the overflow volume can be utilized or disposed of easily; or if a good decanting centrifuge is not available or cannot be set up properly on the system, the perforated rotor machine can be of great assistance in controlling viscosity in a weighted mud system, with worthwhile money savings.

Hydrocyclones for Viscosity Reduction

If the average weighted mud were fed to a balanced hydrocyclone capable of separating most of the inert solids, the underflow opening could not handle the solids load and would rope. Since a very high separation ability is necessary to separate the major part of an API barite grind, and since efficient separation requires spray discharge, weighted muds must be heavily diluted to reduce their total solids content before they enter special barite hydrocyclone units.

TABLE 6-9
Comparative Characteristics

Primary Separation, Partial Flow, Special Colloidal Removal Units of the best designs available in 1973, operated according to manufactures instructions and extraplolated, if necessary, according to published facts or actual tests. Variations are possible by pushing, but at sacrifice of some result.

		Variations by Feed Mud Weight					
	Sp. Gr.	1.44	1.68	1.92	2.16	2.40	2.64
	PPG	12	14	16	18	20	22
FEED MUD CAPACITY (NOT IN-CLUDING DILUTION WATER), GPM:							
Decanting Centrifuge		21.7	15.5	11.9	9.7	8.5	8.0
Perforated Rotor Centrifuge		17.3	14.7	13.2	12.3	11.8	11.6
Hydrocyclone Units		25.0	16.7	12.9	10.8*	**	**
RECOMMENDED DILUTION WATER AT ABOVE FEED MUD RATES, GPM:							
Decanting Centrifuge		3.0	3.0	3.0	3.0	3.0	3.0
Perforated Rotor Centrifuge		12.1	10.3	9.2	8.6	8.3	8.1
Hydrocyclone Units (Average)		53.7	64.3	67.1	68.6*	**	**
UNDERFLOW DENSITY, PPG							
Decanting Centrifuge, Bowl		21 to 25, depending upon feed content					
" " Hopper		10 to 25, depending upon remix adjustment and feed					
Perforated Rotor Centrifuge		12+	14+	16+	18+	20	22
Hydrocyclones, Spray Discharge		14 to 17, depending upon feed content					
" Rope Discharge		17 to 22,** depending upon the cut point					
SEPARATION OF COLLOIDAL SIZE PARTICLES TO OVERFLOW AS PER CENT OF COLLOIDALS FEEDING							
Decanting Centrifuge		98%± at all feed and underflow hopper weights					
Perforated Rotor Centrifuge		50%± at all feed weights					
Hydrocyclone Units		95%± at all feed and underflow weights					
RECOVERY OF FRESH API GRIND BARITES LARGER THAN TWO MICRONS AS A PER CENT OF THE SAME SIZE FEEDING							
Decanting Centrifuge		98%+ at all weights					
Perforated Rotor Centrifuge		92%+ at all feed weights					
Hydrocyclones, Spray Discharge		75% maximum sustainable					
Rope Discharge		65% downward to 20% or plugged					

* Hydrocyclones are not recommended for normal muds this heavily weighted with API barites.
** These hydrocyclone underflow weights should not be attempted without especially ground barites.

This dilution must be sufficient to reduce solids content to between 6 and 12% by volume, or to about 9.8 PPG (1.18 Sp. Gr.) feed slurry density. Such a dilution involves large quantities of water, much more than a perforated rotor centrifuge for example. At high dilution, proper head, and spray discharge conditions the best barite hydrocyclones for field work can separate to the underflow about 80 percent of the barites greater than 2 microns, from an API barite distribution. With a separation this good, the free liquid film on the barite surface area keeps underflow density down to the range 14 to 17 PPG (or 1.7 to 2.05 Specific Gravity) or less.

Hydrocyclones do a fairly good job of separating the colloidal phase to the overflow in spite of the amount of free liquid film in the underflow, because of the high dilution ratio. The high dilution ratio, while not for that purpose, reduces the colloidal content per unit of liquid phase to such a low point that only a small fraction of clay size material is found in the underflow.

In heavier muds, the low underflow weight at maximum separation creates a problem. Either care must be taken to mix new barites all the time the cyclones are operating to keep up mud weight; or the hydrocyclones must be operated with a rope underflow to bring underflow weight up to or above pit mud weight, with a resulting heavy loss of ultrafine barites in the cyclone overflow. Special coarse barite grinds can reduce this problem, but may introduce other problems.

Hydrocyclones for partial flow removal of colloidals from weighted muds are cheaper to buy and to operate than decanting and perforated rotor centrifuges, but they cannot be justified over decanting centrifuges, and perhaps not over perforated rotor machines, on the total mud cost basis. If good decanting centrifuges are not available or cannot be serviced and if dilution water is available in large quantities and at low cost, and if the overflow disposal is not a problem, hydrocyclones can be considered for barite salvage. If mud weights are above 18 PPG, a special coarse grind of barites similar to Figure 6–8 may be helpful, but the effect on drilling and hole must be watched carefully.

Hydrocyclones capable of colloidal removal weighted muds have small dimensions (Pioneer DSC-300C 3″ SOLIDSMASTER used in their CLAY-MASTER, and the Dorr-designed SWACO 50 mm cyclone used in their CLAYJECTOR) and are understandably subject to plugging. Plugging of the cyclone, whether feed or underflow, cannot be tolerated at all on this service. When in use, hydrocyclones on this service must be checked frequently. To facilitate quick, frequent checking, the underflow of each individual cyclone should be always visible without dismantling any part of the unit.

Table 6–9 summarizes the critical separation characteristics of the best designs of the three types centrifugal separators available for colloidal removal, or viscosity reduction, from weighted muds. It may not be possible to always obtain or install one that is functionally best suited, but it is folly not to be aware of the compromise being made. In any mud solids separation problem it is a truth that the greatest and economic variations are in the benefits and savings to be effected, not in the rent or amortization differences between various types and designs of machines.

EQUIPMENT APPLICATION – SECONDARY SEPARATION, UNWEIGHTED MUDS

There are situations in which the full-flow primary solids removal, which always is selectively of the larger particle sizes, creates intolerable problems, in their turn requiring alleviation.

In desilting unweighted muds, the free liquid film associated with the separated ultrafines and fines may create a waste disposal problem or it may involve a very expensive liquid phase loss, or both. An example of the disposal problem would be desilting any drilling fluid in highly populated areas. Another would be desilting an unweighted oil mud in any area. An example of both scarcity and disposal problems would be desilting in the Arctic, where water is extremely expensive and liquid disposal even more so.

In these cases, one solution to the problem would be to confine the removal to coarse material that does not have a high specific surface area and which can be removed by full-flow shale shaker screens. With the best designed screens of sufficient capacity the free liquid draining from the large cuttings should not present a major problem. However, with this solution to the liquid discard problem there is a higher solids content retained in the mud due to the larger particles that would not be present if the mud were being desilted. This means higher mud weight (hydrostatic head on the formation), thicker and coarser filter cake with increased tendency for differential sticking in permeable formations, slower drilling progress, shorter bit footage life, increased pump fluid end abrasion, less hole stability if there is any natural weakness, etc. etc.

Hydroclone — Centrifuge Combination

A better solution has proved to be to desilt the mud with full-flow hydrocyclone desilters (primary separation) and then to use a decanting cen-

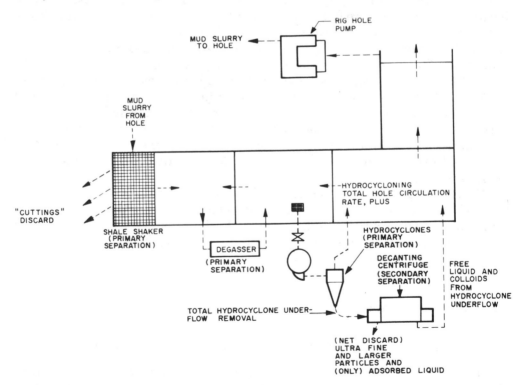

FIG. 6–38. *Decanting centrifuge for secondary separation from primary separation desilters*

trifuge in secondary separation to separate the liquid-colloidal phase and return it to the active system. See Figure 6–38. The underflow of the decanting centrifuge, with no free liquid present, contains the ultrafines, fines, mediums, and larger particles removed by the hydrocyclone desilters. This decanter underflow can be handled by a dirt scoop or blade, but cannot be sucked into a vacuum truck for the lack of free liquid. Disposal of it is proving a very simple problem compared to other alternatives.

In expensive liquid phase muds (water in the Arctic and in the desert; oil anywhere) liquid cost alone is economic justification many times over and continually for the decanting centrifuge in secondary separation. In some areas (the City of Los Angeles and in Europe, for example), reduction of disposal cost alone may justify the decanter. In no case have any of the benefits of primary separation desilting been lost by the addition of secondary separation with the decanting centrifuge.

In unweighted muds there has been no detectable increase in mud weight reported, although the return of a salvaged colloidal phase instead of water or oil makeup obviously must involve some minute greater weight in (less solids removed from) the system. It is conceivable that many of the salvaged colloids (especially in water base muds) are substituting for those that would be purchased and added to the fresh liquid were it not for the salvage. Impressive mud cost reductions in many areas indicate some validity to this reasoning.

Perforated Rotor Machine

The fluidic underflow requirement of the perforated rotor centrifuge precludes its use as a secondary separation unit on the underflow of hydrocyclones. It appears that the inherent characteristics of the separation mechanism prevent its being modified to provide an underflow devoid of free liquid.

Hydroclone-Screen Combinations

Special fine shale shaker screens[23] for secondary recovery from primary separation hydrocyclones are receiving attention and study at this time. See Figure 6–39. They are normally not recommended for water base unweighted muds, because the return to the system of ultrafine and some fine particles removed by the primary separation desilters can decrease the low-silt-size-solids drilling benefits and has been found to increase active system weight (compared to desilting without secondary separation of the underflow) by approximately 0.5 PPG (0.06 Specific Gravity) and more.

The secondary separation screening process definitely diminishes the volume of discard by returning ultrafines, and associated free liquid and suspended colloids to the system. In a situation where the increased mud weight, total solids, and abrasiveness of the returned solids do not harm the drilling operation, the secondary separation screens can be used to advantage. If there are adverse drilling effects, no mud cost savings from secondary separation can be justified, and if in this case the underflow

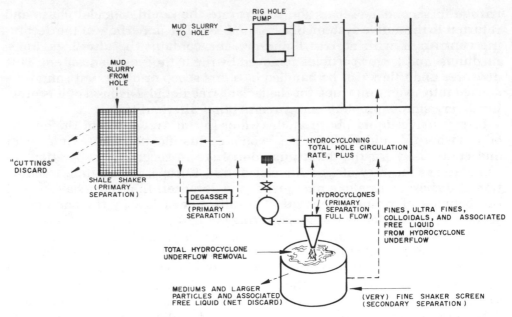

FIG. 6–39. *Special shale shakers for secondary separation from primary separation de-silters*

liquid loss from primary separation desilting cannot be tolerated, the decanting centrifuge becomes the only alternative to stopping desilting.

SECONDARY SEPARATION—WEIGHTED MUDS

Although recorded test data are very scarce, and no published articles are available, hydrocyclone desilters have been run full-flow on weighted muds at least as far back as 1965. Word of good results hole-wise, but no scientific report on which to base cost analyses, has been the common experience. Still, most operators have been reluctant to attempt to improve weighted mud conditioning full-flow hydrocycloning for fear the barite replacement costs would be prohibitive. This has been in spite of the fact that most drilling rigs are equipped with desilters, and it is not difficult to field test the cost and the drilling benefits.

Decanting Centrifuge

In 1969 there were preliminary reports of attempts to apply centrifuges on secondary separation in Europe processing the desilting hydrocyclone underflow on weighted muds to reduce disposal hauling and to investigate effects on the drilling. After problems with improperly designed centrifuges were solved by use of better machines, excellent data were obtained[24] in 1971. Meanwhile, in the Canadian North, centrifuging the underflow of hydrocyclone desilters became fairly commonplace, whether the mud was weighted or not.

Referring to Figure 6–38, the underflow of the hydrocyclones removes only a small fraction of the total solids in the cyclone feed. This small fraction of solids is selectively the coarsest fraction in the mud, and the most detrimental to drilling, to drilling equipment, and to the hole, whether it is barite or drill solids. Pound for pound, it costs 80% as much (or more) to replace the weight effect of light solids as it does to replace barites. The odds are in favor of improving drilling cost by removing coarses rather than compromising the removal by attempting to distinguish closely between fine-size barites and fine-size-plus light solids. The secondary separation centrifuge can be justified only by the elimination of free liquid from the hydrocyclone discard (disposal) and by the salvage of liquid and colloidal material to the system.

The secondary separation and return of colloidal material to a weighted mud system caused some consternation when first reported.[24] The *removal* of this very liquid colloidal phase with decanting centrifuges in partial-flow primary separation for viscosity control was the original, and is still the most common, application of the machine. It is shown in Table 6–3 that the larger size solids in the mud, both barite and low specific gravity, furnish degraded particles to the colloidal size. There is both evidence and logic[3] to indicate that the bigger the particles, the more they contribute to the colloidal phase because of their more forceful collisions caused by their heavy mass.

Since the colloidal materials cause the viscosity problems that plague most weighted muds, it is reasonable that constantly removing the larger particles down even into the fine range, including barite, might reduce the rate of viscosity buildup in a weighted mud by cutting off part of the supply of degraded colloids.

In Table 6–10, Items 1, 3, and 7 represent the screening, desilting, and secondary separation by the decanting centrifuge. It would be surprising if in all formations and muds and drilling conditions this processing provided sufficient viscosity control. It is extremely important to realize that if the viscosity is a problem, a second centrifuge to remove colloidals, as Item 4 or 5, or special cyclones Item 6, is not necessary. The secondary separation centrifuge already on the job is returning colloidal material to the system that would be removed by another centrifuge in primary separation. They would be working against each other. The first remedial step would be to stop the secondary separation and let the desilter underflow pass to waste. If this cannot be tolerated, continue secondary separation, but put the centrifuge liquid overflow to storage, or use it on an unweighted mud job. Either results in colloidal removal from the system.

If the first step of allowing total desilter underflow to leave the system fails to give sufficient viscosity control, and only in that case, the decanting centrifuge should be converted to primary separation, as in Item 4 of Table 10–6. This will result in two effective colloidal discards simultaneously: (1) the colloids in the free liquid film in the desilter solids underflow, and (2) the colloids in the centrifuge liquid overflow. In addition, the larger particles are still being removed by the desilter. This combination should face up to the most severe viscosity problems in weighted mud drilling.

TABLE 6-10

Primary & Secondary Separations Full & Partial Flow Processing

(Normal Pertinent Effects on the Mud System)

Item	Separation	Type Flow	Type of Equipment*	Character of the Discard	Effect On: Unweighted Mud — Total Solids	Unweighted Mud — Mud Weight	Unweighted Mud — Plastic Viscosity	Weighted Mud — Total Solids	Weighted Mud — Plastic Viscosity
1	Primary	Full	Shale Shaker Screens	Wet	Down	Down	Negligible	Down	Slightly Down
2	Primary	Full	Desanding** Hydrocyclones	Wet	Down	Down	Negligible	Down*	Slightly Down
3	Primary	Full	Desilting Hydrocyclones	Very Wet	Down	Down	Up	Down	Down
4	Primary	Partial**	Decanting Centrifuges	Low Vol. – Liquid	M.A.***	M.A.	M.A.	Slightly Down	Down
5	Primary	Partial**	Perf. Rotor Centrifuge	Medium Vol. – Liquid	M.A.	M.A.	M.A.	Slightly Down	Down
6	Primary	Partial**	Special Cyclones	High Vol. – Liquid	M.A.	M.A.	M.A.	Slightly Down	Down
7	Secondary	(Desilter Underflow)	Decanting Centrifuge	Damp	Slightly Up	Slightly Up	Up	Slightly Up	Slightly Up
8	Secondary	(Desilter Underflow)	Special Screens	Wet	Up	Up	Up	Up, but Variable	Slightly Up

* All equipment of each type assumed to be the best design available.
** A misapplication if used on oil muds; use on water base only.
*** M.A. – Misapplication. Full-flow, plus secondary separation if justified, is the only economic approach for unweighted muds.

Special Fine Screens

The special screens[23] mentioned earlier under unweighted muds (see Figure 6–39) have a much greater application in weighted mud situations, for which they were developed. The process is straightforward, and the most important thing is to know or determine the drilling benefits and costs when desilting without secondary separation by the screens. Then when the screens reduce mud cost the savings can be weighed against the drilling benefits, which may be reduced by screening. If drilling benefits are not reduced, the savings are completely valid.

The second thing to keep in mind with secondary separation screening is the viscosity problem. Again, if the viscosity increases, the first remedy is to stop returning colloidal material to the system with the screen. Let it go with the coarses in the desilter underflow. If the viscosity still is a problem, a decanting centrifuge is recommended for primary separation of colloids. At this point, screen salvage may or may not be possible, depending upon the formation.

Two Stage Centrifuging

Occasionally there is word of a "double stage" centrifuging job on weighted mud. This process began on the Gulf Coast of Texas and Louisiana, fell from favor rapidly in the early 1960's, and was not heard of for several years. The problem with this process is that it is based on misconceptions. One is that drill solids, or at least the detrimental ones, are all in a size range just above colloidal size. Examination of Figures 6–5, 6–7, 6–8, 6–9, 6–32, 6–33, and 6–34 quickly indicate the error of this assumption. It is also based on the misconception there are no barites in this range. See Figures 6–7 and 6–8. Another misconception is that colloidal materials (including chemicals) do not cause problems and must not be discarded.

Examine Figure 6–40. Primary-separation centrifuge A is operated at a speed below normal to produce a cut at perhaps 4 to 6 microns (barite equivalent), with the underflow solids particles returning to the system and overflow liquid (with barites smaller than maybe 5 microns, light solids smaller than perhaps $7\frac{1}{2}$ microns) proceeding to the secondary-separation decanting centrifuge B, operated at normal full speed to make a regular cut of 2 microns on barites and 3 microns on light solids particles. The underflow solids of machine B are discarded. This consists, in this typical example, of barites between 2 and 5 microns and light solids between 3 and 7.5 microns. All colloidals are returned to the system by machine B; most ultrafines and all fines and larger are returned by machine A.

In one study of this process[25] more barite than light solids, though not much of either, was being discarded. The mud properties from the remix of machine A underflow and machine B overflow was almost exactly identical to the feed to A from the pit. In a water base mud, if feed dilution were used, that water alone would produce a change. The same water could be added to the system without the machines. The "dual stage," as described here and as pictured in Figure 6–40 cannot be recommended.

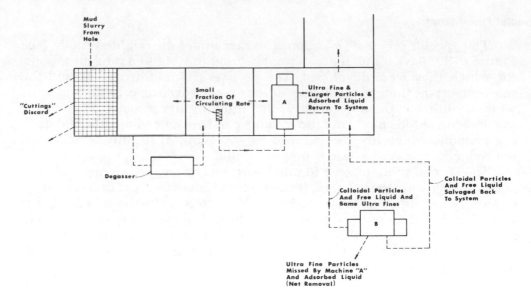

FIG. 6–40. *High speed decanting centrifuge for secondary separation from medium speed decanter*

THE "TWILIGHT ZONE" IN MUDS

It has been discussed that the solids problem and the problem of controlling solids in weighted muds is very different from unweighted muds. This has long posed a dilemma to operators who have reached a depth where they did not need a weighted mud, the formation pore pressure required more hydrostatic head than the desilted (and unweighted) mud in the hole, and the operator did not wish the drilling disadvantages of stopping the desilters. This problem has been classic in the range from a low of 9.0 to 9.5 PPG (1.08 to 1.14 Specific gravity), at the low weight, up to 10.5 to 11 PPG. This problem is reflected in Figure 6–29. It has been approached three basic ways in the past:

1. The most common method has been to stop desilting first, and do basically nothing while mud weight increases from drilled fines and mediums. As higher weight is needed, barites are added, and either a centrifuge is used or discard and makeup is resorted to for viscosity control. This is most adaptable to slower than the fastest drilling rates.

2. In faster drilled holes, with short, rapidly increasing transition zones, a similar except much quicker method has been used often. Intensive desilting is carried on until the transition zone is evident. Then, simultaneously, desilting is stopped, decanting centrifuging for colloidal removal is started, and rapid barite weighting is begun. Fast drilling is continued. Several operators have habitually reported increasing weight 3 PPG (0.36 Specific gravity) per circulation. They also report this can be done only with a thoroughly desilted mud.

3. A third method is very popular with those who have tried it, but the majority are reluctant to test it for fear of barite cost. Intensive de-

silting is continued, but barites are added as more mud is needed. This is usually continued to a mud weight of 11 PPG (1.32 Specific gravity) or more, and often to 13 PPG (1.55 Specific gravity). In this range, desilting usually has been stopped, or the underflow restricted more, or resorted to periodically to thin the filter cake, reduce abrasion, alleviate hole problems, etc. The decanting centrifuge is used as needed for viscosity reduction, or the discard and makeup method is used.

It would seem that the special secondary separation screens may be of benefit for the "Twilight Zone" of neither weighted nor unweighted muds. As shown in Table 6–10, the return of the liquid phase from the screen to the unweighted system will cause some weight increase. If a more rapid increase is needed, barites can be added as necessary. As the transition to a weighted mud is completed additional measures can be taken if needed. This should be tried against Method 3 above to determine comparative drilling advantages.

INSTALLATION AND MAINTENANCE

The subjects of proper mud system design, equipment installation engineering, and maintenance programs are much too lengthy and involved to take up here. Papers[3,16,19] are available. However, some pointers are in order.

On any drilling operation in the world, poorly maintained solids removal equipment reflects the attitude of the management of that operation. No equipment will be kept in operating condition if management does not make it clear that failure to maintain it in top shape and to operate it properly will not be tolerated. If an installation is not as it should be, the chances are that it will be changed soon if proper maintenance is insisted upon.

Equipment inadequate for the job cannot operate properly. The rigs with the most and best solids removal equipment in the world would not part with any of it; the rigs most poorly equipped would gladly give away what they have because it can't work properly. Rigs capable of drilling the fastest need the highest solids removal capacities. The potential drilling benefits justify the best equipment available.

Time and effort spent by a drilling supervisor to understand drilling mud solids, and the machinery used to control them, will return handsome rewards. Besides study, this understanding requires familiarity with the equipment, gained on the drilling rig.

DRILLING FLUIDS SOLIDS CONTROL
SUMMARY

REFERENCES

1. R. A. Bobo and R. S. Hoch, "The Mechanical Treatment of Weighted Drilling Muds," AIME paper No. 290-G.
2. R. L. O'Shields and Don E. Wuth, "New Mud Desander Cuts Drilling Cost," AAODC annual Meeting, Houston, Sept. 1955. *Drilling Contractors,* October 1955.
3. G. S. Ormsby, "Desilting Drilling Muds with Hydrocyclones," AAODC Rotary Drilling Conference, *The Drilling Contractor,* March–April 1965.
4. T. C. Mondshine, "New Fast Drilling Muds also Provide Hole Stability," *Oil and Gas Journal,* March 21, 1966.
5. Official Minutes, API Task Group on Determination of Liquids and Solids, D. B. Anderson, Chairman (set up in 1968 by API Committee on Standardization of Drilling Fluid Materials).
6. "Characteristics of Particles and Particle Dispersoids," Bulletin from Stanford Research Institute Journal, Third Quarter 1961 (Dept. 300, Stanford Res, Institute, Menlo Park, Calif.)
7. John H. Perry, "Chemical Engineers Handbook," Third Edition, McGraw-Hill.
8. Unpublished data sources, including Pioneer Centrifuging Company files.
9. U.S. Dept. of Agriculture, S.C.S., "A Guide for Interpreting Engineering Uses of Soils," 1971 Manual, Director of Documents, Washington, D.C. Also from International Standard Soil Classification System, adopted by International Society of Soil Scientists in 1934.
10. Longwell, Knopf, Flint, "Outlines of Physical Geology," Second Edition, Wiley & Sons.
11. API RP13-C, First Edition, Sept. 1973, API Committee on Standardization of Drilling Fluid Materials, Task Group on Solids Removal Devices, Luther Bartlett, Task Group Head.
12. "Principles of Drilling Fluid Control," Twelfth Edition, Petroleum Extension Service, University of Texas at Austin.
13. R. L. Ritchey, "Effect of Particle Size on $\overline{YP}$ and Viscosity of Drilling Muds," Salt Water Control Co. (SWACO) publication, February 1959.
14. Clyde Orr, Jr. and J. M. Dallavalle, "Fine Particle Measurement," MacMillan.
15. H. G. N. Fitzpatrick, private communication.
16. Louis K. Brandt, "Remarks on Fine Screen Shakers," proceeding of Rotary Drilling Conference of IADC, 1 March 1973, Houston, Texas.
17. J. A. Gill, "Drilling Mud Solids Control—New Look at Techniques," *World Oil,* October 1966.
18. W. F. Roper, leader of screen section of Solids Removal Task Group, verbal statements in Task Group meetings leading to API RP-13C.
19. G. S. Ormsby, "Why Your Removal System Won't Remove," proceedings of Rotary Drilling Conference of IADC, 1 March 1973, Houston. Published also as Part 1 "Proper Rigging Boosts Efficiency of Solids-Removing Equipment," 12 March 1973, and Part 2, "Proper Pumps, Lines Help Cut Costs," 19 March 1973, by *Oil and Gas Journal.*
20. Direct copies of original MOHOLE-calculated data in Pioneer Centrifuging Co. library, from MOHOLE test work conducted near Uvalde, Texas, 1964.

21. Ralph F. Burdyn, D. E. Hawk, and F. D. Patchen, "New Device for Field Recovery of Barite: II Scale-up and Design," *S.P.E. Journal,* June 1965.
22. M. D. Nelson, "Removal of Fine Solids From Weighted Muds," API Paper No. 926–15–H, presented before the Southern District Division of Production, Houston 4–6 March 1970.
23. L. H. Robinson and J. K. Heilhecker, "Solids Control in Weighted Drilling Fluids," Paper No. SPE 4644, 48th Annual Fall Meeting of SPE of AIME, Las Vegas, 1973.
24. Erhard Maikranz, Brigitta-Elwerath Betriebsgermeinshaft, unpublished Intra Company report, 1972.
25. George S. Ormsby, "An Analysis of a Dual Centrifuging Process on a Weighted Inverted Oil Emulsion Mud Containing Asphalt Additives," March 1966, unpublished, in Pioneer Centrifuging Co. Library.

7

Pressure Losses in the Circulating System

T H E determination of pressure losses in the circulating system has been an objective of technology for almost as many years as rotary drilling has been in existence. The first dedicated efforts to determine the pressure losses came when drilling hydraulics were introduced in 1948.

The development of jet bits and hydraulics programs was responsible during the early 1950's for the introduction of graphs, charts and slide-rules, by bit companies, for the determination of pressure losses in turbulent flow. The rotating viscometer was introduced at about the same time to measure mud properties in laminar flow.

In general, the initial calculation of laminar flow pressure losses was performed assuming a Bingham Plastic flow model. The Bingham model has now been replaced generally by the Power Law flow model, which has been shown to be more accurate.

In general, the methods introduced by the bit companies to calculate pressure losses in turbulent flow have been sufficiently accurate in field applications for hydraulics programs. However, efforts are continuing in the drive to improve pressure loss calculations in laminar flow. Operators, at least in geologically young formations, can measure directly the fracture gradients of open hole formations close to the casing seats.

The direct measurement of fracture gradients has emphasized the need to determine, with accuracy, the total pressure loss in the annulus. With accurate measurements of fracture gradients, mud weights and annular circulating pressures it should be possible to eliminate most of the lost circulation problems that are associated with weighted muds.

A schematic diagram of the drilling fluid circulating system is shown

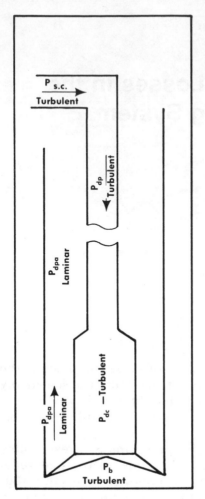

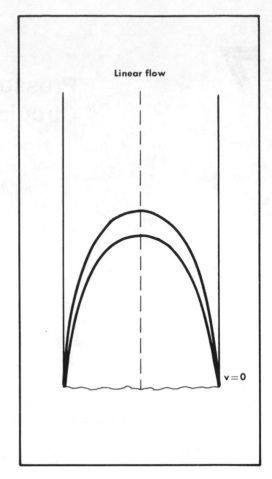

FIG. 7-1. *Circulation system and normal flow patterns*

FIG. 7-2. *Linear flow*

in Figure 7-1. The summation of pressure losses in the entire circulating system is shown on the surface pressure gauge, normally located on the standpipe in a corner of the rig mast. This summation of pressures is shown as Equation 1.

$$P_s = P_{s.c.} + P_{dp} + P_{dc} + P_b + P_{dca} + P_{dpa} \qquad (1)$$

The total pressure gives no indication whether the flow pattern in any part of the system is laminar or turbulent. As noted in Figure 7-1, the flow patterns inside the drill string are generally turbulent. Flow patterns in the annulus may be either laminar or turbulent and methods will be shown to help distinguish the flow pattern.

Laminar flow is distinguished by a smooth flow pattern as shown in Figure 7-2. The velocity of each layer of fluid increases towards the middle of the stream until some maximum velocity is reached. It is not uncommon

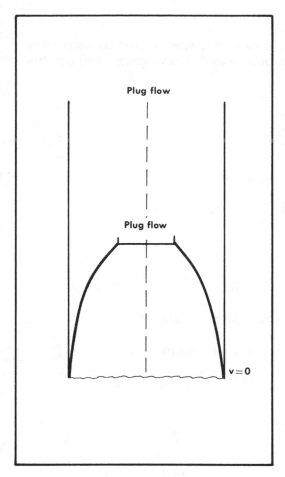

FIG. 7–3. *Plug flow*

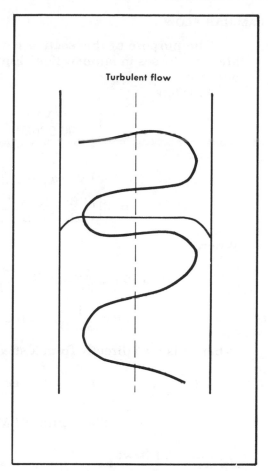

FIG. 7–4. *Turbulent flow*

to encounter special cases of laminar flow, where the center of the flow pattern is flat, as shown in Figure 7–3. In this flat portion of the stream there is no shear of fluid layers and this type flow is called plug flow. In hole cleaning it is often desirable to flatten the velocity profile by increasing the mud thickness; however, this practice would generally increase annular pressure losses.

Turbulent flow can be described as random flow. There is no viscous shear of fluid layers as shown in Figure 7–2. A typical pattern for turbulent flow is shown in Figure 7–4. The fluid velocity at the hole wall is zero; however, the fluid velocity in the core of the stream is essentially flat. There are different degrees of turbulence. Thus, a simple statement that the flow pattern is turbulent is not very definative. As turbulence increases, the pressure losses also increase.

An adverse effect of fluids in turbulent flow is hole erosion. Of significance is the fact that the viscous flow properties of the mud have very little effect on pressure losses in turbulent flow. This fact is used to help determine changes in annulus pressure losses during routine drilling operations.

LAMINAR FLOW

The purpose of this section is to illustrate methods used to determine pressure losses in laminar flow. Equations 2 and 3 have been used for this purpose.

Pipe flow:

$$P = \frac{(PV)\bar{v}l}{90,000D^2} + \frac{yl}{225D} \quad \text{(Bingham)} \qquad (2)$$

Where:

$$PV = \theta_{600} - \theta_{300} \text{ and } y = \theta_{300} - PV$$

$$P = \left[\left(\frac{1.6\bar{v}}{D}\right)\left(\frac{3n+1}{4n}\right)\right]^n \frac{Kl}{300D} \quad \text{(Power Law)} \qquad (3)$$

Where:

$$n = 3.32 \log \frac{\theta_{600}}{\theta_{300}} \text{ and } K = \frac{\theta_{300}}{511^n} \text{ or } \frac{\theta_{600}}{1022^n} \text{ or } n = 3.32 \log \frac{\theta_{2\gamma}}{\theta_{\gamma}} \qquad (4)$$

$$P = \frac{l\tau}{300D} \quad \text{(Direct shear stress reading)} \qquad (5)$$

Where τ is read directly from a shear stress vs. rate diagram

$$\text{Shear rate (sec}^{-1}) = \left(\frac{1.6\bar{v}}{D}\right)\left(\frac{3n+1}{4n}\right) \qquad (6)$$

$$\text{Shear rate (RPM)} = \left(\frac{0.94\bar{v}}{D}\right)\left(\frac{3n+1}{4n}\right) \qquad (7)$$

Annular flow:

$$P = \frac{(PV)\bar{v}l}{60000(D_h - D_p)^2} + \frac{yl}{200(D_h - D_p)} \quad \text{(Bingham)} \qquad (8)$$

$$P = \left[\frac{2.4\bar{v}}{D_h - D_p}\left(\frac{2n+1}{3n}\right)\right]^n \frac{Kl}{300(D_h - D_p)} \quad \text{(Power Law)} \qquad (9)$$

Where n and K are same as for pipe flow

$$P = \frac{l\tau}{300(D_h - D_p)} \quad \text{(Direct shear stress reading)} \qquad (10)$$

Where:

$$\text{Shear Rate (sec}^{-1}) = \left(\frac{2.4\bar{v}}{D_h - D_p}\right)\left(\frac{2n+1}{3n}\right) \qquad (11)$$

$$\text{Shear Rate (RPM)} = \left(\frac{1.41\bar{v}}{D_h - D_p}\right)\left(\frac{2n+1}{3n}\right) \qquad (12)$$

The viscometer readings, θ, for all these equations are obtained from rotating viscometer data. Units are based on normal field usage and are given at the end of this chapter. The normal field model of rotating viscometers has been one with only two speeds. Recent emphasis on accurate determinations of pressure losses has increased the utilization of multi-speed viscometers in field operations.

Bingham Flow

Bingham flow is illustrated by Figure 7–5. The solid line represents a normal behavior pattern for drilling fluids, the dashed line shows the behavior predicted by the Bingham equations. Also shown in Figure 7–5 is

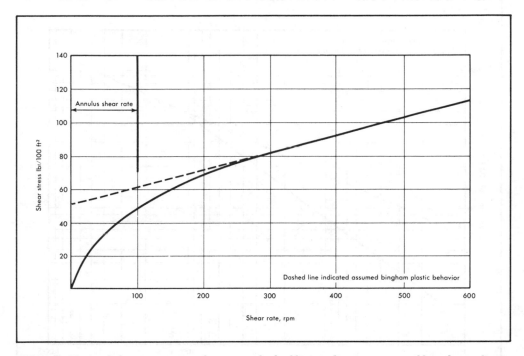

FIG. 7–5. *Normal shear stress vs. shear rate dashed line indicates assumed bingham plastic behavior*

the normal shear rate for an annulus. Thus, it can be seen that the Bingham equation for annular flow would predict shear stresses which are higher than the actual shear stresses in the annulus.

For pipe flow, the pattern is generally turbulent and none of these laminar flow equations would be used. However, if the flow pattern inside the pipe is laminar the power-law equation or the equation using shear stress directly is preferred.

Power-Law Flow

The power-law equations should be used if only a two speed viscometer is available. The term power-law comes from the expression shown in Equation 13 where the shear rate, γ, is raised to some power n.

$$\tau = K\gamma^n \tag{13}$$

Equation 13 plots as a straight line on log paper and the log relationship is shown as Equation 14.

$$\text{Log } \tau = \log K + n \log \gamma \qquad (14)$$

From Equation 14 it can be seen that the slope of the straight line on log paper is n and the intercept at the point the log shear rate equals 1.0 is K. Figure 7–6 shows a log plot of the same data used to plot Figure 7–5 on

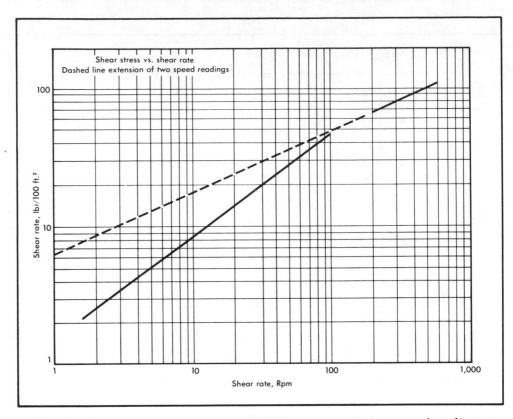

FIG. 7–6. *Shear stress vs. shear rate dashed line extension of two speed readings*

regular coordinate paper. It is noted that the use of only the 600 and 300 readings will still result in a calculated annular pressure loss which is greater than that which would be calculated by data from the multi-speed viscometer. For this reason, a multi-speed viscometer should always be used when annular pressure losses become critical.

Direct Shear Stress Reading

If multi-speed viscometer data are available, pressure losses in laminar flow should be calculated using Equation 5 for pipe flow and Equation 10 for annular flow. The general procedure for this calculation is given as follows:

1. Plot the shear stress versus shear rate from multi-speed viscometer data, preferably on log paper.
2. Determine the slope of the straight line of the log shear stress vs. log shear rate data in the region of interest. This is n or n may be calculated as shown in Equation 4.
3. Calculate the shear rate using Equation 6 or 7 for pipe flow or Equations 11 or 12 for annular flow. In some cases these equations have been simplified by assuming an n = 0.7. This is not a bad assumption in many applications. Using this assumption, the given equations for shear rate are simplified as follows.

Pipe flow:

$$\text{Shear rate (sec}^{-1}) = \frac{1.77\bar{v}}{D} \tag{15}$$

$$\text{Shear rate (RPM)} = \frac{1.04\bar{v}}{D} \tag{16}$$

Annular flow:

$$\text{Shear rate (sec}^{-1}) = \frac{2.74\bar{v}}{D_h - D_p} \tag{17}$$

$$\text{Shear rate (RPM)} = \frac{1.61\bar{v}}{D_h - D_p} \tag{18}$$

4. Using the calculated shear rate, read the shear stress directly from the plot of shear stress vs. shear rate.
5. Calculate the pressure loss using Equation 5 for pipe flow or Equation 10 for annular flow.

Example 1 shows the general procedure for calculating annular pressure losses in laminar flow using Equation 9 for power-law behavior and Equation 10 where shear stress is determined directly.

■ EXAMPLE 1:

Well depth = 10,000 ft.
Hole size = $8\frac{1}{2}$ inch
Drill pipe size = $4\frac{1}{2}$ inch O.D.
Rate of circulation = 300 gpm
Mud weight = 10 ppg

Viscometer Data

$\theta_{600} = 127$
$\theta_{300} = 94$
$\theta_{200} = 78$
$\theta_{100} = 54$
$\theta_6 = 8$
$\theta_3 = 5$

Solution: The viscometer readings versus the viscometer speed in RPM's for Example 1 are plotted in Figure 7–7. It will be noted that the readings at 200, 300 and 600 form a straight line in Figure 7–7 and the readings at 3, 6 and 100 form a straight

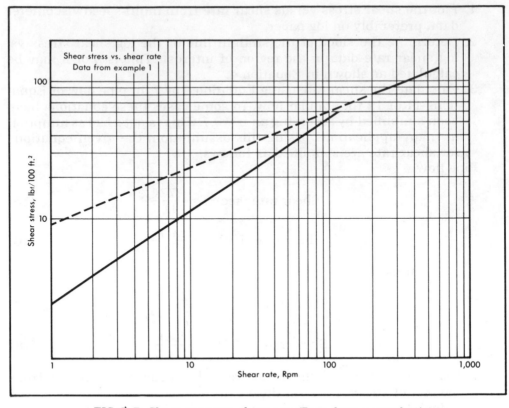

FIG. 7–7. *Shear stress vs. shear rate (Data from example 1)*

line. If only a two speed viscometer were available it can be seen that for any annular shear rate less than a 100 RPM's or 170 sec⁻¹ the calculated shear stress would be higher than that predicted by the actual measurements.

Multi-speed data:

$$\text{Shear rate (RPM)} = \left(\frac{1.41\bar{v}}{D_h - D_p}\right)\left(\frac{2n + 1}{3n}\right)$$

$$n = 3.32 \log \frac{\theta_{100}}{\theta_{50}} = 3.32 \log \frac{54}{34} = .666$$

or n could be obtained by measuring the slope directly. Thus:

$$\frac{2n + 1}{3n} = \frac{2.33}{1.998} = 1.166$$

$$\text{Shear rate} = \frac{(1.41)(138)(1.166)}{4} = 56.7 \text{ RPM}$$

Shear stress (from Figure 7–7) = 37 lb$_f$/100 ft²

$$P = \frac{1T}{300(D_h - D_p)} = \frac{(10^4)(37)}{(300)(4)} = 308 \text{ psi}$$

Two speed data:

$$n = 3.32 \log \frac{127}{94} = 0.433$$

$$\frac{2n + 1}{3n} = \frac{1.866}{1.299} = 1.435$$

$$\text{Shear rate} = \frac{(1.41)(138)(1.435)}{4} = 69.7 \text{ RPM}$$

$$\text{Shear stress} = 52 \text{ lb}_f/100 \text{ ft}^2$$

$$P = \frac{(10^4)(52)}{(300)(4)} = 433 \text{ psi}$$

Note: Equation 9 could have been used to obtain about the same answers as given by either the two speed or multispeed viscometer. In both cases it would have been necessary to calculate the K value. For example, consider the two speed result.

$$n = 0.433$$

$$K = \frac{\theta 300}{\gamma n} = \frac{94}{511^{.433}} = 6.33$$

Then:

$$P = \left[\left(\frac{2.4\bar{v}}{D_h - D_p} \right) \left(\frac{2n + 1}{3n} \right) \right]^n \frac{Kl}{300(D_h - D_p)}$$

$$P = \left[\frac{(2.4)(138)(1.435)}{4} \right]^{.433} \frac{(6.33)(10^4)}{(300)(4)} = 416 \text{ psi}$$

Also, it is not uncommon to assume an n value equal to 0.7, which eliminates one step in the calculation. If this had been done for the two speed readings, the following results would have been obtained:

Two speed data, assuming n = 0.7:

$$\text{Shear rate (RPM's)} = \frac{(1.41)(138)(1.142)}{4} = 55.6$$

$$\text{Shear stress} = 45 \text{ lb}_f/100 \text{ ft}^2$$

$$P = \frac{(10^4)(45)}{(4)(300)} = 375 \text{ psi}$$

For the multispeed viscometer the assumption of n = 0.7 would have given the following results:

Multi-speed data, assuming n = 0.7
Shear rate = 55.6 RPM's
Shear stress = 38 lb$_f$/100 ft^2

$$P = \frac{(10^4)(38)}{(4)(300)} = 318 \text{ psi}$$

The difference in using n = 0.7 in the case of the two speed is an error of about 10 per cent as compared with the use of the actual n of 0.433. While in the case of the multispeed viscometer the error is about 3 per cent.

Example 1 has been used to illustrate the basic procedures to determine annular pressure losses from either a two speed or multi-speed viscometer. It should be remembered that Example 1 is a specific example and the results cannot be used to estimate the differences in two speed and multispeed pressure loss calculations on a general basis.

Another procedure that may be used to determine annular pressure losses involves a re-arrangement of Equation 1 as shown in Equation 19.

$$P_{dca} + P_{dpa} = P_a = P_s - P_{s.c.} - P_{dp} - P_{dc} - P_b \qquad (19)$$

The annular pressure loss, P_a, may be determined by reading the standpipe pressure directly and subtracting the calculated pressure losses inside the drill string and bit.

The advantages of this procedure are: (1) the inside measurements of the drill string are more precise than hole size estimates; (2) pressure losses through the bit are considered very accurate; (3) flow patterns inside the drill string and through the bit are generally turbulent and viscous flow properties have a minor effect on the total pressure loss in this part of the string; and (4) flow patterns in the annulus are generally laminar and viscous flow properties of the mud affect pressure losses substantially. Thus, down-hole changes in annulus mud properties would result in higher pressure readings on the standpipe pressure gauge.

These same changes in mud properties would have only a minor effect on pressure losses inside the drill string and through the bit. This means that in most cases if the standpipe pressure at a given flow rate increases above normal levels, that most of the increase in pressure loss is in the annulus. This, of course, assumes no mechanical problems or bit plugging.

The disadvantages of using Equation 19 are: (1) Most of the pressure loss in the circulating system occurs inside the drill string and through the bit. Thus, a five per cent error in these calculations would result in a potential 30 to 40 per cent error in the determination of annular pressure losses. (2) The procedure is limited to a determination of total annular pressure losses and the point of interest is very often just below the last casing seat to the surface, not from total depth to the surface.

Considering the advantages and disadvantages of Equation 19, it is believed that the greatest advantage in using Equation 19 is to determine changes in annular pressure and to check for substantial variations between calculated annular pressure losses and those obtained using Equation 19.

The accuracy of annular pressure loss equations is limited because the flow properties of the mud must be measured at a given surface temperature and these measurements are used in the equations. Because the flow properties of the mud are sensitive to temperature changes, this practice of using surface flow properties introduces a potential error. However, down-hole measurements of actual annular pressure losses as compared with calculated values have at times been surprisingly accurate.

   Measurements which have not been very accurate are those where the down-hole mud is very thick due to flocculation. Also, calculated values immediately following trips are not generally accurate because down-hole properties may be substantially different from those measured at the surface. Example 2 shows the application of Equation 19 to determine the annular pressure losses.

■ EXAMPLE 2:

Well depth = 15,000 ft.
$9^5/_8$ inch casing set at 12,000 ft.
First sand below casing seat is at 12,100 ft.
Leak-off test after first sand below casing seat is exposed shows a fracture gradient = 0.91 psi/ft
Hole size = $7^7/_8$ inch
Drill pipe size = $4^1/_2$ inch O.D., 3.78 inch I.D.
Drill collars = 700 ft, $6^1/_4$ inch O.D., 2.75 inch I.D.
Bit nozzles = Two, $^3/_8$ inch
Mud weight = 16.5 ppg = 0.86 psi/ft
Mud properties, two speed viscometer:
$\theta_{600} = 65$
$\theta_{300} = 40$
Circulation rate, Q = 250 gpm
Before making a round trip, the following standpipe pressures are recorded.

Q, gpm	P, psi
250	2900
200	2050
170	1550

Determine the annular pressure losses and decide if it is safe to make an additional 500 feet of hole.

Solution:

$$n = 3.32 \log \frac{65}{40} = 0.70$$

$$K = \frac{40}{(511)^{.7}} = 0.52$$

$\bar{v}$ (average annular velocity) $= \dfrac{24.5Q}{D_h^2 - D_p^2}$

Around drill collars

$$\bar{v} = \frac{(24.5)(250)}{62 - 39} = 266 \text{ fpm}$$

Around drill pipe (open hole)

$$\bar{v} = \frac{(24.5)(250)}{62 - 20} = 146 \text{ fpm}$$

Around drill pipe (cased hole)

$$\bar{v} = \frac{(24.5)(250)}{72 - 20} = 118 \text{ fpm}$$

Use Equation 9:

Around drill collars

$$P = \left[\frac{(2.4)(266)(1.142)}{1.625}\right]^{0.70}\frac{(0.52)(700)}{(300)(1.625)}$$

$$P = \frac{(72)(0.52)(700)}{(300)(1.625)} = 54 \text{ psi}$$

Around drill pipe (open-hole)

$$P = \left[\frac{(2.4)(146)(1.142)}{(3.375)}\right]^{0.70}\frac{(0.52)(2300)}{(300)(3.375)}$$

$$P = \frac{(28.5)(0.52)(2300)}{(300)(3.375)} = 34 \text{ psi}$$

Around drill pipe (cased-hole)

$$P = \left[\frac{(2.4)(118)(1.142)}{(4.0)}\right]^{.70}\frac{(0.52)(12000)}{(300)(4.0)}$$

$$P = \frac{(21.6)(0.52)(12000)}{(300)(4.0)} = 112 \text{ psi}$$

The total calculated annulus pressure loss is as follows:

$$P_a = P_{dca} + P_{dpa_1} + P_{dpa_2} = 54 + 34 + 112 = 200 \text{ psi}$$

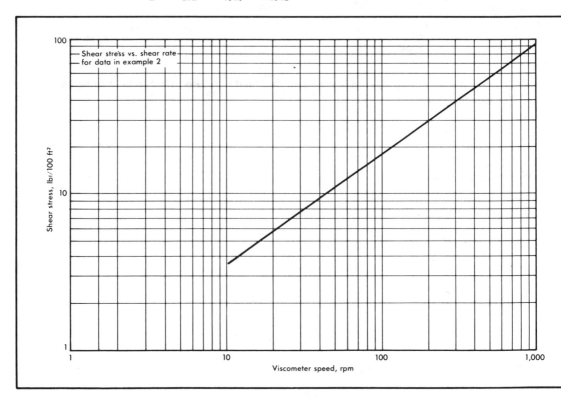

FIG. 7–8. *Shear stress vs. shear rate for data in example 2*

The same calculations could have been made by plotting the data as shown in Figure 7–8, calculating the shear rate and reading the shear stress directly from Figure 7–8.

Pressure loss around drill collars:

$$\text{Shear rate} = \left(\frac{(1.41)\bar{v}}{D_h - D_p}\right)\left(\frac{2n + 1}{3n}\right)$$

$$= \frac{(1.41)(266)(1.142)}{(1.625)} = 264 \text{ RPM}$$

Shear stress = 37 lb$_f$/100 ft^2

$$P = \frac{(700)(37)}{(300)(1.625)} = 53 \text{ psi}$$

Pressure loss around drill pipe and open hole:

$$\text{Shear rate} = \frac{(1.61)(146)}{3.375} = 69.7 \text{ RPM}$$

Shear stress = 14 lb$_f$/100 ft^2

$$P = \frac{(2300)(14)}{(300)(3.375)} = 32 \text{ psi}$$

Pressure loss around drill pipe and cased hole:

$$\text{Shear rate} = \frac{(1.61)(118)}{4.0} = 47.5 \text{ RPM}$$

Shear stress = 10.8 lb$_f$/100 ft^2

$$P = \frac{(12000)(10.8)}{(300)(4.0)} = 108 \text{ psi}$$

Total annulus pressure loss:

$$P_a = 53 + 32 + 108 = 193 \text{ psi}$$

Pressure loss in annulus using standpipe pressure:

$$P_a = P_s - P_{s.c.} - P_{dp} - P_{dc} - P_b$$

The pressure losses inside the drill string and through the bit were determined using a Hydraulic slide-rule. Table 7–1 shows these results and the calculated annulus pressure loss using Equation 19.

TABLE 7–1
Determination of Annulus Pressure Loss

Q	P_s	$P_{s.c.}$	P_{dp}	P_{dc}	P_b	P_a
250	2900	21	673	150	1770	286
200	2050	16	500	114	1250	170
170	1550	12	372	84	900	182

It should be remembered that the indicated fracture gradient at 12,100 feet is 0.91 psi/ft. Thus the circulating pressure from 12,100 feet to the surface, using a 16.5 ppg mud, is limited to a maximum of 630 psi. The annular pressure losses shown in Table 7–1 are for the total annular circulating pressure. Considering the calculated pressure losses, the circulating pressure from 12,100 feet to the surface should be about 60 per cent of the total annular losses. Because 40 per cent of the annular pressure drop is below 12,100 feet and will not affect the formation at 12,100 feet, this will be used as a safety factor.

It will be assumed that any increase in surface pressure at the given circulation rates shown in Table 7–1 will be caused by increases in the annular pressure. This assumption is believed valid because the annular flow is laminar and will be affected by the viscous flow properties of the mud. Flow inside the drill string and through the bit is generally turbulent. Using this assumption, the maximum permissible standpipe pressure can be determined and this is shown in Table 7–2

TABLE 7–2
Maximum Permissible Standpipe Pressure

Q	P_s	Actual P_a	Permissible P_a	Additional Increase Permitted	Maximum Permissible P_s
250	2900	286	630	344	3244
200	2050	170	630	460	2510
170	1550	182	630	448	1998

These data are shown on Figure 7–9 and cover all ranges of circulation rates. If the actual surface pressure exceeds the maximum permissible surface pressure, the operator faces the danger of lost circulation and an underground blowout if high pressure permeable formations are open to the well-bore. A reasonable question is what action does the operator take if the surface pressure exceeds the maximum permissible surface pressure.

If the increase is unavoidable, he should try to condition mud and set a protective liner. If the increase occurs immediately after a round trip, he may need to pull back to say 8000 feet to circulate and condition mud and in this manner work the pipe back to bottom. Specific situations which could increase annular pressures and which should be watched closely by the operator include: (1) surge pressures when running pipe back to bottom after a round trip; (2) pressure that remains trapped in the mud by the gel strength of the mud; (3) pressures caused by impact forces when breaking circulation; (4) flocculation of the downhole mud caused by contamination or insufficient chemical treatment; and (5) thickening of the bottom-hole mud, caused by cooling, as it nears the surface.

The surge pressures can be estimated and the procedures used to do this are shown in the chapter on Pressure Surges. Pressure trapped in the mud by the gel strength may be the aforementioned surge pressure or surge pressures created while making pipe connections. If pressure is trapped

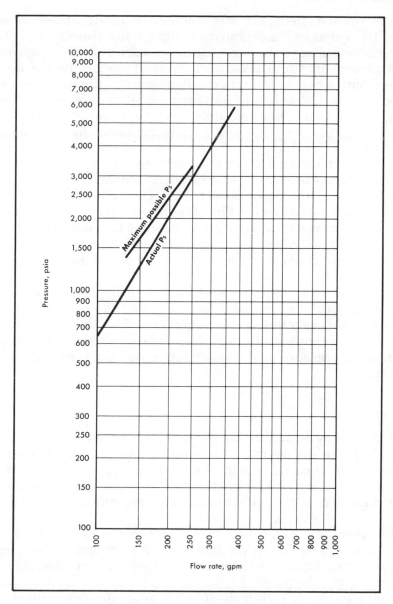

FIG. 7–9. *Standpipe pressure vs. circulation rate for example 2*

in the mud it will be additive to the normal circulation pressure losses and will increase the standpipe pressure.

Downhole flocculation of the mud will also be reflected by an increase in standpipe pressure. It may not be possible to determine immediately the cause of the increase in standpipe pressure; however, the important thing is to not continue pumping if all the parameters indicate the possibility of fracturing the formation.

The control of annular pressures may be difficult. When running pipe there is no way other than by using equations that the annular pressure can be determined. Instantaneous changes in annular pressure might be predicted from changes in drill string weight, when running pipe; however, the operator would never be sure that the change in drill string weight was due to a change in surge pressure or to wall contact by the drill string.

Some control can be exercised when starting pumps. One procedure would be to rotate pipe before initiating circulation, then lift the pipe as pumping is initiated. This procedure would minimize the effect of mud gelation by stirring the mud and reduce the surge by creating pressure reductions due to swabbing. If the procedure is not successful, the pump should be stopped, the pipe lowered and the procedure repeated.

Another alternative is to use a standpipe choke. The choke should be open while starting the pump slowly. Then close choke slowly while maintaining the same pump speed. The standpipe choke provides a method of minimizing the surge created when pumping is first initiated. Most pumps cannot be started and run at very low speeds thus the surge associated with starting the pump is unavoidable unless some of the fluid can be bypassed initially.

Some operators, to minimize the initial surge pressures, have developed the practice of breaking circulation several times when running pipe. There is a danger in this practice if circulation is broken just as the bit reaches the casing seat, because with the larger collars just above the casing shoe the circulating pressure at the casing shoe would be more than with the bit on bottom.

TURBULENT FLOW

Turbulent flow is defined by its name. There are different degrees of turbulence and as the degree of turbulence is increased, the pressure loss for a given set of conditions is increased. Thus, there is no way to be completely accurate using any set of equations for pressure losses in turbulent flow, because there is no reliable method to determine the degree of turbulence.

Fluid properties which affect flow patterns substantially in laminar flow have a minor affect on the flow pattern in turbulent flow. However, fluid properties do affect the fluid velocity at which transition from laminar to turbulent flow is initiated.

Pressure losses in turbulent flow are generally determined by using empirical friction factors. A common correlation of this type is shown in Figure 7–10, which shows a lot plot of the Fanning friction factor versus Reynold's number. The Fanning friction factor is defined in Equation 20 and the Reynolds number is defined in Equation 21.

$$f = \frac{9.3(10^4)PD}{v^2 1} \qquad (20)$$

$$Re = \frac{15.47\rho vD}{\mu} \qquad (21)$$

Equation 20 may be solved for pressure and the result is shown in Equation 22.

$$P = \frac{\rho v^2 l f}{9.3(10^4)D} \tag{22}$$

Equation 22 is the most commonly used equation to determine pressure losses in turbulent flow. In general, the equation is modified by relating the friction factor to the Reynolds number in the turbulent flow region. This is where many of the final equations used by various companies and individuals are slightly different. The most common procedure is to use a linear

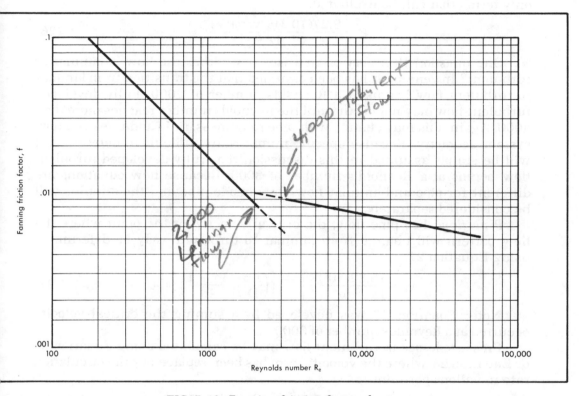

FIG. 7-10. *Fanning friction factor chart*

relationship between the friction factor and Reynolds number. This procedure generally results in small differences in the slope of the line and small differences in the intercept on the ordinate. In the turbulent flow region of Figure 7-10, Equation 23 defines a straight line relationship between the friction factor and Reynolds number.

$$f = \frac{0.046}{R_e^{0.2}} \tag{23}$$

Using Equation 23 for the friction factor in Equation 22 results in Equation 24 for pressure loss in turbulent flow.

$$P = \frac{2.86(10^{-7})\rho^{.8}v^{1.8}\mu^{.2}l}{D^{1.2}} \tag{24}$$

One further alteration is needed in Equation 24 to permit its routine use. The fluid viscosity term, μ, is practically impossible to determine in the turbulent flow region. For this reason, it is a common practice to relate μ to the plastic viscosity. One procedure that has proven to be accurate enough for general use is shown in Equation 25.

$$\mu = \frac{(PV)}{3.2} \quad \textit{turbulent flow only} \tag{25}$$

Using Equation 25 in Equation 24 results in Equation 26, which includes only terms that can be evaluated.

$$P = \frac{2.27(10^{-7})\rho^{.8}v^{1.8}(PV)^{.2}l}{D^{1.2}} \tag{26}$$

One question that very often arises is whether the flow is laminar or turbulent? In general, it has been assumed that the transition from laminar to turbulent flow begins at a Reynolds number of 2000. Fully developed turbulent flow may not occur until the Reynolds number reaches more than 4000. Again, this may change with pipe roughness, fluid additives and flow system geometry. Because specific numbers are difficult to determine, it will be simply assumed in this discussion that fully developed turbulent flow begins at a Reynold's number of 3000. Because flow equations are different for pipe and annular flow, flow inside pipe and the annulus will be considered separately.

Pipe flow: For pipe flow, the critical velocity, or the point where the flow pattern will change from laminar to turbulent, may be determined using Equation 27.

$$v_c = \left[\frac{5.82(10^4)K}{\rho}\right]^{\frac{1}{2-n}} \left[\left(\frac{1.6}{D}\right)\left(\frac{3n+1}{4n}\right)\right]^{\frac{n}{2-n}} \tag{27}$$

Note: Equation 27 was developed by assuming the critical velocity occurred at a Reynolds number of 3000.

Pressure losses in turbulent flow may be calculated using Equation 26 or Equation 28, where the velocity term has been replaced by the circulation rate in gallons per minute.

$$P = \frac{7.7(10^{-5})\rho^{.8}Q^{1.8}(PV)^{.2}(L)}{D^{4.8}} \tag{28}$$

The use of Equation 28 is shown in Example 3.

■ EXAMPLE 3:

Given Data: Drill pipe size = $4\frac{1}{2}$ inch O.D. and 3.78 inch I.D.
Drill pipe length = 1000 feet
Mud weight = 10 ppg
Q = 300 gpm
PV = 25 cp

Determine: The pressure loss in psi.
Solution:

$$P = \frac{7.7(10^{-5})(10)^{.8}300^{1.8}25^{.8}}{3.78^{4.8}}$$

$$P = \frac{7.7(10^{-5})(6.31)(2.892)(10^4)(1.902)(10^3)}{6(10^2)} = 44.6 \text{ psi}$$

A hydraulic slide-rule manufactured and distributed by a service company shows a pressure loss of 43.5 psi for Example 3.

These equations for pipe flow are considered accurate enough for general use. Charts and hydraulic slide-rules available from major service companies are generally reliable for the determination of turbulent flow pressure losses inside pipe. There are small variations in calculated pressures from these sources because of variations in the friction factor relationship used by each. Because the pressure losses in turbulent flow are dependent on the degree of turbulence which is affected by fluid type and pipe roughness, it is difficult to develop a general solution for pressure losses in turbulent flow.

Many of the polymers suppress the degree of turbulence and thus reduce pressure losses in turbulent flow. To determine the degree to which these polymers are effective would require experimental testing and a wide range of results from this testing. Many companies selling polymers claim to have done this type testing and have published results of these tests. It is suggested that these published results be checked in field operations where possible for accuracy.

Annular flow: It is very difficult to determine accurately annular pressure losses in turbulent flow. It is also difficult to determine if flow in the annulus is laminar or turbulent. In laminar flow pressure loss calculations, the fluid velocity affects the pressure loss substantially less than it does in turbulent flow. Because the annular velocity is affected substantially by hole size, it may be very difficult to determine annular velocities in open-hole sections with any degree of accuracy.

Caliper surveys of open-hole sections are helpful, but unless a four point caliper is being used the accuracy of caliper surveys are open to question. These questions are not being introduced to discourage the determination of flow patterns and turbulent flow pressure losses in the annulus, but rather to emphasize the importance of a careful analysis when these pressures become critical to further drilling.

Assuming the flow pattern changes from laminar to fully developed turbulent flow at a Reynolds number of 3000, the critical velocity may be defined as shown in Equation 29 for annular flow.

$$v_c = \left[\frac{3.878(10^4)K}{\rho}\right]^{\frac{1}{2-n}} \left[\frac{2.4}{D_h - D_p}\left(\frac{2n+1}{3n}\right)\right]^{\frac{n}{2-n}} \tag{29}$$

If the flow pattern in the annulus is determined to be turbulent, then Equation 26 may be modified as shown in Equation 30 for the determination of annular pressure losses.

$$P = \frac{7.7(10^{-5})\rho^{0.8}Q^{1.8}(PV)^{0.2}l}{(D_h - D_p)^3(D_h + D_p)^{1.8}} \tag{30}$$

Again, solutions to equations similar to Equation 30 are available, using charts and hydraulic slide-rules with the same limitations already mentioned.

Other methods have been suggested to determine pressure losses in turbulent flow in the annulus. One procedure suggested by Forbes[1] includes construction of a regular rheogram chart of shear stress versus shear rate. The next step is to determine whether the flow pattern in the annulus is laminar or turbulent and Forbes uses a Reynolds number of 2000 for this purpose. A straight line with a slope of 2.0 is then constructed from the critical velocity point on a regular rheogram chart.

A procedure based on the same approach as used by Forbes, but assuming fully developed laminar flow begins at a Reynolds number of 3,000, will be illustrated in Example 4.

■   EXAMPLE 4:

Given Data: Hole size $= 8\frac{1}{2}$ inch
Drill pipe size $= 4\frac{1}{2}$ inch O.D.
Mud weight $= 16$ ppg
Viscometer readings: $\theta_{600} = 90$
$\theta_{300} = 50$
Circulation rate $= 450$ gpm
Well Depth $= 10,000$ feet

Determine: Pressure loss in the annulus

Solution:

$$n = 3.32 \log \frac{90}{50} = 0.848$$

$$K = \frac{50}{511^{.848}} = 0.254$$

$$v_c = \left[\frac{3.878(10^4)6.254}{16}\right]^{0.869} \left[\frac{2.4(1.06)}{4}\right]^{0.736}$$

$$v_c = (260)(0.716) = 186 \text{ fpm}$$

$$\text{Annular velocity} = \frac{(24.55)(450)}{72 - 20} = 212 \text{ fpm}$$

The annular velocity of 212 fpm is greater than the critical velocity of 186 fpm, thus the annular flow pattern is assumed to be turbulent. Next, the viscometer readings are plotted on Figure 7–11. Then determine the shear rate in RPM's at the critical and annular velocities.

Critical velocity:

$$\text{Shear rate} = \frac{(1.41)(186)(1.06)}{4} = 69.5 \text{ RPM}$$

Annular velocity:

$$\text{Shear rate} = \frac{(1.41)(186)(1.06)}{4} = 79.1 \text{ RPM}$$

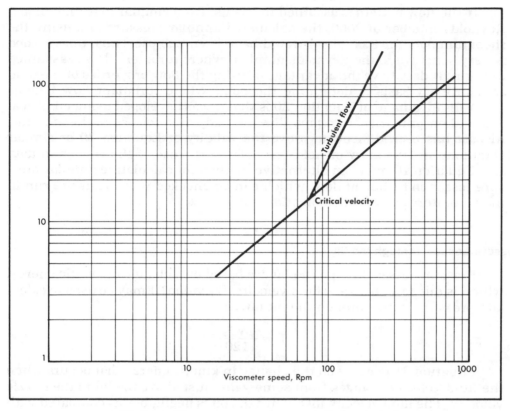

FIG. 7–11. *Shear stress vs. shear rate for example 3*

On Figure 7–11, construct a straight line with a slope of 2.0, beginning at the critical shear rate of 69.5 RPM's. From the turbulent flow line on Figure 7–11, the shear stress at the annular shear rate of 79.1 RPM's is 18.5 lb$_f$/100 ft^2.

Calculate the pressure loss using Equation 10.

$$P = \frac{(10^4)(19.0)}{(300)(4)} = 158 \text{ psi}$$

This result can be compared with the pressure loss calculated by Equation 30.

$$P = \frac{7.7(10^{-5})16^{0.8}450^{1.8}40^{0.2}10^4}{(4^3)(13)^{1.8}}$$

$$P = \frac{7.7(10^{-5})(9.2)(5.98)(10^4)(2.09)(10^4)}{(64)(101)} = 137 \text{ psi}$$

If would be very difficult to determine which procedure of calculating annular pressure losses is the most accurate. If the flow pattern actually changes from laminar to turbulent at a Reynolds number of 3000, then the calculated pressure of 154 psi using the plotted data on Figure 7–11 would probably be the most accurate solution to Example 4.

If the flow pattern is assumed to change from laminar to turbulent at a Reynolds number of 2000, the calculated annular pressure loss using the rheogram data for Example 4 would have been 229 psi. Thus, this method is very sensitive to the Reynolds number, where turbulent flow is assumed to begin. It does have the advantage of using the flow properties of the mud as a primary basis to determine the point where turbulent flow begins.

A disadvantage of Equation 30 is that it was developed using empirical data, taken primarily from pipe flow, and adjusted for use in the annulus. Also, it has been difficult to prove the validity of Equation 30 or similar equations in laboratory or field tests. Thus, the use of the rheogram data and Equation 10 offers an alternative method for calculating annular pressure losses in turbulent flow, which can be checked when tests are run in the laboratory or in field applications.

Pressure Losses Through the Bit

Pressure losses through the bit are based on a change in kinetic energy, which is due to a change in fluid velocity. Equation 31 may be used to calculate pressure losses through the bit nozzle.

$$P = \frac{\rho v_n^2}{1120} \tag{31}$$

Equation 31 is based on the change in kinetic energy that occurs when the fluid velocity changes from some value just above the bit to the nozzle velocity. The fluid velocity just above the bit is negligible, as compared with the nozzle velocity, and this is the reason it does not appear in Equation 31. Also, a nozzle coefficient of 0.95 was assumed for Equation 31.

The nozzle velocity in feet per second may be calculated using Equation 32.

$$v_n = \frac{0.32Q}{A_n} \tag{32}$$

In general, bit nozzle pressure losses are determined accurately using Equation 31 or similar relationships. The nozzle velocity, and thus bit nozzle pressure losses, are very sensitive to nozzle size. Thus, nozzle sizes should be determined accurately.

Summary of Pressure Losses in the Circulating System

Emphasis in this discussion has been placed on an accurate determination of annular pressure losses. It is difficult to determine annular pressure losses accurately because of (1) hole size variations, (2) changes in fluid behavior, (3) changes in mud properties because of changes in temperature, and (4) a general lack of concern with the problem. Reason 4 is changing because methods have been developed to determine pore pressures and fracture gradients more accurately, which places a premium on the determination of the total annulus pressure imposed at any point.

More accurate equations have been developed for calculating annular pressure losses; however, these do not eliminate reasons one through three for the lack of accuracy. To combat this, Equation 19 is being used to determine annular pressures. Advantages and disadvantages of this procedure were discussed. Accuracy, using Equation 19, requires an accurate surface pressure gauge, accurate determinations of pressure losses inside the drill string and accurate determinations of bit nozzle pressure losses. This need for accuracy places a premium on actual field measurements of these pressure losses in some critical wells.

The success or failure of a specific well may be determined by what appears to be small changes in operating practices. Accurate pressure loss determinations may be the key to reaching objectives that heretofore could not be reached or to reducing the cost of reaching objectives such that more prospects become attractive economically.

REFERENCES

1. Forbes, Gene. Personal communication concerning fluid circulation.

NOMENCLATURE

A_n = nozzle area, inches
D = diameter, inches
D_h = hole diameter, inches
D_p = pipe diameter, inches
l = length, feet
P = pressure loss for any section of length, l, in the circulating system
P_b = pressure loss through the bit, psi
P_{dp} = pressure loss through drill pipe, psi
P_{dc} = pressure loss through drill collars, psi
P_{dpa} = pressure loss through hole and drill pipe annulus, psi
P_{dca} = pressure loss through hole and drill collar annulus, psi
$P_{s.c.}$ = pressure loss through surface connections, psi
P_s = total pressure loss in circulating system, psi
P_a = total pressure loss in annulus, psi
Q = circulation rate, gpm
$\bar{v}$ = fluid velocity, fpm
$\bar{v}_n$ = nozzle velocity, fps
PV = plastic viscosity, centipoise
Y = yield point, $lb_f/100\ ft^2$
θ = viscometer readings, $lb_f/100\ ft^2$
n = slope of the straight line on a plot of log shear stress versus log shear rate
K = ordinate intercept on a plot of log shear stress versus log shear rate, $lb_f/(sec^{-1})^n$
γ = shear rate, sec^{-1}
f = Fanning friction factor, dimensionless
ρ = mud weight, ppg
μ = mud viscosity, centipoise
R_e = Reynolds number, dimensionless

Lifting Capacity of Drilling Fluids

INITIALLY, the primary purpose of the drilling fluid was to remove formation solids continuously. No time was spent on a scientific evaluation of the carrying capacity of the fluid and very little effort was made to control the fluid properties. Even with the advancement of science in mud treating, operators ignored the lifting capacity of muds.

If an operator felt that the quantity of solids being removed was less than that being generated, he increased the fluid velocity or made the mud thicker. Thus, the days of high annular velocities and thick muds were introduced with an almost painless effort.

The introduction of jet bits in 1948 reversed the trend towards high circulation rates. Jet bits introduced the need for increased pressure losses through the bit nozzles for adequate bottom-hole cleaning. One way of meeting this need was to reduce pressure losses in other parts of the drill string by reducing circulation rates. Pressure losses were then increased through the bit by using small nozzle sizes.

This philosophy did not occur suddenly, but the needed direction was recognized and in 1950 Williams and Bruce[1] published a paper on the lifting capacity of fluids that has become a classic of the technical literature. Based on many months of controlled research on the lifting capacity of fluids, Williams and Bruce concluded the following:
1. Turbulent flow in the well annulus was the most desirable flow pattern for removing formation cuttings.
2. Low viscosity or thin fluids were, in general, more desirable than thick fluids for good hole cleaning.
3. Pipe rotation provides an aid to removing formation cuttings.

4. Annular fluid velocities of 100 to 125 fpm using water are adequate for removing formation solids.

These conclusions were based on tests performed in an actual well and in plastic pipe in the laboratory. Figure 8–1 shows the results of tests conducted in the laboratory where the actual behavior of the transported solids could be photographed. Parts A, B, and C show the reason for Conclusion 1 – the solids were being removed continuously while the fluid was in turbulent flow, as shown in part A. In parts B and C, the solids next to the center pipe and the wall of the hole were falling back. Part E shows the effect of pipe rotation and the advantages gained for lifting solids next to the rotating pipe. Part D simply illustrates the lifting of large discs which are too thick to be turned on their side next to the wall of the pipe and slip back as shown in parts B and C.

Formation cuttings weigh about 21.0 ppg, which means the solids will tend to slip downward through any fluid which weighs less than 21.0 ppm. In Figure 8–1, the solids weighed in a range comparable to formation solids and the fluid was water, with thickening agents, which weighed 8.33 ppg.

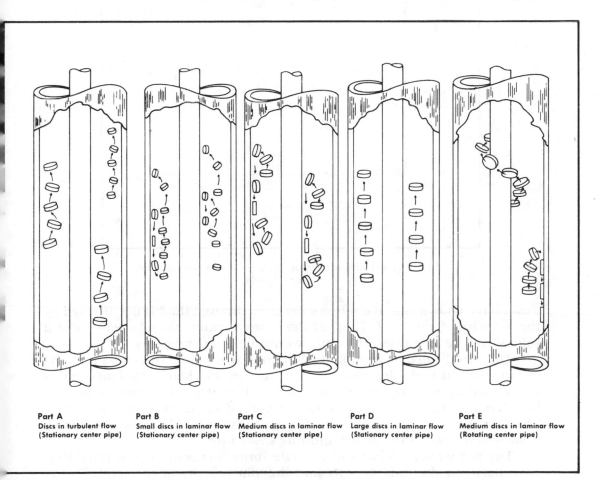

Part A	Part B	Part C	Part D	Part E
Discs in turbulent flow (Stationary center pipe)	Small discs in laminar flow (Stationary center pipe)	Medium discs in laminar flow (Stationary center pipe)	Large discs in laminar flow (Stationary center pipe)	Medium discs in laminar flow (Rotating center pipe)

FIG. 8–1. *Lift of solids in liquid solutions*

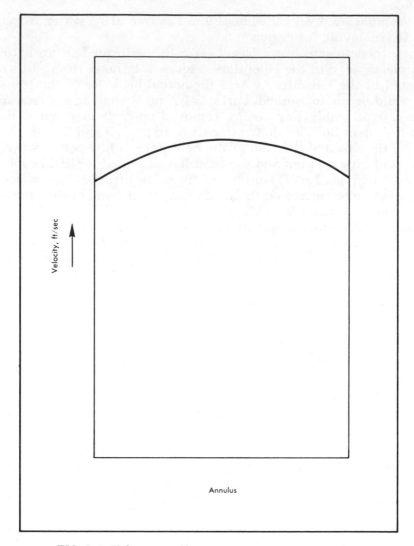

FIG. 8-2. *Velocity profile in annulus for turbulent flow*

Tests have shown that the rate a solid slips through fluid is not affected by the velocity of the fluid. Thus, in the case of water, the slip velocity of a solid would be the same if the water were quiescent or flowing.

This would also be true for drilling mud if the mud thickness remains the same at different flow rates. In laminar flow, it is known that most muds shear thin as velocity increases, while in turbulent flow there would be little change in mud thickness in the normal range of circulation rates. Thus, in any precise analysis of lifting capacity, mud thickness would have to be determined at the actual annulus shear rate.

The net upward velocity of a single formation solid will be the difference between the fluid velocity and the slip velocity of the solid. This is expressed mathematically in Equation 1.

$$V_p = V_f - V_s \qquad\qquad (1)$$

In turbulent flow, the velocity distribution of the fluid is almost flat, as shown in Figure 8–2. Thus, if the fluid velocity exceeds the solids slip velocity, the solids will be removed continuously, as shown in part A of Figure 8–1. In laminar flow, the velocity distribution is affected by the mud properties; however, a typical velocity distribution is shown in Figure 8–3. Referring to Figure 8–3, the average annular velocity would be at point A. This means that part of the fluid is traveling at a velocity which is higher, and a part at a velocity which is lower, than the average fluid velocity. Thus, in the middle of the stream, the recovery of particles may be faster than anticipated and next to the pipe walls some particles may never reach the surface.

The practice of injecting dye or oats in the mud to determine travel time

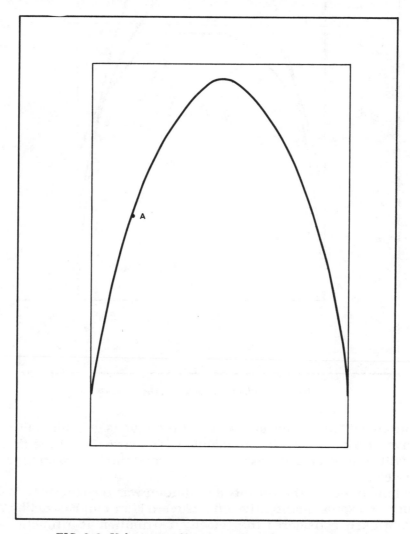

FIG. 8–3. *Velocity profile in annulus for laminar flow*

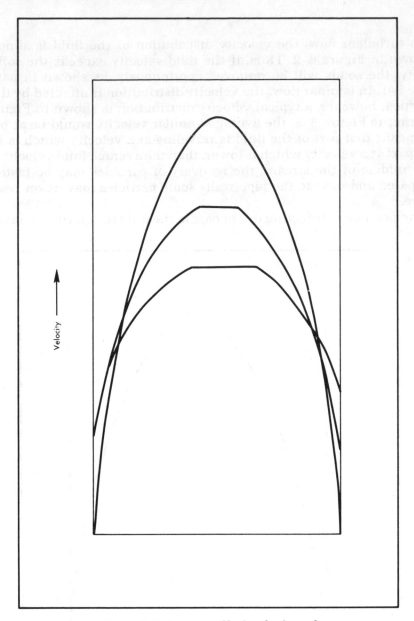

FIG. 8–4. *Velocity profile for thick muds*

from bottom will give erroneous results if the fluid is in laminar flow. Some of the dye or oats will be in the middle of the stream and the time from bottom will be substantially less than the normal time based on the average annular velocity.

The importance of the velocity distribution was recognized by Williams and Bruce and some quantitative effects were later emphasized by Walker in 1963. Walker[2] considered the velocity distribution as a function of the mud properties. In general, an increase in the ratio of yield point to plastic viscosity or a decrease of the line slope, n, for the power law behavior of fluids results in a flattening of the velocity profile as shown in Figure 8–4.

The flat part of the velocity profile shown in Figure 8–4 is referred to as plug flow. In the plug flow region there is no shearing of the fluid layers. Thus, as the width of plug flow region increases, the low velocity parts of the stream decreases and the cleaning action of the fluid is increased. Equations may be developed to determine velocity profiles, using either the Power-law or Bingham Plastic assumptions of flow behavior.

However, an estimate of the shape of an annulus profile would not be possible because pipe rotation disturbs the profile, the drill string is not concentric in the hole and the shape of the hole is irregular. It is enough to know that increases in the yield point or a decrease in the n factor will flatten the velocity profile and provide an assist in hole cleaning.

The actual lifting capacity of a fluid is related directly to the rate a solid will slip through the fluid. The slip velocity of a particle may be estimated using Equation 2.

$$v_s = 113.4 \left[\frac{D_p(\rho_p - \rho_f)}{C_D \rho_f} \right]^{1/2} \tag{2}$$

The particle diameter may be estimated from a visual inspection or, if more precision is needed, the equivalent diameter may be determined by a screen analysis. The particle density is generally considered constant at 21.0 ppg for Equation 2. The drag coefficient is the frictional drag between the fluid and the particle.

There are no methods available to determine this frictional drag precisely; however, Figure 8–5 shows the drag coefficient curve versus the Particle Reynolds number. Figure 8–5 was prepared using limestone and shale cuttings from field drilling operations. Equation 2 was used to calculate the drag coefficient after the slip velocity had been determined experimentally. The particle Reynolds number was calculated using Equation 3.

$$R_p = \frac{15.47 \rho v_s D_p}{\mu} \tag{3}$$

Water, and glycerin mixed with water, was used as the base fluid for the determination of Figure 8–5. Above a particle Reynolds number of about 2000 the drag coefficient remains constant at about 1.50. Thus, when the flow pattern around the particle is turbulent, a drag coefficient of 1.50 can be used in Equation 2 and the slip velocity calculated directly. When the fluid flow pattern is laminar around the particle, the drag coefficient varies with the particle Reynolds number. For a particle Reynolds number of 1.0 or less, the drag coefficient can be determined using Equation 4.

$$C_D = \frac{40}{R_p} \tag{4}$$

Substituting this value of the drag coefficient in Equation 2 gives Equation 5 for the slip velocity.

$$v_s = \frac{4980 D_p^2 (\rho_p - \rho_f)}{\mu} \tag{5}$$

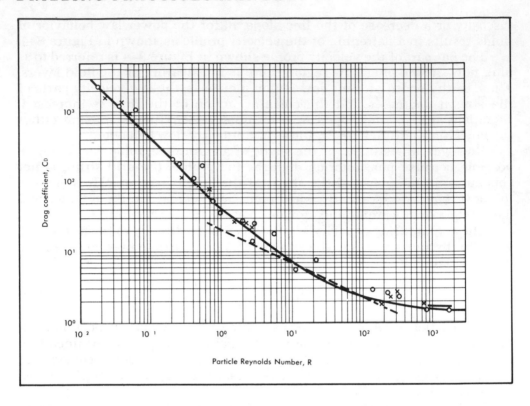

Equation 5 has limited application, because in most cases the particle Reynolds number will exceed 1.0 when the flow around the particle is laminar. Drawing the best approximate straight line between particle Reynolds number of 10 and 100 gives Equation 6 for the drag coefficient.

$$C_D = \frac{22}{R_p^{0.5}} \tag{6}$$

Using this value of drag coefficient in Equation 2 gives Equation 7 for the slip velocity.

$$v_s = \frac{175 D_p (\rho_p - \rho_f)^{.667}}{\rho_f^{.333} \mu^{.333}} \tag{7}$$

Equation 7 is the equation that should be used in routine solutions. After the determination of slip velocity, the particle Reynolds number may be calculated. If the particle Reynolds number is more than 2000, Equation 2 should be used with a drag coefficient of 1.50. If the operator is in doubt about whether the flow around the particle is turbulent or laminar, the equation giving the lowest slip velocity for the specific problem should be used. This procedure will always provide the most nearly correct solution.

The use of these equations will be illustrated by Examples 2, 3, and 4. Example 2 shows the slip velocity when the fluid around the particle is in turbulent flow. Example 3 shows the particle slip velocity in a thick low weight mud and Example 4 shows the particle slip velocity in a weighted mud.

■ EXAMPLE 2:

Well depth = 10,000 ft
Hole diameter = $7\frac{7}{8}$ in.
Drill Pipe diameter = $4\frac{1}{2}$ in.
Fluid = water, weight = 8.33 ppg
Shale cuttings
Weight of formation solids = 21 ppg
Average particle size 0.3 in.
Solution: Using Equation 2

$$v_s = 113.4 \left[\frac{0.3(21 - 8.33)}{(1.5)(8.33)} \right]^{1/2} = 63 \text{ fpm}$$

■ EXAMPLE 3:

Well Depth = 10,000 ft
Hole diameter = $7\frac{7}{8}$ in.
Enlarged sections of hole = 20 in.
Drill pipe diameter = $4\frac{1}{2}$ in.
Fluid, weight = 9.5 ppg
viscometer = $\theta_{600} = 100$
          $\theta_{300} = 60$
Weight of formation solids = 21 ppg
Average particle size = 0.3 in.
Circulation rate = 300 gpm
Solution: Using Equations 2, 5, and 7.
Equation 2:

$$v_s = 113.4 \left[\frac{(0.3)(21 - 9.5)}{(1.5)(9.5)} \right]^{1/2} = 56 \text{ fpm}$$

Equation 5:

$$v_s = \frac{(4980)(.09)(21 - 95)}{92\ 89} = \frac{(4980)(.09)(11.5)}{92\ 89} = 57 \text{ fpm}$$

$$\mu = \left[\left(\frac{2.4\bar{v}}{D_h - D_p} \right) \left(\frac{2n + 1}{3n} \right) \right]^n \frac{200K(D_h - D_p)}{v} \qquad p.107$$

$$\bar{v} = \text{(for } 7\frac{7}{8} \text{ in hole)} = 175 \text{ ft/min}$$

$$n = 3.32 \log \frac{100}{60} = (3.32)(.223) = 0.74$$

$$K = \frac{60}{511^{.74}} = \frac{60}{101} = 0.595$$

$$\mu = \left[\frac{(2.4)(175)(1.12)}{3.375} \right]^{.74} \frac{(200)(.595)(3.375)}{175}$$

$$\mu = (39)(2.29) = 89 \, c_p$$

Equation 7:

$$v_s = \frac{(175)(.3)(21 - 9.5)^{.667}}{(9.5)^{.333}(89)^{.333}} = \frac{268}{9.4} = 28.5 \text{ fpm} \qquad correct\ answer$$

Equation 7 represents the most nearly correct solution. It is noted that the slip velocity is about one-half that calculated by Equations 5 and 2. This shows the fluid to be in laminar flow around the particle. These solutions would be changed because of the change in equivalent viscosity in the enlarged hole section. Considering the 20-inch hole, the slip velocity is determined as follows using Equation 7:

$$\mu = \left[\frac{(2.4)(19)(1.12)}{15.5} \right]^{.74} \frac{(200)(.595)(15.5)}{19}$$

$$\mu = (2.42)(97.1) = 235 \; c_p$$

$$v_s = \frac{(175)(.3)(11.5)^{.667}}{(9.5)^{.333}(235)^{.333}} = \frac{(175)(.3)(5.1)}{(2.12)(6.16)} = 20.5 \text{ fpm}$$

The mud is substantially thicker, 235 c_p, compared with 89 c_p, in the larger hole because of the reduction in shear rate. However, in the gauge hole section the average annular velocity of 175 fpm is substantially above the slip velocity of 28.5 fpm and there should be no problem cleaning the hole. In the enlarged hole section, the average annular velocity of 19 fpm is just below the slip velocity of 20.5 fpm, thus the mud would have to be thickened to remove particles which are 0.3 inch in diameter or larger. The average annular velocity is the velocity at only one point in the flow stream, part of the fluid is moving at a higher velocity and part of the fluid is moving at a lower velocity.

This introduces the question, what is the required average annular velocity for adequate hole cleaning? To give a complete answer the velocity profile would have to be determined, the size of the cuttings to be removed would have to be known and downhole mud thickness would have to be predicted. In most practical applications there is no way to make these precise determinations, thus the required lifting capacity is generally based on observations of hole cleaning while the drilling operation is under way.

Again consider the case of the 20 inch hole. In this case the slip velocity was slightly greater than the annular velocity. This condition could result in bridges in the hole, and plugging of the annulus and regardless of the calculations made the operator would be aware of the problem. If corrections were not made quickly he might suffer the potential consequences of stuck pipe or other associated problems. The useful part of the calculation is that the operator is able to define the precise problem and is aware of the required treatment.

It is assumed hole problems were being experienced in Example 3. The operator knows he must either increase the annular velocity or thicken the mud. Considering the problem to be associated directly with the enlarged hole section, the slip velocity is also calculated using Equation 5.

$$v_s = \frac{(4980)(.09)(21 - 9.5)}{235} = 21.9 \text{ fpm}$$

The particle Reynolds number is determined using Equation 3.

$$R_p = \frac{(15.47)(9.5)(21.9)(.3)}{235} = 4.1$$

If the mud is thickened further, the slip velocity should be calculated, using Equation 5. It can be seen by observation that the slip velocity in Equation 5 is inversely proportional to viscosity while in Equation 7 the slip velocity is inversely proportional to viscosity to the 0.333 power. The required viscosity assuming the slip velocity must be at least 25 per cent less than the average annular velocity is calculated as follows:

$$v_s = (19)(.75) = 14 \text{ fpm}$$

Using Equation 5:

$$\mu = \frac{(4980)(.09)(11.5)}{14} = 368 \text{ c}_p$$

This shows that for a 32 per cent reduction in slip velocity, the mud thickness must be increased 57 per cent. These calculations have been based on particle sizes of 0.3 inch in diameter. If the formations are sloughing, the cuttings may be substantially larger and the mud may have to be thickened even more. In any event, such calculations provide a basis for solving the problem so that the knowledge gained may be used to solve future problems of a similar nature.

Lifting capacity with weighted muds is examined in Example 4.

■ EXAMPLE 4:

Well depth = 15,000 ft
Hole diameter = $7\frac{7}{8}$ in.
Drill pipe diameter = $4\frac{1}{2}$ in.
Maximum hole enlargement = 10 in.
Fluid weight = 16 ppg
Weight of formation solids = 21.0 ppg
Average particle size = 0.3 in.
Circulation rate = 250 gpm
Viscometer = $\theta_{600} = 90$
$\qquad\qquad\quad \theta_{300} = 50$
Solution: for the gauge hole section
v = 143 fpm

$$n = 3.32 \log \frac{90}{50} = (3.32)(.255) = .85$$

$$K = \frac{50}{(511)^{.85}} = \frac{50}{200} = .25$$

$$\mu = \left[\frac{(2.4)(143)(1.06)}{3.375} \right]^{.85} \frac{(200)(.25)(3.375)}{143}$$

$$\mu = (53)(1.179) = 62.4 \text{ c}_p$$

Using Equation 5:

$$v_s = \frac{(175)(0.3)(21 - 16)^{.667}}{(16)^{.333}(62.4)^{.333}}$$

$$v_s = \frac{(175)(0.3)(2.93)}{(2.5)(3.96)} = 15.5 \text{ fpm}$$

For the enlarged hole section:
$\bar{v}$ = 75 ft/min

$$\mu = \left[\frac{(2.4)(75)(1.06)}{5.5} \right]^{.85} \frac{(200)(.25)(5.5)}{75}$$

$$\mu = (20.38)(3.67) = 74.7 \; c_p$$

$$v_s = \frac{(175)(0.3)(21 - 16)^{.667}}{(16)^{.333}(75)^{.33}}$$

$$v_s = \frac{(175)(0.3)(2.93)}{(2.5)(4.2)} = 14.65 \; fpm$$

The slip velocity of the cuttings changed very little from the gauged to the enlarged hole section. The average annular velocity in the enlarged hole section is 75 fpm, which is still five times higher than the calculated slip velocity. From this analysis it can be seen that the mud can be thinned substantially and still be thick enough to remove cuttings. In fact, the required viscosity is only that required to support barite. This is the normal case for weighted muds of 15 ppg or heavier, because (1) the density of the mud provides lifting capacity and (2) hole enlargement in deeper hole sections where weighted muds are normally used is less than in the normal pressure upper hole sections.

It would be impossible to cover all the potential hole problems associated with hole cleaning; however, some specific problems will be discussed that seem to be reoccurring situations in all drilling areas.

DOWNHOLE MUD PROPERTIES

Surface measurements of mud properties are not always indicative of downhole mud properties. Muds may thin with increased temperature or they may thicken by flocculation. Unweighted muds are more likely to thin as the temperature increases, thus what appears to be an adequate thickness at the surface may not be sufficient in the enlarged downhole sections. An example was noted while drilling in 600 feet of water offshore.

The water temperature cooled the mud to about 60°F and the measured thickness of the mud was considered adequate to remove cuttings. When the mud thickness was corrected for downhole temperature, calculations showed the mud was too thin to clean the hole. This corresponded with the actual problem at the well, where problems of hole cleaning were occurring daily. Fortunately, before some expensive unnecessary corrective procedure was initiated, the real problem was discovered. After thickening the mud, at a small additional cost, the problems of hole sloughing and hole cleaning were corrected.

HOLE CLEANING ASSOCIATED WITH LOST CIRCULATION

When more hole cleaning capacity is needed, it is difficult to achieve the desired result without increasing the annular pressure losses. If lost circulation is also a potential problem, the operator is often placed in a position of trying to achieve the best compromise solution. One method might be to reduce the mud weight and increase mud thickness keeping the annular pressure loss constant. If mud weight cannot be reduced because of formation pore pressures, then the next best alternative is to use a slug of very viscous mud. This procedure involves the mixing of about 25 barrels of very thick mud, which would occupy only a small portion of the entire annulus.

The thick mud cleans out the cuttings and the increase in total annulus pressure loss is very small. This procedure would have to be repeated periodically if drilling was extended in the same open hole section.

A serious hole problem in the East Texas area was solved in this manner. On round trips, several bridges were encountered when running back to bottom and in addition over 100 feet of fill on bottom was a common occurrence. Small gas kicks had been noted and circulation had already been lost in the hole. On subsequent days, about 25 barrels of thick mud was circulated through the hole before trips. Circulation was not lost and no further trouble was experienced with fill on bottom or bridges in the hole.

This was a case where the problem was diagnosed properly and the treatment was successful. The basis for treating the problem was conceived on a sound basis and did not involve trial and error experimentation.

RATE OF MUD THICKENING

For most drilling operations, the velocity of the mud through the bit nozzles is several hundred feet per second and the flow pattern is in a high degree of turbulence. The energy imparted to the mud through the bit nozzles may prevent some muds from thickening rapidly at low rates of shear when lifting capacity is needed. It has been noted that the rate of thickening for various muds is different and this may be important when drilling in fractured sloughing shale zones.

In all wells, where potential problems exist, the operator should know the total hydrostatic pressure imposed on the underground formations, the approximate hole size, and the lifting capacity of the mud in use. If high temperature properties are needed, then laboratory tests should be run on the mud thickness. Many expensive alternatives have been used to cure relatively simple hole problems, because requirements for hole cleaning were not recognized.

REFERENCES

1. Williams, C. E., Jr., and Bruce, G. H., "Carrying Capacity of Drilling Muds," Petroleum Transactions Reprint Series, Drilling No. 6.
2. Walker, R. E., "Practical Oil-Field Rheology," Southern District, API Division of Production, San Antonio, March, 1964.

NOMENCLATURE

v_p = Particle velocity, fpm
v_f = Fluid velocity, fpm
v_s = Slip velocity of particle, fpm

D_p = Particle diameter, inches
ρ_p = Weight of particles in ppg
ρ_f = Weight of drilling fluid, ppg
C_D = Drag coefficient dimensionless
R_p = Particle Reynold's number, dimensionless
μ = Mud viscosity, centipoise
v   = Annular velocity, fpm
n   = Slope of shear stress versus shear rate straight line on log paper
K   = Intercept of Shear Stress versus shear rate straight line on log paper

Surge and
Swab Pressures

SURGE pressures are associated with
fluid flow, caused by running equipment into a liquid filled bore-hole. Swab
pressures are associated with fluid flow, caused by pulling equipment out
of a liquid filled bore-hole. Procedures used for estimating the magnitude
of these pressures are similar to estimating the pressure losses in conventional fluid circulation. To reduce problems of calculation, the swab pressure is estimated by calculating the surge pressure and assuming that this
is equal to the swab pressure for the same rate of pipe movement.

The magnitude of surge and swab pressures are important to the operator for the following reasons:

1. More than 25 per cent of the blowouts result from pressure reductions in the bore-hole due directly to swabbing when pulling pipe.
2. Excessive surge pressures have initiated lost circulation problems both during the drilling operation and during the running of casing into the hole.
3. Pressure changes caused by alternating between surge and swab pressures due to pipe movements, such as those made on connections cause hole sloughing and generally promote other unstable hole conditions such as solids bridges and solids fill on bottom.
4. Swab pressure reductions may result in contamination of the mud by the entry of formation fluids. This may result in expensive mud treating costs and cause other hole problems.

The adverse effects of surge and swab pressures were recognized very
early in rotary drilling. Cannon[1] in 1934 became concerned with blowouts
that were occurring in normal pressure wells. The mud weights being used
were substantially in excess of measured formation pore pressures, yet

blowouts were occurring. To investigate the problem, Cannon initiated a series of tests to measure the actual surge and swab pressures, table 1 shows the results from one series of his tests.

These series of tests were run at different depths and hole sizes, with mud thicknesses measured as a function of gel strength. At that time it was difficult to obtain any type of quantitative evaluation of mud thickness. The magnitude of the pressure surges was surprising. At 7,000 feet, with a gel strength of 36, the surge pressure in the 7-inch hole using 3½ inch drill pipe was 462 psi. This is comparable to an increase in mud weight of about

TABLE 9-1

Annulus Size	Depth, Ft.	Gel Strength	Pressure Surge
10¾ inch casing:			
4½ inch drill pipe	7,000	36	275
	7,000	12	125
	3,000	36	125
	3,000	12	62
7 inch casing:			
3½ inch drill pipe	7,000	60	487
	7,000	36	462
	7,000	6	362
	3,000	60	212
	3,000	36	200
	3,000	6	160

1.3 ppg and if considered when pulling pipe easily shows the potential hazard of a blowout.

Under these conditions, a normal weight mud of 10.0 ppg would not be sufficient to control normal formation pore pressures of 9.0 ppg, which are common in the coastal areas. These tests also show that the surge pressures are directly proportional to depth. For example, it is noted that the surge pressure in the 7-inch cased hole was 200 psi at 3,000 feet, as compared with 462 psi at 7,000 feet for a gel strength of 36. This would be expected because the surge pressure is simply the pressure required to overcome friction at the displacement rate of the fluid.

In 1951, Goins, Weichert, Burba, Dawson and Teplitz[2] related the problems of lost circulation in coastal area wells directly to surge pressures associated with pipe movement. They also noted that the surge pressures often exceeded equivalent mud weights of 1.0 ppg. They noted that surge pressures could be reduced significantly by reducing the pipe running speed.

In 1953, Cardwell[3] presented information on surge pressures and pulling suction and provided a chart for estimating the magnitude of these pressure changes. This chart was related to hole and pipe geometry and pipe running speed. He assumed an equivalent viscosity of 300 centipoise for the mud. The chart is not included because subsequent work has shown

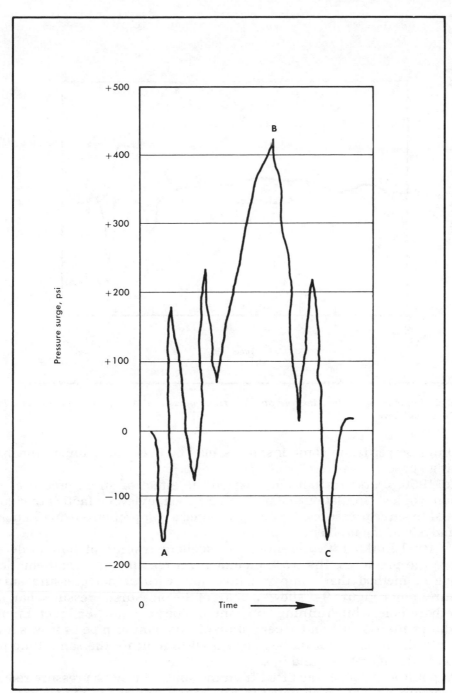

FIG. 9-1. *Typical pressure surge pattern measured as a joint of casing was lowered into well bore*

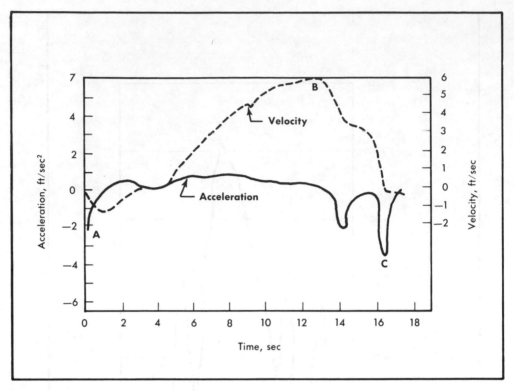

FIG. 9–2. *Typical pipe velocities and accelerations measured as a joint of casing was lowered into well bore*

that fluid properties in some instances may be even more important than running speed.

In 1956, Clark[4] further emphasized the effect of surge pressures on problems of lost circulation and presented a good review on factors that contributed to surge pressures. He also presented a comprehensive mathematical analysis of the problem.

In 1960 Burkhardt[5] presented an excellent resume of how to determine surge pressures. His work included test results and a mathematical prediction method that compared favorably with actual pressure surge measurements. Figure 9–1 shows a chart of the measured pressure changes in the bore-hole, while running one joint of pipe at about 1,850 feet. Figure 9–2 shows the velocity and acceleration of this joint of pipe as it was lowered into the hole. Lettering has been used to indicate the same time period on both Figures 9–1 and 9–2.

At point A, the pipe was lifted from the slips. The swab pressure reduction was almost 200 psi as noted in Figure 9–1 and the fluid velocity was negative as shown in Figure 9–2. At point B, the surge pressure was over 400 psi and it is noted that this is the point of maximum pipe velocity. At point C, swabbing again occurred and it is noted that this point corresponds to the maximum deceleration of the pipe.

Of interest was the fact that the swab pressure reduction at point C was almost 200 psi and this occurred when running pipe into the hole. This deceleration pressure indicates a well can be swabbed when running pipe into the hole. Other changes in pressure are probably due to the changes in pipe speed.

Figures 9–1 and 9–2 show that (1) the pressure can change drastically when running pipe, (2) the running speed changes substantially when running pipe, and (3) the pressure surges can be very high at shallow depths.

Surge pressures or swab pressure reductions may be caused by breaking the gel strength, viscous shear, fluid displacement in turbulence or by pipe acceleration and deceleration. These effects will be considered separately.

Surge–Gel Factors

The surge pressure required to break the gel strength may be estimated using the definition of shear stress as a function of pressure, this relationship is shown in Equation 1.

$$P = \frac{l\tau}{300(D_h - D_p)} \tag{1}$$

The gel strength in Equation 1 will be gel strength of the mud in $lb_f/100$ ft^2 as determined from the rotating viscometer. The surge pressure recorded when breaking the gel strength is generally less than that experienced at the maximum pipe running speed as indicated in Figure 9–1. However, this may not always be the case and the operator should be careful when initiating fluid circulation either by starting the pump or by lowering the drill string. In the deep high temperature well, gelation may be severe, as shown in the chapter on drilling muds.

Equations have been presented for calculating pressure losses in the normal circulation of drilling muds. These same equations can be used to determine surge pressures providing the velocity of the displaced fluid can be determined.

The displacement velocity of the fluid will depend on whether the pipe is open or closed and for this reason the two cases will be considered separately. Also, the operator must determine whether the flow pattern is laminar or turbulent. This may be done in exactly the same manner as shown in the chapter on pressure losses during normal fluid circulation. Probably the best way is to calculate the surge pressures, assuming both laminar and turbulent flow and to use the answer which gives the highest surge pressure.

CLOSED PIPE

The first consideration of surge pressures will be made assuming closed pipe. The pipe would actually be closed only when using drill pipe floats or when using conventional float collars when running casing. It may be considered almost closed when using differential float collars in casing. The

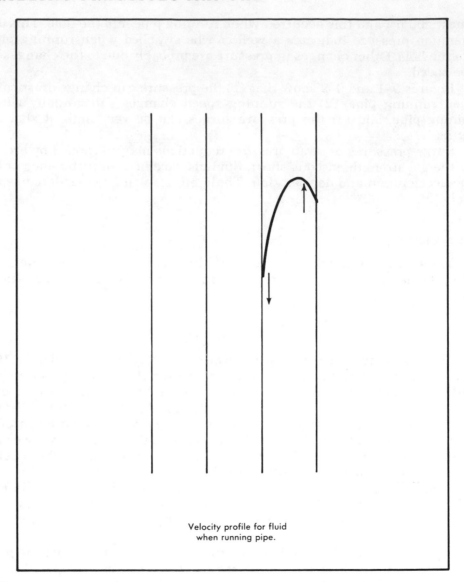

Velocity profile for fluid
when running pipe.

FIG. 9–3. *Velocity profile for*
fluid when running pipe

differential float collars are excellent tools for the continuous filling of the casing; however, most of the fillup generally occurs after the joint of casing has been lowered into the hole. Thus, in general, even with differential fill-up tools, the casing can be considered closed when calculating surge pressures.

The fluid displaced by the pipe volume is easily determined; however, friction losses are determined by the relative movement of the fluid to the pipe wall and thus the pipe movement into the hole must also be considered. The rate of fluid movement by the pipe is increased by the rate the pipe is lowered into the hole. The assumed velocity profile is shown by Figure 9–3.

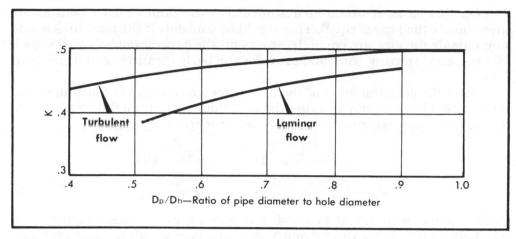

FIG. 9–4. *Mud "clinging" constant*

Equation 2 shows the estimated annular fluid velocity for plugged pipe.

$$v = \left[K + \frac{D_p^2}{D_h^2 - D_p^2} \right] v_p \tag{2}$$

Burkhardt determined the value of K in Equation 2 by the use of Figure 9–4. A good general average for K is 0.45. One other problem arises in the use of Equation 2. The pipe velocity, v_p, changes as pipe is being run; this is shown in Figure 9–2. The maximum surge pressure will generally occur at the maximum pipe velocity. Thus, the maximum fluid velocity may be estimated, using Equation 3.

$$v_m = 1.5v \tag{3}$$

Equation 3 is a rough estimate of the maximum fluid velocity and the 1.5 is primarily an experience factor. When running a joint or stand of pipe, it is common practice to accelerate to some point and then decelerate to set the pipe in the slips. In general, the maximum pipe speed would be about 1.5 times the average pipe speed, which is determined by dividing the length of pipe lowered into the hole by the total time required in minutes.

OPEN PIPE

Pressure surges with open pipe are difficult to define. The relative flow in the annulus and inside the drill string would be dependent on the relative pressure forces and cross-sectional areas. Pressure forces in the annulus will depend many times on mud properties, while those through the bit and inside the drill string will, in general, be independent of mud properties, because the flow would be turbulent. As a first step, it can be assumed that the pipe is completely open and for this purpose Equation 4 can be used to calculate the displacement velocity.

$$V = \left[K + \frac{D_p^2 - D_i^2}{D_h^2 - D_p^2 + D_i^2} \right] V_p \tag{4}$$

On this basis, it would be assumed that the fluid velocity both inside and outside the pipe is equal. This would be true only if the pressures inside and outside the pipe are equal. In any event the assumption of equal velocities is a good starting point and provides the basis for further trial and error steps.

After the determination of the fluid velocity, the calculation of pressure surges may be determined, using the equations for normal fluid circulation. These equations are given again for convenient reference.

$$P_s = \frac{(PV)vml}{60,000(D_h - D_p)^2} + \frac{yl}{200(D_h - D_p)} \qquad (5)$$

$$P_s = \left[\left(\frac{2.4v_m}{D_h - D_p}\right)\left(\frac{2n + 1}{3n}\right)\right]^n \frac{Kl}{300(D_h - D_p)} \qquad (6)$$

Note: Equation 5 should be used only when a multispeed viscometer is available to determine the PV and Y at viscometer speeds of 30 and 60 rpm. If only a two speed viscometer is available, Equation 6 should always be used.

Turbulent Flow

$$P_s = \frac{7.7(10^{-5})\rho^{.8}Q^{1.8}(PV)^{.2}l}{(D_h - D_p)^3(D_h + D_p)^{1.8}} \qquad (7)$$

The use of these equations is shown in Examples 1 and 2.

■ EXAMPLE 1:

Well Depth = 15,000 feet
Hole Size = $7\frac{7}{8}$ in.
Drill pipe size = $4\frac{1}{2}$ in.
Drill collar length = 700 feet
Drill collar size = $6\frac{1}{4}$ inch O.D.
$\quad\quad\quad\quad\quad\quad\quad\quad\quad$ $2\frac{3}{4}$ inch I.D.
Mud properties: Mud weight = 15 ppg
$\quad\quad\quad\quad\quad\quad$ Viscometer readings, $\theta_{600} = 140$
$\quad\quad\quad\quad\quad\quad\quad\quad\quad\quad$ $\theta_{300} = \;\;80$
Average pipe running speed, $v_p = 270$ fpm
Bit nozzle size = $3\frac{11}{32}$ in.
Determine: The surge pressure at 15,000 feet assuming plugged pipe.
Solution:

$$P_s = \left[\left(\frac{2.4v_m}{D_h - D_p}\right)\left(\frac{2n + 1}{3n}\right)\right]^n \frac{Kl}{300(D_h - D_p)}$$

$$n = 3.32 \log \frac{140}{80} = (3.32)(.243) = 0.805$$

$$K = \frac{80}{(511)^{.805}} = \frac{80}{152} = 0.527$$

$$v = \left[0.45 + \frac{20}{42}\right]270 = 250 \text{ fpm}$$

$$v_m = (1.5)(250) = 375 \text{ fpm (maximum fluid velocity)}$$

Around Drill Pipe (14,300 Feet)

Assume laminar flow:

$$P_s = \left[\frac{(2.4)(375)(1.08)}{(3.375)}\right]^{.805}\frac{(.527)(14.3)(10^3)}{.3(10^3)(3.375)} = (95)(7.44) = 707 \text{ psi}$$

Assume turbulent flow:

$$Q = 642 \text{ gpm (equivalent)}$$
$$P_s = 704 \text{ psi}$$

Because P_s is about the same, the flow pattern is not distinguishable.

Around Drill Collars (700 Feet)

$$v = \left[0.45 + \frac{39}{23}\right]270 = 580 \text{ fpm}$$
$$v_m = (1.5)(580) = 870 \text{ fpm}$$

At the higher velocity, the flow around the drill collars can be assumed to be turbulent.

$$P_s = \frac{7.7(10^{-5})(15)^{.8}(816)^{1.8}(60)^{.2}(700)}{(1.625)^3(14.125)^{1.8}} = 375 \text{ psi}$$

The total annular surge pressure at 15,000 feet:

$$707 + 375 = 1082 \text{ psi}$$

■ EXAMPLE 2:

Same conditions as shown in Example 1, except pipe is open.
Determine the surge pressure at 15,000 feet.
Solution: Annular surge around Drill Pipe:

$$V = \left[.45 + \frac{5.9}{56.1}\right)270 = 150 \text{ fpm}$$

$$V_m = 150 \times 1.5 = 225 \text{ fpm}$$

$$P_s = \left[\frac{(2.4)(225)(1.08)}{(3.375)}\right]^{.805}\frac{(.527)(14.3)(10^3)}{(300)(3.375)} = (64)(7.44) = 476 \text{ psi}$$

Around Drill Collar:

$$V_m = \frac{(225)(42)}{23} = 410 \text{ fpm}$$

Assume turbulent flow:

$$P_s = 140 \times .7 = 98.0 \text{ psi}$$

Total annular surge $= 476 + 98 = 574$ psi

Surge inside drill string through drill pipe. Assume laminar flow:

$$P_s = \left[\frac{(1.6)(225)(1.06)}{(3.75)}\right]^{.805}\frac{(.527)(14.3)(10^3)}{(300)(3.75)} = (41.5)(6.7) = 278 \text{ psi}$$

Through Drill Collars assuming turbulent flow:

$$Q = \frac{VA}{19.2} = \frac{(410)(.785)(7.56)}{19.2} = 127 \text{ gpm}$$

$P_s = 76 \times .7 = 53.2$ psi
Through bit
$P_s = 288$ psi
Total surge inside drill string $= 278 + 53 + 288 = 619$ psi.

These calculations show the annular surge will be between the 574 psi considering the pipe completely open and 619 psi which represents the pressure required to flow fluid inside the pipe at the indicated rate. A reasonable estimate of annular surge in Example 1, with the pipe open, would be 600 psi. This is about 55 per cent of the surge with the pipe plugged, and indicates a substantial potential reduction in annular surge pressure by omitting float collars even with small jet nozzles.

The calculation assuming open pipe was a rough estimate and the relative effects of having open pipe will change with mud properties and drill string geometry. In any event, these calculations do show a reduction in annular surge pressures by omitting the use of drill string floats.

In addition to the surge or swab pressures associated directly with pipe velocity, it is noted from Figure 9–1 that acceleration and deceleration effects may need to be considered. Of particular interest is the swab pressure reduction noted at point C in Figure 9–1, when the pipe was stopped suddenly. This simply shows that a well could be swabbed-in while running pipe and emphasizes the need for the operator to be conscious of this effect. Equation 8 may be used to estimate acceleration or deceleration pressures associated with pipe movement if pipe is plugged and Equation 9 may be used for the same purpose if the pipe is open.

Plugged pipe:

$$P_s = \frac{.00162 \rho l D_p^2 a_p}{D_h^2 - D_p^2} \tag{8}$$

Open pipe:

$$P_s = \frac{.00162 \rho (D_p^2 - D_i^2) a_p l}{D_h^2 - D_p^2 + D_i^2} \tag{9}$$

Example 3 illustrates the use of these equations.

■ EXAMPLE 3:

Same conditions as given in Example 1.
Assume $a_p = -4.5$ fps,² which is comparable to stopping the pipe in one second while running the pipe into the hole at a velocity of 4.5 fps.
Determine the surge pressure assuming plugged and open pipe.
Plugged pipe around drill pipe:

$$P_s = \frac{1.62(10^{-3})(15)(14.3)(10^3)(20)(-4.5)}{42} = -745 \text{ psi}$$

Around drill collars:

$$P_s = \frac{1.62(10^{-3})(15)(.7)(10^3)(39)(-4.5)}{23} = -130 \text{ psi}$$

Total negative surge = −875 psi.
Open pipe around drill pipe:

$$P_s = \frac{1.62(10^{-3})(15)(14.3)(10^3)(6)(-4.5)}{56} = -168 \text{ psi}$$

Around drill collars:

$$P_s = \frac{1.62(10^{-3})(15)(.7)(10^3)(31.4)(-4.5)}{(30.6)} = -79 \text{ psi}$$

Total negative surge = −168 + (−79) = −247 psi.

The negative surge would be higher than −247 psi and to be precise, the actual quantities of fluid entering the pipe would have to be determined.

The deceleration losses noted in Example 3 are rough estimates and at best simply show why the negative surge pressures occur and provide a warning signal on their potential order of magnitude. This warning becomes even more important as mud weights are reduced to levels close to pore pressure.

SPECIAL CONSIDERATIONS

In deep high temperature wells, the mud may gel considerably on bottom. The surge pressures may be trapped in the mud and have an additive effect to circulating pressure losses when circulation is commenced. For this reason, it is always important to pick-up pipe slowly when pumping is commenced. Also, surface pressures should be watched carefully when breaking circulation to avoid excess pressure on the formation.

If problems should be experienced with lost circulation, it should be remembered that the crucial point in slowing down pipe running speed is when the bit first reaches the weak zone. In coastal areas, this will generally be just below the protective casing seat. This is true because at this point the drill collars are above the weak zone and this restriction is annulus size as compared with drill pipe increases fluid velocity over the collar interval.

It can be seen that pressure fluctuations when handling pipe can be very high. This places a premium on careful pipe handling practices in areas where hole stability is a problem.

REFERENCES

1. Cannon, George E., "Changes in Hydrostatic Pressure Due to Withdrawing Drill Pipe from the Hole," Drilling and Production Practice, 42 (1934).
2. Goins, W. C., Jr., Weichert, J. P., Burba, J. L., Jr., Dawson, D. D., Jr., and Teplitz, A. J., "Down-The-Hole Pressure Surges and Their Effect on Loss of Circulation," Southwestern District, Division of Production, Beaumont, Texas, 1951.
3. Cardwell, W. T., Jr., "Pressure, Changes in Drilling Wells Caused by Pipe Movement," Drilling and Production Practices API (1953).
4. Clark, E. H., Jr., "A Graphic View of Pressure Surges and Lost Circulation," Drilling and Production Practices, API 1956.
5. Burkhardt, J. A., "Wellbore Pressure Surges Produced by Pipe Movement," SPE of AIME paper No. 1546-G, Denver, Colorado, 1960.

NOMENCLATURE

P_s = Surge pressure, psi
l   = Well depth, feet
τ = Gel strength of mud, $lb_f/100 \ ft^2$
D_h = Hole diameter, inches
D_p = Outside diameter of pipe, inches
v   = Fluid velocity, fpm
v_p = Pipe Velocity, fpm
D_i = Inside pipe diameter, inches
PV = Plastic viscosity, centipoise
v_m = Maximum fluid velocity, fpm
Y   = Yield point, $lb_f/100 \ ft^2$
Q   = Circulation rate, gpm
ρ = Mud weight ppg
a_p = Pipe acceleration, fps^2

10 Hydraulics in Rotary Drilling

TH E use of hydraulic horsepower is associated with the utilization of jet bits. Very little attention was directed towards fluid-circulation programs prior to the introduction of jet nozzles in bits in 1948. Conclusions on the value of jet bits were not made until 1949 and very limited industry acceptance was noted for several more years.

Even today many operators do not recognize the importance of bottom-hole cleaning. They may use jet bits, but the circulation program is so poorly designed that bottom-hole cleaning is not much better than that achieved with conventional bits. We need to go back to the analogy of the child cleaning mud off the driveway with a garden hose. First he turns on the water full pressure, directs the water discharge on the mud, and, if the hose is equipped with a nozzle, he reduces the nozzle size until he blasts away the mud. No engineering is involved; by simple observation he has noted that this is the fastest way to remove the mud.

For some reason, maybe because the nozzle is hidden, the industry first resisted turning the nozzle on the mud and then it resisted the nozzle adjustment necessary to remove the mud in the shortest time. Maybe we have not been in as big a hurry as the child, or perhaps intuition is better than scientific conclusions. In any event, 25 years have elapsed since jet bits were proven to be superior to regular bits for bottom-hole cleaning. Numerous articles have been written on hydraulics and all conclusions have shown the importance of adequate bottom-hole cleaning.

Basically there are two steps in designing a hydraulics program: (1) the determination of how much bottom-hole cleaning is needed to maximize drilling rate and (2) the maximizing of bottom-hole cleaning based on the surface hydraulic horsepower available.

OPTIMUM BOTTOM-HOLE CLEANING

In very soft formations, it may be very difficult to determine the bottom-hole cleaning required for maximum penetration rates. It is possible in soft formations, such as those found in some coastal areas, to generate hole by the jetting action of the bits. Thus, in areas such as this, maximum penetration rates will be achieved with maximum bottom-hole jetting action. Thus, the problem is actually one of using the maximum jetting action that is economically feasible. Economic feasibility will depend on the maximum penetration rates possible, based on hole conditions and other activities that have an equalizing effect, such as connection time and the maintenance of support equipment.

In formations of normal hardness, and there is no specific breaking point, the amount of bottom-hole cleaning required may be determined directly in field operations.

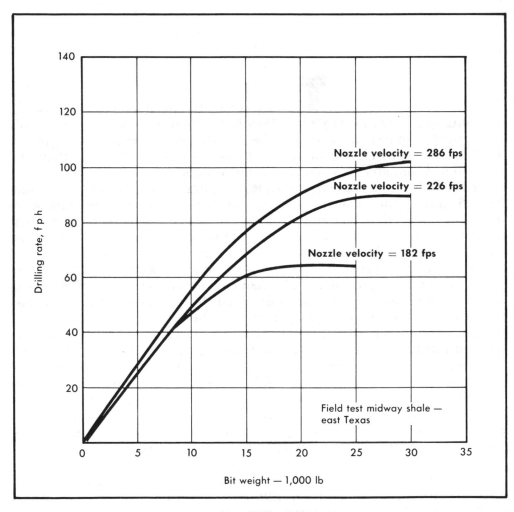

FIG. 10–1. *Effect of bottom-hole cleaning*

A method of determining the amount of bottom cleaning needed is indicated by the field tests shown in Figure 10–1. In this series of tests, the nozzle velocity was increased while holding circulation rate constant, and by this method both bit hydraulic horsepower and hydraulic impact were increased. However, the surface horsepower was also increased.

Figure 10–1 shows at 20,000 lb. of bit weight that drilling rate was increased 25% when nozzle velocity was increased by 24.2%. A further 26.5% increase in nozzle velocity resulted in a 12% increase in drilling rate. This test shows that bottom-hole cleaning was not adequate at the highest nozzle velocity. Note that the top curve is a straight line starting at zero only to a bit weight of 7,500 lb. This simply indicates that more bottom-hole cleaning was needed.

The drilling rate should be directly proportional to bit weight if bottom-hole cleaning is adequate. Exceptions would be in those cases when low bit weights are not high enough to fracture the rock. Thus, this field test suggests one of several methods that could be used to determine bottom-hole cleaning requirements.

Tests such as those shown in Figure 10–1 could be run where nozzle velocity was increased in increments until all of the available hydraulic horsepower had been utilized. However, it seems that as much accuracy would be obtained by designing for maximum bottom-hole cleaning with the horsepower available, then increasing bit weight to some maximum in increments. If the relationship between drilling rate and bit weight is a straight line, the results indicate that all of the bottom-hole cleaning is not needed, unless you were lucky enough to be exactly right the first time.

The bottom-hole cleaning power should then be reduced and the incremental changes in bit weight continued. This practice should establish the cleaning required for a given formation at a given depth using a given drilling program.

Tests of the type described would follow the pattern shown in Figure 10–2. Consider the Points, 1, 2 and 3 as levels of bottom-hole cleaning action. If the operator intends to use a bit weight at the level indicated by A, then he needs a level of cleaning action between Points 2 and 3.

In all probability, any series of field tests would not be this simple. Formations are not generally homogeneous, bits dull as drilling proceeds and it may be difficult to maintain precise levels of bit weight and rotary speed. For these reasons, extended tests may be required to establish the level of cleaning required at point A.

A suggested procedure is to determine drilling rates over five foot intervals at different bit weights on an alternating basis during one 24 hour period. The specific problem may be simplified, using insert bits with friction bearings. This is true because the maximum loading on these bits is generally fixed rather than selected on an economic basis.

As is the case with most applications of technology, rules-of-the-thumb appear. In coastal operations using mill-tooth bits, a rule-of-the-thumb number is 5.2 Bit Hydraulic Horsepower per square inch of hole. Using diamond bits, a number of 3.1 Bit Hydraulic Horsepower per square inch of hole is frequently used. In the West Texas area, where formations are hard,

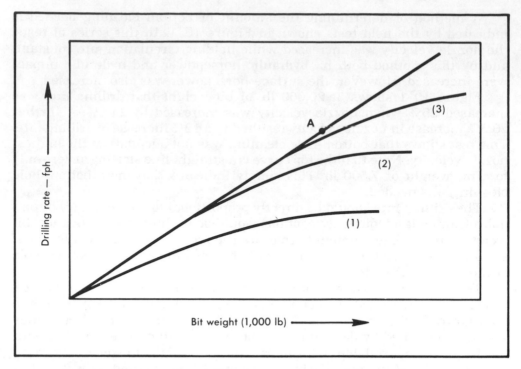

FIG. 10-2. *How drilling rate increases with bit weight*

less bottom-hole cleaning is required than in the soft formations of the coastal areas.

Operators are cautioned to consider rules-of-the-thumb as starting points only. There are far too many variables and drilling costs are too high to permit guessing on a continued basis.

DESIGN OF THE HYDRAULICS PROGRAM

The design of a hydraulics program is based on maximizing bottom-hole cleaning. Methods of design that are being used include: (1) hydraulic impact, (2) bit hydraulic horsepower, (3) nozzle velocity, and (4) combinations of these methods. Regardless of the design method used, the first step in the program is to determine the available surface hydraulic horsepower.

Hydraulic horsepower is defined mathematically as shown in Equation 1.

$$Hp = \frac{PQ}{1714} \qquad (1)$$

The amount of surface hydraulic horsepower available is based on the maximum permissible discharge pressure and the maximum flow rate. Example 1 illustrates the calculation of maximum surface hydraulic horsepower.

■ EXAMPLE 1:

Maximum permissible surface pressure = 3000 psi
Maximum flow rate = 500 gpm

Available surface horsepower: $Hp_s = \dfrac{(3000)(500)}{1714} = 875$

It should be noted that the size of drilling rig pumps is based on input horsepower to the pump. Example 2 illustrates a typical field situation.

■ EXAMPLE 2:

Rig pump size = 1500 Hp
Using 6½ inch liner, maximum surface pressure = 4118 psi.
Recommended maximum pumping rate = 500 gpm
Available surface horsepower = 1200 Hp

In actual practice, the maximum surface pressure is usually limited to some value less than maximum. In this case, if the pressure is limited to 3000 psi, a typical practice, the surface horsepower available is 875. Thus, the 1500 horsepower pump is capable, under these conditions, of delivering only 875 horsepower. It is also possible that many operators, or just the design of the hydraulics program, will reduce the circulation rate below the maximum. Assume the maximum circulation rate is reduced to 350 gpm at a reduced pressure of 3000 psi. The available surface horsepower is now reduced to 613.

From Example 2, it can be seen that the selection of a pump is very important in the design of a hydraulics program. Also, of equal importance, is the maximum pressure and circulating rates that can or will be used. In addition, it should be remembered that the prime movers available must be capable of supplying the required input horsepower. Thus, the operating practices of the drilling rig operator is of equal importance to the type equipment included in the rig inventory.

When the available surface horsepower has been determined, the next step in designing the hydraulics program is to select the design method. Two methods for designing hydraulics programs will be considered: (1) hydraulic impact, and (2) bit hydraulic horsepower. In both methods it is necessary to use empirical charts or tables to determine pressure losses at given rates of circulation. Also, in each method, there are three separate design criteria: (1) the situation where there is no limit on surface pressure; (2) the case where surface pressure is limited, and reductions in surface horsepower are necessary, to maximize bottom-hole cleaning; and (3) the intermediate case which falls in between the two maximums stated in (1) and (2).

In both methods of design, it will be assumed that pressure relates to circulation rate as shown in Equations 2 and 3. Any corrections necessary in actual field operations will be shown in a later example.

$$P_c = KQ^{1.80} \tag{2}$$

$$P_b = \frac{K'Q^2}{A_N^2} \tag{3}$$

If information is required considering the origin of Equations 2 and 3, please refer to the chapter on fluid circulation.

The horsepower distributions are shown in Equation 4.

$$Hp_s = Hp_c + Hp_b \tag{4}$$

In this case the surface horsepower, Hp_s, is distributed to horsepower in the circulating system excluding the bit, Hp_c, and to the horsepower expended through the bit, Hp_b.

The object of the design program is to maximize bottom-hole cleaning. The manner in which this is done will depend on the design method used.

Hydraulic impact is defined mathematically in Equation 5.

$$\text{I.F.} = \frac{\rho Q V_N}{1932} \tag{5}$$

Bit hydraulic horsepower is defined mathematically in Equation 6.

$$Hp_b = \frac{P_b Q}{1714} \tag{6}$$

Either hydraulic impact or bit hydraulic horsepower can be maximized, based on the use of a certain per cent of the surface horsepower at the bit. These required percentages can be determined through mathematical derivations. While the derivations will not be shown in this text, the percentages to maximize each will be shown for each design criteria.

Consider first the unlimited surface pressure case. Hydraulic impact would be maximized in this case, as shown in Equation 7.

$$Hp_b = 0.74 \, Hp_s \tag{7}$$

This equation simply states that the maximum impact force would be obtained for a given surface horsepower when 74 per cent of the surface horsepower is used at the bit. The maximum bit hydraulic horsepower is obtained as shown in Equation 8.

$$Hp_b \rightarrow Hp_s \quad \text{and} \quad Hp_c \rightarrow 0 \tag{8}$$

This equation simply states that the minimum circulation rate and the maximum surface pressure should always be used to obtain maximum bit hydraulic horsepower in the unlimited pressure case. In the design program, this would mean using a minimum permissible circulation rate or the minimum size pump liner.

The second criteria of limited surface pressure changes substantially the percentages of power used at the bit. Hydraulic impact in this case would be maximized, according to Equation 9.

$$Hp_b = 0.48 \, Hp_s \tag{9}$$

Bit hydraulic horsepower would be maximized in the limited surface pressure case, as shown in Equation 10.

$$Hp_b = 0.65 \, Hp_s \tag{10}$$

It should be remembered that one definition of the limited surface pressure case was that the surface horsepower had to be reduced to maintain the required percentages shown in Equations 9 and 10.

The third criterion was called the intermediate case, which cannot be described mathematically. This will be the condition where surface pressure and circulation rate are maintained at constant levels until the horsepower through the bit has been reduced to that shown in either Equation 9 or 10, depending on which method of design is being used.

Example 3 shows the design of a hydraulics program, using both hydraulic impact and bit hydraulic horsepower methods.

■ EXAMPLE 3:

Projected well depth = 16,000 ft.
Average hole size = $7\frac{7}{8}$ in.
Drill pipe = $4\frac{1}{2}$ in. O.D., 3.75 in. I.D.
Drill collars = 700 ft., $6\frac{1}{4}$ in. O.D., $2\frac{3}{4}$ in. I.D.
Pump = 1500 Hp, $7\frac{1}{2} \times 18$ in., Duplex
Maximum surface pressure = 3500 psi or 90% of liner rating, whichever is less.
mud weight = 15 ppg
PV = 20
Y = 15

Design a hydraulics program using both hydraulic impact and bit hydraulic horsepower design methods for well depths of 4000, 8000, 12000 and 16000 feet using 7, $6\frac{1}{2}$ and 6 inch liners.

Solution: Determine pressure losses at 300 gpm for conditions shown in Example 3. These are shown in tabular form in Table 10–1.

TABLE 10–1

Well Depth, Ft.	$P_{dp} + P_{dpa}$	$P_{dc} + P_{dca}$	$P_{s.c.}$	P_c
4000	248	255	29	532
8000	548	255	29	832
12000	848	255	29	1132
16000	1148	255	29	1432

Construct Figure 10–3, using the P_c data from Table 10–1 and the relationship between pressure and circulation rate shown in Equation 2. Expanding Equation 2 as a log equation gives the following:

$$\text{Log } P_c = \log K + 1.80 \log Q$$

This shows that a plot of P_c versus Q on log paper will give a straight line with a slope of 1.80. Pump data for the three liner sizes is shown in Table 10–2.

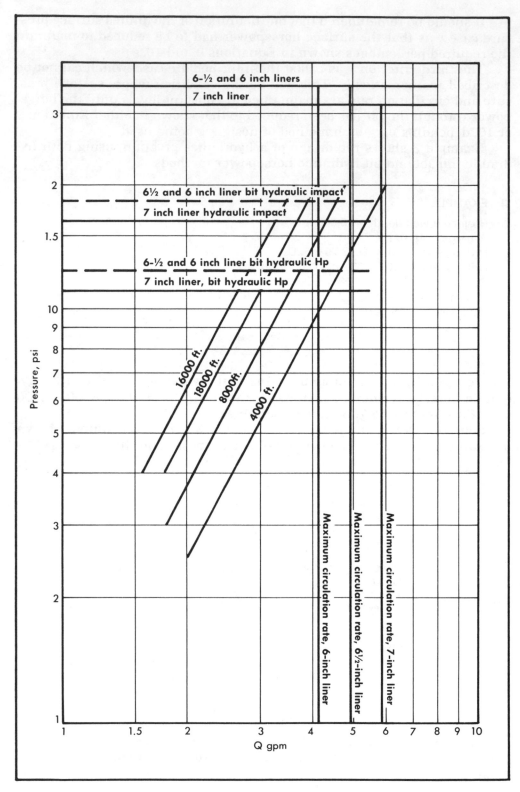

FIG. 10-3. *Pressure loss vs. circulation rate for example 3*

TABLE 10-2

Liner Size	Q at 60 SPM	Liner Rating	Maximum Surface Pressure
7	585	3470	3125
6½	495	4118	3500
6	415	4972	3500

Maximum surface pressure was determined by taking 90 per cent of the liner rating for the 7 inch liner and by accepting the maximum operator limitations of 3500 psi for the 6½ and 6 inch liners, this pressure being less than 90 per cent of the liner ratings.

Circulation rates are taken from Figure 10-3 and the hydraulics program is designed as shown in Table 10-3.

The circulation rates in Table 10-3 were selected based on the pressure limitations shown in Equation 9 and the use of design criteria number 3 for the intermediate case. For example, with the 6-inch liner, the maximum circulation rate of 415 gpm was used, until the depth of 12000 feet was reached. At 12000 the circulation rate was reduced to 390 gpm, which was selected at the point the 1820 psi pressure line crosses the circulating pressure line for the 12000 foot depth. The 1820 psi represents 52 per cent of the maximum surface pressure of 3500 psi.

The circulation rate of 345 gpm was selected at 16000 feet where the 1820 psi pressure line crossed the circulating pressure line for 16000 feet. The nozzle sizes are based on the circulation rate and the pressure drop shown for the bit. The actual pressure loss through the bit is that actually

TABLE 10-3
Hydraulic Impact

Depth, Ft.	Q	Pb	Nozzles, In.	Actual Pb	V_N	Impact Force, Lbs.
			7-Inch Liner			
4000	540	1500	Three $^{15}/_{16}$	1500	335	1400
8000	430	1500	Two $^{13}/_{32}$, one $^{14}/_{32}$	1520	337	1125
12000	365	1500	Two $^{12}/_{32}$, one $^{13}/_{32}$	1500	335	950
16000	325	1500	Two $^{12}/_{32}$, one $^{11}/_{32}$	1480	333	840
			6½-Inch Liner			
4000	495	2125	Two $^{13}/_{32}$, one $^{14}/_{32}$	2025	380	1460
8000	460	1680	Two $^{14}/_{32}$, one $^{13}/_{32}$	1580	345	1235
12000	390	1680	Two $^{12}/_{32}$, one $^{13}/_{32}$	1700	357	1080
16000	345	1680	Two $^{12}/_{32}$, one $^{11}/_{32}$	1670	353	945
			6-Inch Liner			
4000	415	2510	Two $^{12}/_{32}$, one $^{11}/_{32}$	2420	425	1370
8000	415	2000	Two $^{12}/_{32}$, one $^{13}/_{32}$	1940	382	1230
12000	390	1680	Two $^{12}/_{32}$, one $^{13}/_{32}$	1700	357	1080
16000	345	1680	Two $^{12}/_{32}$, one $^{11}/_{32}$	1670	353	945

obtained from the nozzles selected. The nozzle velocity in feet per second is determined from charts, but may also be calculated using the following equation:

$$P_b = \frac{\rho V_N^2}{1120} \qquad\qquad V_N = \frac{0.32Q}{A_N}$$

Normally, the 6-inch liner would have given the maximum impact for the conditions shown. However, the limitation on surface pressure to 3500 psi shows little difference between the $6\frac{1}{2}$ and 6-inch liner. The hydraulics program with the 7-inch liner was inferior because of the lower permissible surface pressure of 3125 psi.

It will be noted that at 16000 feet, with the 7-inch liner, the surface horsepower being used was only 592 and with the $6\frac{1}{2}$ and 6-inch liners, the output surface horsepower was only 705. This is less than one-half of the stated pump horsepower of 1500. As the well is drilled deeper, the surface horsepower will be reduced even further.

Another problem arises when considering these programs. The annular velocity in all cases will be very high. At a circulation rate of 345 gpm, the annular velocity around the drill pipe is 200 fpm and the annular velocity around the drill collars is 370 fpm. Any type of hole problem would preclude the use of annular velocities this high.

Before considering limitations, a bit hydraulic horsepower program will be designed for Example 3. The maximum permissible pressure in the circulating system for maximum bit hydraulic horsepower is 1095 psi [3125 × .35] for the 7-inch liner and 1225 psi [3500 × .35] for the $6\frac{1}{2}$ and 6-inch

TABLE 10–4
Bit Hydraulic Horsepower

Depth	Q	P_b	Nozzles, In.	Actual P_b	Bit Horsepower
			7-Inch Liner		
4000	440	2030	Two $^{13}/_{32}$, one $^{12}/_{32}$	1960	503
8000	350	2030	Two $^{11}/_{32}$, one $^{12}/_{32}$	1940	396
12000	300	2030	Two $^{10}/_{32}$, one $^{11}/_{32}$	2030	355
16000	265	2030	Two $^{10}/_{32}$, one $^{9}/_{32}$	2100	302
			$6\frac{1}{2}$-Inch Liner		
4000	470	2275	Two $^{13}/_{32}$, one $^{12}/_{32}$	2250	617
8000	375	2275	Two $^{11}/_{32}$, one $^{12}/_{32}$	2200	481
12000	320	2275	Two $^{10}/_{32}$, one $^{11}/_{32}$	2330	434
16000	280	2275	Three $^{10}/_{32}$	2050	335
			6-Inch Liner		
4000	415	2500	Two $^{12}/_{32}$, one $^{11}/_{32}$	2410	584
8000	375	2275	Two $^{11}/_{32}$, one $^{12}/_{32}$	2200	481
12000	320	2275	Two $^{10}/_{32}$, one $^{11}/_{32}$	2330	434
16000	280	2275	Three $^{10}/_{32}$	2050	335

liners. These limitation pressure lines are shown on Figure 10–3 and form part of the basis for selecting circulation rates.

The maximum pump rates are used until the pressure limitation lines are reached and the circulation rates are reduced along these lines to limit the pressure loss in the circulating system to this maximum. The bit hydraulic horsepower program using 7, 6½ and 6-inch liners is shown in Table 10–4.

With the limitation on surface pressure set at 3500 psi for both the 6½ and 6-inch liners, either liner may be used. The 6-inch liner would definitely be preferable if higher surface pressures were permitted. At 16000 feet the surface horsepower being used is 575. Again, the most severe limitation being the limits placed on surface pressure. The circula-

TABLE 10–4
Limited Pressure Case
Pressure Loss Versus Circulation Rate

	Hydraulic Impact		Bit Hydraulic Horsepower	
Slope	$P_c = \% \, P_s$	$P_b = \% \, P_s$	$P_c = \% \, P_s$	$P_b = \% \, P_s$
2.00	0.50	0.50	0.33	0.67
1.86	0.52	0.48	0.35	0.65
1.60	0.56	0.44	0.38	0.62
1.50	0.57	0.43	0.40	0.60
1.25	0.61	0.39	0.44	0.56
1.00	0.67	0.33	0.50	0.50

tion rate of 280 gpm results in an annular velocity around the drill collars of 280 fpm. Again, circulation rates this high may not be permissible if there are any potential hole problems.

These programs emphasize the requirement for smaller liners and shorter stroke pumps and create an argument for the single acting short stroke multiplex pumps.

As mentioned, the hydraulic programs may have to be changed if other hole conditions dictate. Any changes and corrections in the hydraulic programs, to conform with actual field operations, may be done very simply by following a method proposed by Scott:[1] The field procedure proposed by Scott is summarized generally as follows:

1. Measure the standpipe pressure at three or more rates of flow and plot these pressures versus flow rate on log paper.
2. Determine the pressure drop through the bit at the three flow rates and subtract bit pressure losses from the standpipe pipe pressures. This pressure difference represents the system circulating pressure.
3. Determine the slope of the system pressure losses versus flow rate. From this slope, refer to Table 10–4 for the percentage of pressure loss that should be used through the circulating system and the percentage of the pressure loss through the bit.
4. From step 2, determine the flow rate that will produce the pressure loss in the circulating system determined in step 3.

5. Using the circulation rate selected from step 4 and the pressure loss for the bit from step 3, pick the appropriate nozzle size for the next bit run.

Example 4 has been prepared to illustrate this procedure.

■ EXAMPLE 4:

Well depth = 15000 ft.
Mud Wt. = 15 ppg.
PV = 30
Y = 15
Average hole size = $8\frac{1}{2}$ inch
Drill pipe = $4\frac{1}{2}$ inch O.D., 3.75 inch I.D.
Drill collars = 700 ft., $6\frac{1}{4}$ inch O.D., $2\frac{3}{4}$ inch I.D.
Pump = 1600 Hp at 65 spm, $7\frac{1}{4} \times 16$ inch
Maximum permissible surface pressure = 3700 psi
Liner size = $6\frac{1}{2}$ inch
spm = 60, pump rate = 480 gpm
Bit nozzle size = three $\frac{9}{16}$ inch
Pressure loss @ 480 gpm = 3700 psi
Pump is slowed down and these additional pressure measurements are made:
Q = 350 gpm,   P = 2230 psi
Q = 200 gpm,   P = 910 psi
Determine the circulation rate and nozzle size for the next bit run to obtain maximum bit horsepower and hydraulic impact.
Solution: The three measured pressures versus circulation rate are plotted in Figure 10–4. The bit nozzle losses at the three circulation rates are determined as follows:
Q = 480 gpm, P_b = 575 psi
Q = 350 gpm, P_b = 305 psi
Q = 200 gpm, P_b = 100 psi
Pressure loss in circulating system is as follows:
P_c at 480 gpm = 3700 − 575 = 3125 psi
P_c at 350 gpm = 2230 − 305 = 1925 psi
P_c at 200 gpm = 910 − 100 = 810 psi

The pressure loss in the circulating system is the bottom curve on Figure 10–4. The measured slope of the circulating pressure versus the circulation rate is 1.55. Table 10–5 shows the bit nozzle program for the next bit run.

The circulating pressures, P_c, were based on a slope of 1.55 from Figure 10–4 and the per cent of P_s shown in Table 10–4. The circulating rates were selected to provide the required P_c. The circulation rate and P_b were used to select the nozzle sizes. The per cent improvement in hydraulic impact was 26.2, the per cent improvement in bit hydraulic horsepower was 136.

Before making this correction the surface horsepower output was 1035. After making the correction in the case of hydraulic impact, the surface horsepower output was 800. In the case of bit hydraulic horsepower, the surface horsepower output was 627. These are substantial reductions in surface horsepower output and substantial increases in bottom hole clean-

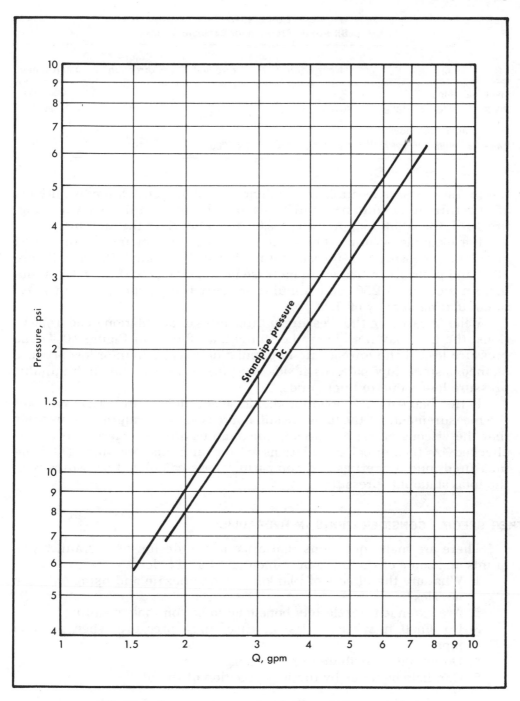

FIG. 10–4. *Pressure loss vs. circulation rate for example 4*

ing. Even the hole conditions would be improved by the lower circulation rates.

This field procedure is very easy to use and the circulating pressure loss curve in Figure 10–4 has other uses. Remember, this curve represents

TABLE 10–5
Bit Nozzle Program for Example 4

P_c	Q	P_b	Nozzle Size	Corrected	Before Correction	Improvement
Hydraulic impact						
2072	370	1628	Two $^3/_8$, one $^{13}/_{32}$	975	773	26.2
Bit hydraulic horsepower						
1443	290	2257	Three $^{10}/_{32}$	380	161	136.0

the pressure losses assuming turbulent flow. When circulation is initiated after returning to bottom and if P_c is above the straight line shown in Figure 10–4, the additional pressure loss is probably all in the annulus.

For example, assume the pumping rate is 250 gpm and the measured circulating pressure is 1400 psi. This is 250 psi above the 1150 psi on Figure 10–4 and probably indicates an increase in pressure of 250 psi in the annulus. An increase of 250 psi in annulus pressure is equivalent to an increase in circulating density of 0.32 ppg.

When measuring the pressure at three rates of circulation in field operations, the line may not always be straight, as shown in Figure 10–4. The pressure loss at the lowest rate of circulation may be high or low because of inconsistent flow patterns inside the pipe or because of high annulus pressure losses due to thick muds.

If this happens, the operator should select higher circulation rates for the measurements. If the lines are still not reasonably straight, it is probable that the viscous pressure losses in the annulus are very high. This can be checked by theoretical calculations of the annulus pressure loss from measured mud properties and comparing these calculated pressures with the total standpipe pressure.

OTHER SPECIAL CONSIDERATIONS IN HYDRAULICS

There are many questions that arise in the design of hydraulics programs and some of the more prominent are listed below.

1. What are the effects of blanking-off one nozzle and using only two nozzles?
2. Are extended nozzle bits beneficial to bottom-hole cleaning?
3. Do small nozzles in bits increase surge pressure when running pipe?
4. Do jet bits contribute to hole enlargement?
5. Can hole be made by the jetting action of the bit?

Blanking-Off One Nozzle

Using one blank nozzle is a common industry practice. The practice was started to prevent the need for very small jets in some hydraulic programs. For example, two $^3/_8$-inch jets are about equal to three $^5/_{16}$-inch jets and in many cases might be preferable because of the potential plugging of the smaller jets. The question arises, will bottom-hole cleaning be affected by using two instead of three jets?

A study by Sutko and Myers,[2] indicated bottom-hole cleaning would be improved at constant power levels by reducing the number of nozzles. Unpublished field work showed improvements in drilling rates, at constant power levels, when using two instead of three bit nozzles. There have been reports of overheating in bearings using only one jet and some isolated reports of the same problem using two jets.

In general, there seems to be no disadvantage to using two instead of three nozzles in three cone bits. The largest advantage of two nozzles appears to be a reduction in the danger of plugging when compared to the same area divided among three nozzles.

Extended Nozzles

The concept of extending the nozzles on three cone bits is an old one. However, the practice is still experimental. Laboratory results by Sutko and Myers,[2] and unpublished work, shows that improvements in bottom-hole cleaning have resulted from extending nozzles. Field tests have confirmed the laboratory tests. However, it is difficult to prevent the breaking of nozzles in field operations and the bits because of the custom work necessary cost more.

The economic feasibility of extended nozzles will have to be determined in field operations. Extended nozzles will increase bottom-hole cleaning and this advantage will have to offset occasional nozzle breakage and higher bit costs.

Small Nozzle Effects on Pressure Surges

Pressure surges are increased by using jet nozzles, but much less than generally believed. An example of this effect is shown in the chapter on surge and swab pressures.

Jet Bit Effects on Hole Enlargement

This is a general myth introduced by anti-jet bit proponents in the early part of the 1950–1960 decade. There is still a carry over in this type thinking, although there has never been any type of confirmation in controlled laboratory or field tests.

Hole-Making Action of Jet Bits

Hole can be made by the jetting action of jet bits. This is confirmed in soft formations in coastal areas where hole making is achieved in many cases by practically no rotation of the bit. Also, new erosion drilling techniques with very high surface pressures (up to 20,000 psi) have resulted in drilling rates of three times normal rock bit rates in very hard formations. Economic feasibility of the erosion drilling is being studied.

REFERENCES

1. Scott, Kenneth F., "A New Practical Approach to Rotary Drilling Hydraulics," SPE paper No. 3530, New Orleans, Louisiana, October 1971.
2. Sutko, A. A., and Myers, G. M., "The Effect of Nozzle Size, Number and Extension on the Pressure Distribution under a Tricone Bit," SPE paper No. 3109, Fall meeting of SPE of AIME, Houston, Texas, October 1970.

NOMENCLATURE

Hp = Hydraulic horsepower

P = Pressure, psi

Q = Circulation rate, gpm

Hp_s = Surface hydraulic horsepower

P_s = Pressure at surface, psi

P_b = Pressure drop through bit, psi

P_c = Pressure loss in circulating system, excluding bit, psi

A = Nozzle area, inches

Hp_b = Bit hydraulic horsepower

Hp_c = Horsepower required in circulating system, excluding bit

ρ = Mud weight, ppg

V_N = Nozzle velocity, fps

K and K^1 = Proportionality constants

11

Prediction of Pore Pressures and Fracture Gradients

TH E accurate prediction of pore pressures and fracture gradients has become almost essential to the drilling of deep wells with higher than normal pore pressures. Costs and drilling problems can be reduced substantially by the early recognition of abnormally high pore pressures. The accurate determination of fracture gradients defines the need for protective casing and the limits that must be placed on total annulus pressures during the drilling operation.

Some of the primary methods used to predict pore pressures are enumerated as follows:
1. Seismic data *only in Gulf Coast*
2. Drilling rate
3. Sloughing shale
4. Shale density
5. Gas units in mud
6. Chloride increases
7. Mud properties
8. Temperature measurements
9. Bentonite content in shale
10. Paleo information
11. Wire-line logs

SEISMIC DATA

A detailed explanation of using seismic field data to locate the existence and depth of formations with abnormally high pressures was given in a paper by Pennebaker.[1] The technique relies primarily on normal formation

269

compaction with depth. When this normal compaction trend is not followed, the velocity of sound waves is reduced. These velocity changes can be detected and converted to the degree of abnormal pore pressures.

To be accurate, precise sound velocity data are needed and this may determine the successful application of this method. In coastal areas, seismic data have been used to predict the top of abnormal pressure formations within a few hundred feet. In some new areas, seismic data have given erroneous results. Actually the use of seismic data has more utility in exploratory areas. However, in these new areas there is generally a lack of sound velocity data and underground formations vary considerably.

This lack of velocity information limits substantially the use of seismic data for pore pressure predictions in new exploratory areas. Specific interpretations of seismic data are difficult and experienced geophysicists will be needed if results are to be meaningful. Any method of this type improves as field data are collected. This means that drilling personnel and geophysicists can profit by working closely with each other.

DRILLING RATE

Drilling rate is a very useful tool in the detection of changes in pore pressure; however, it should also be remembered that drilling rate is effected by litology changes, bottom-hole cleaning, bit weight, rotary speed, fluid properties, and bit type and condition. Differentials between the hydrostatic and pore pressures have a large effect on drilling rate.

These effects are documented by many and are generally accepted facts in the oil industry. To review the principle in general, Figure 11–1 shows the bottom of the hole with the hydrostatic pressure shown as, P_2, and the pore pressure shown as P_1. The key to changes in penetration rate assuming

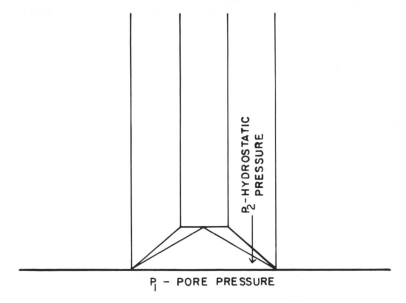

FIG. 11–1. *Differential pressure*

all other factors affecting drilling rate remain constant, is the magnitude of the differential pressure $(P_2 - P_1)$. Thus, if mud weight remains constant and the pore pressure increases penetration rate will generally increase.

The range of these effects is shown by Figure 11-2. It is noted that as pressure differentials increase the rate of change in penetration rate is decreased. This latter affect is well recognized in noting changes in pore pressures in the transition zone. An increase in mud weight to a 12 to 13 ppg level as soon as the pressure transition zone is encountered will often partially mask the formation pressure increases with depth and make it more difficult to select a protective casing seat.

For this reason it has become more common to increase the mud weight as drilling progresses in the transition zone. Exceptions are made in areas where the pressure transition zone is short and improvements in selecting casing seats are still being made.

To use drilling rate properly, all the other factors that affect drilling rate should be held constant. Of course, changes in lithology cannot be controlled and this represents one of the most serious changes. A sudden large increase in penetration rate may be indicative of a change in lithology. However, specific examples can be used to prove or disprove any conclusion. At times there is no way to tell unless other symptoms of abnormal pore pressure are also used. Bit dulling is also uncontrollable but often times it is predictable. A discussion on drilling trends is given in the chapter on cost control.

Changes in bit weight and rotary speed are items the operator can

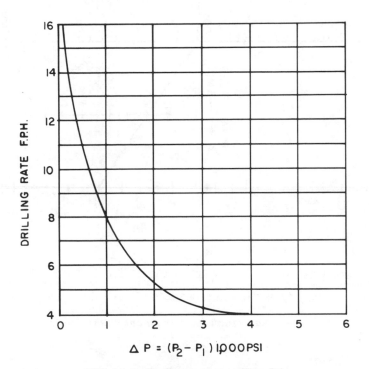

FIG. 11-2. *Drilling rate vs.* $(P_2 - P_1)$

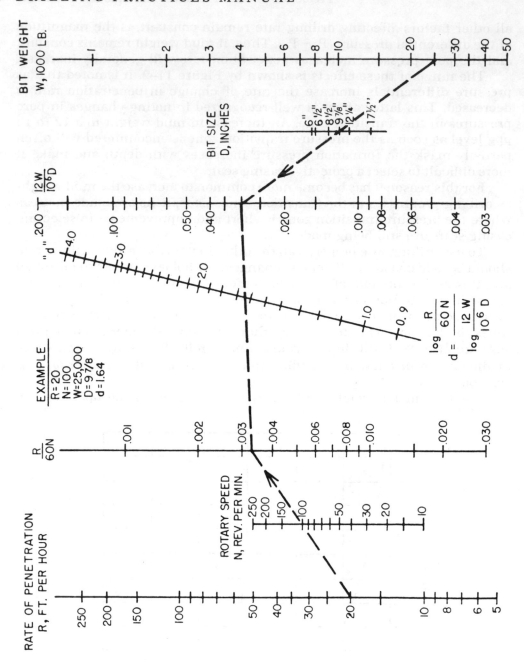

FIG. 11-3. *Nomogram for d exponent determination*

not for d_c

"Extracted from an article entitled, Application of Drilling Performance Data to Overpressure Detection, and written by J. R. Jorden and O. J. Shirley. Article appeared in November, 1966 of the JOURNAL OF PETROLEUM TECHNOLOGY."

control and they should be kept constant in the pressure transition zone. In unknown areas or in evaluating past practices, where bit weight and rotary speed are being changed, Jordan and Shirley,[2] developed a useful method of evaluating drilling rate, known as the "d" exponent. The theoretical base for this exponent is derived from a standard drilling rate equation shown as Equation 1.

$$R = K \, N \left(\frac{W}{D_b}\right)^d \tag{1}$$

With no changes in either rotary speed, bit weight, or lithology, the only factor in Equation 1 that would change the "d" exponent is drilling rate, thus under these conditions drilling rate instead of the "d" exponent should be used. It is noted that drilling rate is shown as being directly proportional to rotary speed. This is true in soft formations such as the soft shales in coastal areas but is not true in hard rock areas. Thus the use of this correlation is limited to soft compactible type rocks. If the "d" exponent is used in hard rock areas, the corrections for rotary speed are not valid and only the corrections for bit weight should be used.

In the Gulf Coast areas of the United States the "d" exponent has proven to be a valuable tool in general field applications. For this reason, the nomograph solution offered by Jorden and Shirley has been included as Figure 11–3. This nomograph is a solution of Equation 2.

$$d = \frac{\log \dfrac{R}{60N}}{\log \dfrac{12W}{10^6 D_b}} \times \frac{9}{MW} \quad \text{for } d_c \tag{2}$$

In addition to the standard "d" exponent a correction for mud weight has been used. One procedure includes multiplying the standard "d" exponent by the inverse ratio of the mud weight increase. There does not appear to be any theoretical justification for this practice; however, it is claimed the correction improves results in the field.

Methods of estimating the pore pressure from changes in drilling rate and the "d" exponent have been developed by several companies and service organizations. All of these methods depend on empirical correlations. The most common procedure is to establish a normal drilling rate trend and then establish the pore pressure from drilling rate increases above the normal trend line.

Another procedure that has been used includes increases in mud weight to keep the drilling rate on a predicted normal trend. From this procedure, the increases in pore pressure are equated to the increases in mud weight required to control the drilling rate. Any procedure of value will depend heavily on a knowledge of drilling rates in normal pore pressure formations. This places a premium on an accurate record of drilling parameters and rates.

SLOUGHING SHALE

Sloughing shale may be the result of the following hole conditions.
1. Formation fluid pressures in excess of the hydrostatic pressure.
2. Hydration or swelling of shale.
3. Erosion caused by fluid circulation, surge pressures or pipe movement.

In some cases the sloughing problem may be a combination of more than one of the above causes. For this reason operators should always try to diagnose the causes. In the coastal areas of the United States items 1 and 2 are the primary reasons for sloughing shale. Actually most of the serious sloughing problems in coastal areas are caused by item 1.

Thus this symptom is watched carefully when penetration rate begins to increase. If penetration rate increases and sloughing shale are noted about the same time, the operator is alerted to the problem of increasing pore pressures. Again this is a phenomenon associated primarily with soft compactible shales. In the mid-continent, Permian Basin or Rocky Mountains, shale sloughing is associated primarily with item 3.

SHALE DENSITY Reduces

The normal trend is for the density of compactible shales to increase with depth. If this trend is reversed as shown in Figure 11–4 by the dashed lines, it is assumed that a pore pressure increase within the shales has

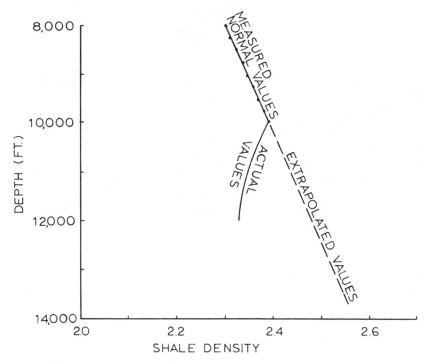

FIG. 11–4. *Shale density vs. depth*

prevented compaction. In theory this is an excellent tool; however, its actual field value is clouded by the difficulty of making precise measurements and selecting the shale that represents drill cuttings.

Two methods of measuring shale density are commonly used in the field. (1) the variable density liquid column and (2) the mud balance density. The variable density liquid column would theoretically be an accurate procedure if all the shale cuttings represented bottom-hole cuttings from the depth assumed and if the shale cuttings were prepared exactly the same way each time. The mud balance method is based on Equation 3.

$$\text{Specific gravity} = \frac{8.33}{16.66 - W_s} \tag{3}$$

The mud cup is filled with shale until the weight is 8.33 ppg and then filled with water. The shale weight plus the added water is W_s. Again the problem is to equate depth with the shale cuttings and to prepare the cuttings in the same manner each time.

It is not easy to determine the source depth for the shale cuttings. The velocity profile of the mud in laminar flow is not flat and shale cuttings from the same depth may arrive at the surface at time intervals that vary as much as several hours. This variation may be minimized by close observation of the size and shape of cuttings used in the analysis, which means expertise is an important part of the procedures.

Washing and drying the shale samples is important. It can be readily seen that the water content will influence the results of any method of measurement. Precise methods of shale preparation are shown in service company manuals. In general, preparation methods call for washing carefully and towel drying the cuttings.

GAS IN MUD

Gas cut mud has always been considered a warning signal, but not necessarily a serious problem. There are multiple sources of gas. Gas may enter the mud as a result of the following:

1. Gas in shales that form a base line for a continuous gas unit level.
2. Gas from sands that may cause sudden changes in the gas concentration level.
3. Connection gas associated with swabbing on connections.
4. Trip gas associated with swabbing following round trips with the drill string.
5. Gas that enters the mud because of insufficient mud weight to control formation fluids.

Because gas is a compressible material it often gives the appearance of being a more serious problem at the surface than actually exists. Many shales contain gas in the pore spaces and furnish a continuous level of gas in the mud. This continuous level of gas forms a base line reference and for given areas it is predictable. Very little attention is paid to this source of gas.

Some gas sands may increase the gas in the mud substantially and may result in severe reductions in surface mud weights. This always concerns the operator and may cause trouble if it comes unexpectedly. To illustrate the effect of this gas, Example 1 shows an assumed set of conditions.

■ EXAMPLE 1:

Well Depth = 15000 ft
Hole Size = $7^7/_8$ inches
Drill pipe size = $4^1/_2$ inches
Mud Weight = 15 ppg
Drilling rate = 20 fph
Sand: gas saturation = 70%
          porosity = 25%
Circulation rate = 7.0 bpm
z_s = 1.0
z_b = 1.35
T_b = 250°F
T_s ≈ 100 F

Bottom-hole ratio of mud volume to gas volume:

$$\frac{Mud}{gas} = \frac{(7)(60)^{min/hr}}{(.062)(20)(.25)(.70)^{bbls/ft}} = 1{,}935 \; Bbls\,mud/bbls\,gas$$

Ratio of Surface Volume of Gas to bottom-hole volume of gas:

$$\frac{V_s}{V_b} = \frac{P_b\, z_s\, T_s}{P_s\, z_b\, T_b} = \frac{11{,}700(1.0)(560)}{14.7(1.35)(700)} = 466$$ or 1 bbl gas will expand to 466 bbls @ surface

$$\frac{1935}{466} = 4.15 \; \frac{mud\; vol}{gas\; vol} \; \text{at surface}$$

Mud Weight

$$\frac{4.15(15)}{4.15 + 1} = \frac{62.3}{5.15} = 12.1 \; ppg$$

A reduction in mud weight from 15 to 12.1 ppg would be frightening if the operator was not aware of the source. In actual practice all the gas would probably not be retained in the mud when it was weighed and most probably the observed reduction in mud weight would be less than that shown in this example. In any event any mud weight reduction at all is a basis for concern and the symptom needs to be associated with other symptoms in diagnosing the problem and the future course of action.

For this specific example the reduction in mud weight in the annulus would not be very high if the gas occupied the entire annulus. The reduction in hydrostatic pressure in Example 1 can be estimated by using Equation 4.

$$\rho D - P = \frac{CP_s Z_a T_a}{(100 - c)Z_s T_s} \; \ln\left(\frac{P + P_s}{P_s}\right) \tag{4}$$

$$\rho D = .78 \, \frac{psi}{ft} \, 15000 \; ft = 11{,}700 \; psi$$

$$c = \frac{(1)(100)}{1 + 4.15} = 19.4$$

Assume, $P = \rho D$ and find the answer by solving the right hand side of Equation 4.

$$\frac{cP_s Z_a T_a}{(100-c)Z_a T_s} \ln\left(\frac{P+P_s}{P_s}\right) = \frac{(19.4)(14.7)(1.18)(635)}{(80.6)(1)(560)} \ln\left(\frac{11,700+14.7}{14.7}\right)$$

$$\rho D - P = 4.74 \ln 798 = (4.74)(6.7) = 31.8 \text{ psi}$$

$$P = 11,700 - 31.8 = 11,668.2 \text{ psi}$$

Thus in this 15,000 foot well, a continuous column of mud cut back in surface weight as shown, would result in only a reduction of about 32 psi in hydrostatic pressure. The primary emphasis with gas cuts of this type is placed on recognizing the source. Sudden increases in mud weight to control the influx of gas might result in excessive hydrostatic pressure, cause a loss of circulation and result in substantially more hole trouble.

On the other hand, failure to recognize a well kick could be disastrous. In a 15,000 foot well if by the time the gas is seen at the surface there are no other indications of a well kick or blowout, then the problem can be associated with gas entry due to drilling the sand or a high pressure low productivity gas zone. In either case the operator generally has time to determine the source of the problem.

Again this emphasizes the need to be familiar with the lithology of the area. In a shallow well, very little decision time is available regardless of the source. Familiarity with the area is desirable. In new areas, the operator may need to close-in the well and observe bottom-hole pressures. If doubt continues the operator should stop drilling and circulate under controlled conditions until the gas source is determined. If the gas disappears after two or three circulations, it is probably due to gas from the sand, if it continues then the problem is probably a low productivity gas sand.

Connection gas and trip gas are introduced into the mud by swabbing or just by the reduction in total annulus back pressure when the pump is stopped. Any increasing trend in either connection or trip gas should be watched carefully. Both sources of gas will be affected by the practices of specific drillers.

Thus where conditions are critical some effort should be expended to standardize practices of making connections and pulling pipe on round trips. With standard practices, an increasing trend in either connection or trip gas may be a warning signal that pore pressures are increasing.

CHLORIDE TRENDS

Chloride trends in the mud are not as easily recognizable as changes in gas concentration. Methods of measurement make it more difficult to obtain information on chloride changes. Also in many cases the water used in mud is a brackish or sea water type with a high level of chloride. However, the operator cannot afford to ignore the symptom and should keep regular checks on the chloride content in the mud both going into and coming out of the hole. A comparison of trends may provide a warning or confirmation signal of increasing pore pressures.

An alternative to measuring the chloride content of the filtrate, is a continuous method of measuring the mud resistivity both into and out of the hole. Such resistivity measurements may be a part of a regular mud logging service or could be added at a small additional cost.

MUD PROPERTIES

The measurement of mud properties into and out of the hole may provide the first warning signal of gas or chloride changes. A sudden increase in yield point or a sudden decrease in the n value will signal potential flocculation of the mud. While the flocculation may be caused by excessive solids or high temperatures it may also be caused by contamination. In any event the operator should immediately check for the cause.

This places a premium on the regular measurement of mud properties into and out of the hole and emphasizes the need to record the precise time that each measurement is made. A drilling mud is normally not a homogenous mixture thus in making comparisons of mud properties it is desirable to know what part of the mud system is being checked. A plot of the mud property trends should be maintained and any addition of additives should be noted.

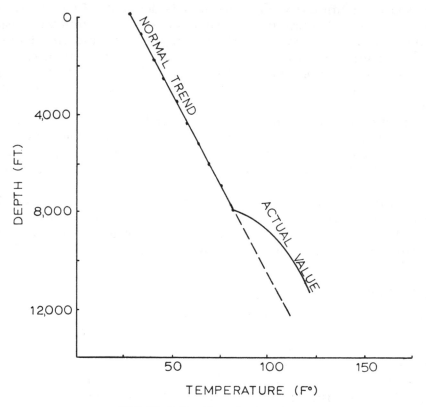

FIG. 11–5. *Depth vs. temperature*

TEMPERATURE MEASUREMENTS

A continuous recording of flow line temperature measurements may or may not be an aid to detecting increases in pore pressures. Figure 11–5 shows a typical flow line temperature profile and point A shows a temperature increase above the normal trend. This increase in temperature above the normal trend may be caused by an increase in circulation rate, a change in solids content in the mud, a change in drilling practices, increased bit torque, or an increase in pore pressure. Thus the temperature curve alone is not definitive.

However, in many areas such as the Anadarko Basin, the temperature log may be the most definitive tool available. In coastal areas, where drilling rates and other symptoms of increasing pore pressure are more definitive the use of flow line temperature has played a secondary role. Many specific examples can be cited where the use of flow line temperature was very useful in Coastal operations and where equipment permits an accurate record of the flow line temperature should be maintained.

BENTONITE CONTENT IN SHALE CUTTINGS

It is normal for the bentonite content in shales to decrease with depth. If the bentonite content does not decrease or if it increases in deeper shales, this is a warning signal that normal shale compaction trends are not being followed. The accuracy of this method is hampered by the inaccuracy of the methylene blue test used to determine bentonite content. The utility of this method compares with the utility of using shale densities. Both procedures have theoretical merit and are hampered by the available methods of measurement.

PALEO INFORMATION

Abnormally high pore pressures are frequently related to certain environmental conditions within a given geologic time period. For example in the coastal areas of South Louisiana if deposition occured in water depths greater than 1,500 feet, the shale to sand ratio increased introducing permeability barriers to sand continuity. This isolation of sands resulted in traps for entrained fluids and the later overburden deposition resulted in part of the overburden being supported by the trapped fluids.

The depth of the water during deposition is marked by the presence of certain fossils. If an examination of cuttings reveal these fossils then the operator should be alerted to the potential problem of encountering abnormally high pore pressures.

WIRELINE LOGS

The use of wireline logs for determining pore pressures is well documented. All of the methods for calculating formation pore pressures from logs are similar. A base line is established from normal pressure levels and

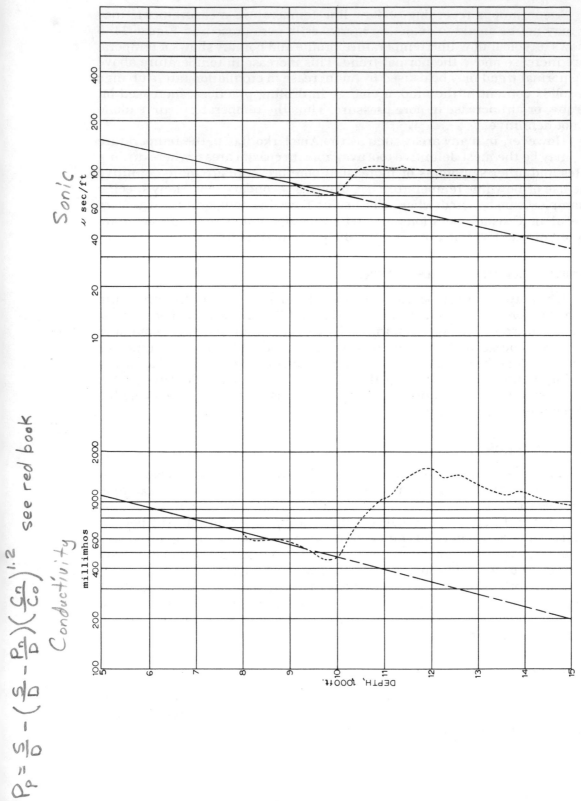

$$P_a = \frac{S}{D} - \left(\frac{S}{D} - \frac{P_o}{D}\right)\left(\frac{C_n}{C_o}\right)^{1.2} \quad \text{see red book}$$

$Conductivity$

$Sonic$

FIG. 11-6. Log data from frio formation—South Texas

extrapolated into the abnormal pressure region. Information in this discussion will be based on the method presented by Matthews and Kelly.[4]

Their technique for predicting pore pressures included the following steps:

1. Establish a straight base line from normal pressure shale readings.
2. Extrapolate this straight line into the abnormal pressure region.
3. In the abnormal pressure region relate the actual log values to the normal extrapolated values.
4. Determine formation pore pressures from empirical data that have been developed for the area or in lieu of data for the specific area relate the information to similar areas.

Figure 11–6 shows a classic example of how abnormal pressure affects both conductivity and sonic logs. The departure of actual log data, shown by the dashed lines, indicates that shale compaction has been reduced. This means that fluid has been trapped within the pore spaces and this fluid is maintaining the porosity by supporting a part of the overburden weight. With additional fluid in the pore spaces electrical conductivity is increased and the travel time in seconds per foot for sound waves has been increased. Although a density log will not be considered specifically, this log would show a decrease in density in the abnormal pressure region.

Figure 11–7 taken from Matthews and Kelly,[4] shows a plot of pressure gradients versus the log ratio of observed conductivity to normal conductivity data, for wells in the South Texas Gulf Coast. These data are developed from experience in an area and represent known data from the

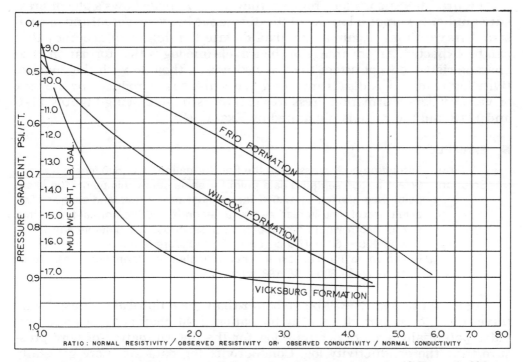

FIG. 11–7. *Pressure gradients vs. resistivity or conductivity ratios, South Texas Gulf Coast*

area. Actually each well drilled is a test well and as new data are obtained curves such as Figure 11–7 are being modified. Pressure data are available from oil companies, service companies and drilling contractors in many areas and can be accumulated to improve accuracy. Using Figures 11–6 and 11–7, the formation pore pressure can be estimated. This is illustrated by Example 2.

■ EXAMPLE 2:

Find: Formation pore pressure in the Frio formation at 11,600 feet which is a porous zone.

Solution: Select a representative data point for shale below this depth. This point is selected at 12,000 feet, where the actual conductivity is 1600 millimhos. A normal conductivity for this depth is 330 millimhos. The ratio of the actual conductivity to the normal conductivity is 1600/330 = 4.85. Using this ratio and figure 11–7 the pressure gradient is found to be 0.84 psi/ft. Thus the formation pore pressure at 11,600 feet is 11,600 × .84 = 9744 psi.

This calculated pore pressure should be compared with actual measurement data, if available. If not isoboric data for the area can be used to estimate the degree of accuracy. A large deviation from anticipated pressures may mean, (1) the empirical data needs to be adjusted, or (2) log readings are inaccurate. Inaccurate log readings may be obtained because of changes in fluid resistivity, undetected changes in lithology, and personal interpretation.

Figure 11–8, also taken from Matthews and Kelly,[4] shows a plot of pressure gradients versus a log plot of the difference between actual and normal sound travel times. Again actual data dictate the degree of accuracy that can be expected. An advantage of the travel time log is that formation fluid characteristics do not affect the travel times as they do with the conductivity log readings. A disadvantage in many areas is introduced by the lack of velocity data. Formation pressures can be estimated using Figures 11–6 and 11–8 and their use is illustrated by Example 3:

■ EXAMPLE 3:

Find: Formation pore pressure in the Frio formation at 11,600 feet.

Solution: Select a representative data point for shale below this point. The shale point was chosen at 12,000 feet or just below the sand section. The actual travel time is read as 100 μ seconds and the normal travel time at this depth would have been 53 μ seconds. The difference in travel time is 100 − 53 = 47 μ seconds. From Figure 8, the pressure gradient is shown to be 0.86 psi/ft. This indicates the formation pressure at 1,600 feet is 11,600 × .86 = 9876 psi.

Again these results will be no better than the empirical data used. The travel time log is not affected by formation fluid characteristics and provides potentially a more accurate log for the determination of pore pressure than the conductivity log. Conductivity log data are more generally available from past operations and empirical correlations such as those

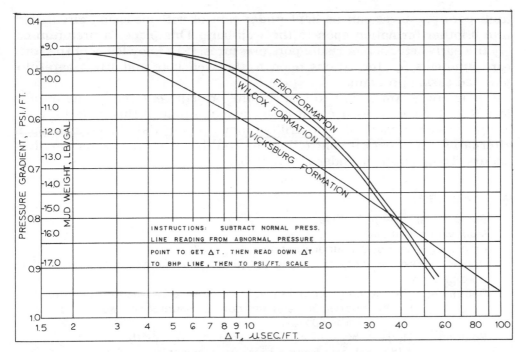

FIG. 11-8. *Pressure gradients vs. shale travel time difference, South Texas Gulf Coast*

shown in Figure 11-7 are thus prepared more easily than those shown for the travel time log in Figure 11-8. The density log is also an excellent method to determine the presence of abnormally high pore pressures; however, again data from this log are not available in many areas.

This method of predicting pore pressures is a simple straight forward approach. There are other more complicated methods which include several correlating parameters for the extrapolation of the normal pore pressure line into the abnormal pressure region. With experience and an abundant amount of lithology and pore pressure data, a greater degree of accuracy might be obtained by a more sophisticated approach. However, in general, this simple approach has proven to be as accurate as some of the more complicated methods and it is easy to understand.

Methods of presenting empirical data such as that shown by Figures 11-7 and 11-8, which were taken from Matthews and Kelly,[4] may be varied. For example, a ratio of travel times could be used in Figure 11-8 instead of the difference as shown. In addition to simplify the determination of pore pressures, log overlays can be prepared from these data where formation pressures could be determined directly. The engineering and art of presentation is in a state of development in this area and improvements in accuracy and utility will depend on the interest assumed by drilling personnel.

SUMMARY

It is common to encounter abnormally high pore pressures while drilling in all parts of the world. The greatest danger is generally whether these high pore pressures can be kept under control without losing circulation into another formation open to the well-bore. This places a premium on (1) the early detection of rising pore pressures, (2) a knowledge of the fracture gradients for formations open to the well bore and (3) a carefully planned casing program.

It is always important to know if the pressure exerted by the mud is less than the pore pressure and if so the magnitude of this underbalance. Some risks are always taken in drilling; however any risk should leave open alternatives that permit controlling the well if some unexpected problem occurs. To illustrate consider Example 5.

■ EXAMPLE 5:

Surface casing set at 2,500 feet
Mud weight = 10 ppg
Fracture gradient below surface casing = 0.73 psi/ft
Drilling in pressure transition zone at 10,000 feet and mud weight may be less than pore pressure.

Determine: The maximum permissable underbalance between pressure exerted by the mud weight and the pore pressure if the well kicks from a formation at 10,000 feet.

Solution: Mud weight exerts a pressure gradient = 0.52 psi/ft
Additional permissable pressure gradient at 2,500 feet is = 0.73 − 0.52 = 0.21 psi/ft

Total permissable underbalance at 10,000 feet is = $(0.21) \left(\dfrac{2,500}{10,000} \right) =$ .0525 psi/ft $\approx$ 1.0 ppg

Example 5 simply shows that if the operator is drilling at more than 1.0 ppg underbalance at 10,000 feet, he may lose the well completely if the drill bit encounters a permeable zone which necessitates closing the blowout preventers. The same analogy is true for any area at any depth where it would be impossible to pump at a rate in excess of the formation fluid influx. If formation productivity permits only a very low rate of fluid influx into the well bore, it may be possible to leave the choke line open and circulate a heavier mud to kill the well.

It is very important in any drilling operation to recognize the symptoms of increasing pore pressures, to be able to estimate the actual magnitude of pore pressures, to know the fracture gradients of exposed formations, and to maintain the drilling practices within controllable limits. Any one symptom of increasing pore pressure may not be definitive enough to provide the basis for precise conclusions.

For this reason it is suggested that all of the methods used to detect and predict pore pressures be plotted on one chart as shown in Figure 11–9. If most of these detection methods indicate increasing pore pressure, the operator should either accept the indicators or run a confirmation wireline log.

There are methods to calculate fracture gradients and open-hole fracture tests are becoming more common. These procedures are discussed in the following section on fracture gradients.

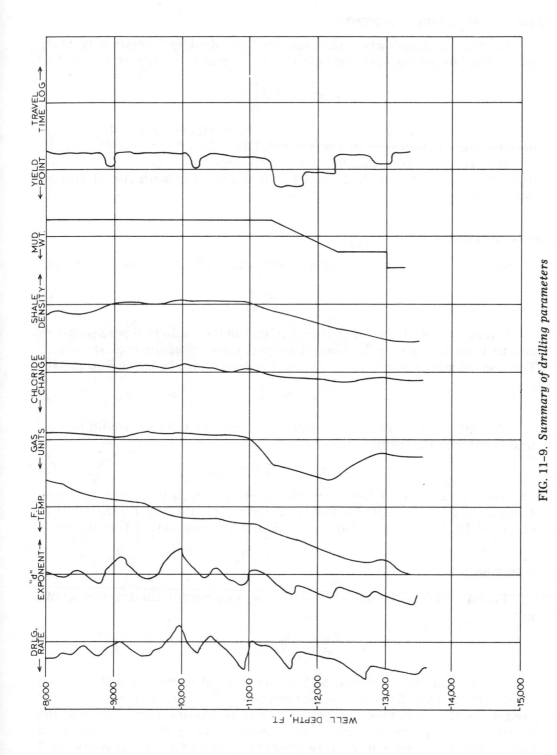

FIG. 11-9. Summary of drilling parameters

FORMATION FRACTURE GRADIENT

Several methods have been presented to calculate formation fracture gradients. One of the first was equation 5, presented by Hubert and Willis.[5]

$$F = \frac{1}{3}\left(1 + 2\frac{P_f}{D}\right) \quad \text{(5)}$$

not good for Soft Rock

Another useful method was presented by Matthews and Kelly[4] and this method will be discussed in some detail. Later Eaton[6] presented a method to calculate fracture gradients and applications have shown this method to be accurate. Both methods will be shown starting with the Matthews and Kelly method.

Matthews and Kelly Method

Matthews and Kelly determined the fracture gradient using Equation 6.

$$F = \frac{P_f}{D} + Ki\,\frac{\sigma}{D} \quad \text{(6)}$$

Equation 6 relates the fracture gradient to the formation pore pressure and to some fraction of the normal support capacity of the rock structure. The normal support capacity of the rock is determined by using Equation 7.

$$\sigma = S - P \quad \text{(7)}$$

The depth, D_i, at which the support capacity of the rock would be normal is determined using Equation 8.

$$\sigma = 0.535\,D_i \quad \text{(8)}$$

Then D_i is used to determine K_i. A plot of depth D_i versus K_i, taken from Matthews and Kelly,[4] is shown in Figure 11–10. Data for this plot are obtained by a solution of Equations 6 and 8 using actual fracture data.

Eaton Method

Eaton proposes the use of Equation 9 to determine the fracture gradient.

$$F = \left[\frac{S}{D} - \frac{P_f}{D}\right]\left(\frac{\gamma}{1 - \gamma}\right) + \frac{P_f}{D} \quad \text{(9)}$$

The overburden gradient, S/D, is determined using a density log and Figure 11–11 from Eaton's presentation shows S/D versus depth for coastal area wells. The pore pressure gradient, P/D, is determined as shown in the section on pore pressure predictions. Poisson's ratio, γ, must be determined empirically using known fracture gradient information. If S/D is assumed to be 0.25 both common assumptions, equation 9 reduces to equation 5 introduced by Hubert and Willis.

$\frac{S}{D} = 1 \neq$

Figure 11–12, taken from Eaton, shows a plot of poisson's ratio versus

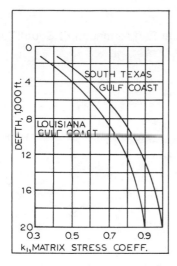

FIG. 11-10. *Matrix stress coefficient*

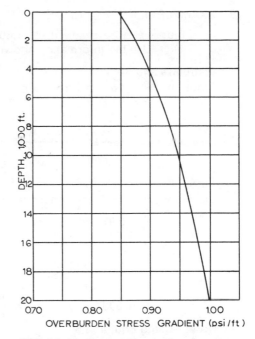

FIG. 11-11. *Composite overburden stress gradient for all normally compacted Gulf Coast formations*

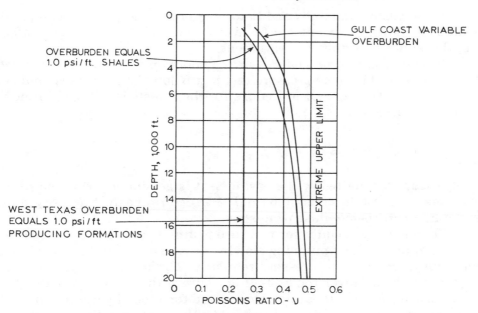

FIG. 11-12. *Variations of poissons ratio with depth*

depth for coastal area formations. Figures 11–11 and 11–12 furnished the necessary information for solving Equation 9 for fracture gradient.

Example 4 shows how both the Matthews and Kelly and Eaton methods may be used to calculate fracture gradients.

■ EXAMPLE 4:

Find: Fracture gradient at 11,600 feet in the Frio formation of South Texas. Use the information shown in Figure 11–6.

Matthews and Kelly
Solution: $P_f = 9876$ psi from sonic log data
$\sigma = 11,600 - 9876 = 1724$ psi
$D_i = 1724/0.535 = 3222$ feet
$K_i - 0.57$ (from Figure 11–9)
$$F = \frac{9876}{11,600} + \frac{0.57(1724)}{11,600} = 0.945 \text{ psi/ft}$$

Eaton
Solution: $P_f = 9876$ psi from sonic log data
$$\frac{S}{D} = 0.97 \text{ (from Fig. 11–10)}$$
$$\gamma = 0.46 \text{ (from Fig. 11–11)}$$
$$F = [.97 - .86]\frac{.46}{.54} + .86$$

$$F = .0937 + .86 \approx 0.954 \frac{\text{psi}}{\text{ft}}$$

Note: Using Equation 5
$$F = \left[\frac{1}{3}\Big(1 + 2(.86)\Big)\right] = 0.907 \text{ psi/ft}$$

The Matthews and Kelly and Eaton methods generally give comparable results as shown in Example 4 for wells below 10,000 feet. Above 10,000 feet there may be significant variations.

These calculation procedures are in common use; however, the most definitive procedure is to pressure test just below the casing seat and open hole sections. The general procedure for these tests is outlined in the following discussion.

Pressure Testing

Pressure testing below the casing seat is generally performed for two reasons: (1) to test the cement job and (2) to determine the fracture gradient in the first sand below the casing shoe.

The general procedure for pressure testing below the casing seat is to close-in annulus, start pumping slowly at a rate of about 0.3 to 0.5 barrels per minute. Record the pressure regularly. The maximum pressure may be at a preselected level, the leak-off pressure or the rupture pressure. If for some reason the operator anticipates a need for a total hydrostatic pressure in excess of the test pressure, he should either re-test with the drill string back in the last casing string or set a liner before drilling ahead.

Testing initially to leak-off is an arbitrary decision that will depend on the operator's objectives. If experience shows the hydrostatic pressure plus circulating pressures will not exceed 16.0 ppg, then there is no need to increase the pressure to the leak-off point that may reach 18.0 ppg. On the other hand if the next casing string depends on what can be contained with

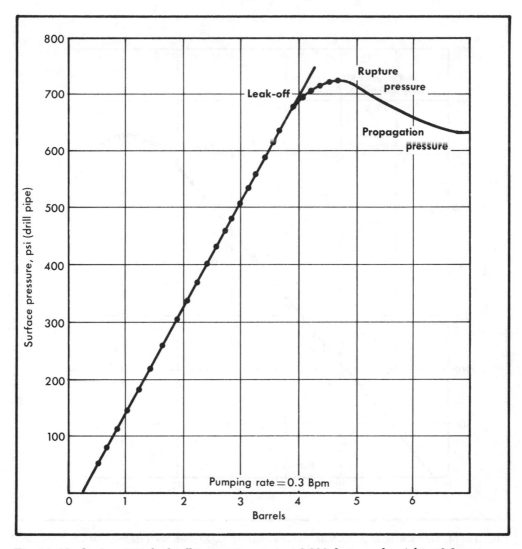

Fig. 11–13. *Casing seat, leak off test casing seat at 3,000 feet, mud weight = 9.6 ppg*

the last casing string then maximum advantage can be obtained from a specific knowledge of the maximum permissable well-bore pressure that can be imposed on the formation. Thus in this case the operator can profit from a knowledge of the maximum leak-off and rupture pressure.

Some examples of leak-off tests are shown in Figures 11–13, 11–14, and 11–15. Figure 11–13 shows a leak-off test below surface casing at 3,000 feet. In this case 20 feet of hole was made below the surface casing. Thus this test is primarily a test of the cement job. In this example the leak-off occurred at 660 psi, the rupture pressure was 725 psi and the propogation pressure was 635 psi. Considering the leak-off pressure as the maximum permissable, results in a maximum mud weight of 13.8 ppg as shown below:

$$\frac{9.6}{} + \frac{660}{(3000)(.052)} = 13.8 \text{ ppg}$$

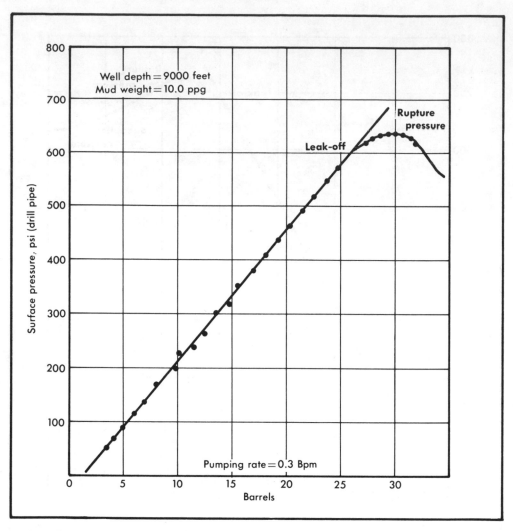

FIG. 11–14. *Open hole leak-off test, casing seat at 3,000 feet, well depth = 9,000 feet, mud weight = 10.0 ppg*

This does not prove that all formations below the surface casing will hold 13.8 ppg mud. It does show that no more than 13.8 ppg mud should be used unless a retest shows an increase in the fracture resistance. An increase in formation strength has been noted in many cases after several days of drilling. It is not uncommon for a zone which held only 13.8 ppg as shown in Figure 11–13 to later hold 15.0 ppg.

A knowledge of this increase in strength might be very helpful in cases when drilling into a pressure transition zone. The increase in strength, when it occurs, is probably due to a plugging of pore spaces by drill solids. It should be emphasized that this strength increase may or may not occur and it is not something the operator can assume will happen.

Figure 11–14 is a leak-off test where the drill string is back in the surface casing, but there is 6,000 feet of open hole. It is noted that the leak-off occurred at about 600 psi. Also it is noted that 26 barrels of mud were required to reach this point, while in Figure 11–13 only 3.8 barrels of mud were required for the casing seat test.

The primary difference is the amount of open hole. In Figure 11–13

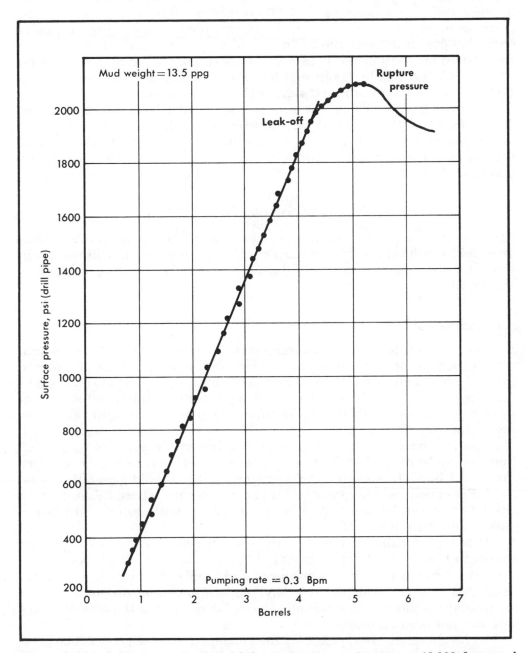

FIG. 11–15. *Leak-off test at first sand below protective casing seat at 10,000 feet, mud weight—13.5 ppg*

only 20 feet of hole had been opened below the casing seat. In Figure 11–14, 6,000 feet of hole had been opened. The additional mud was required because of filtration and loss of mud to very permeable sands.

The leak-off pressure of 600 psi in Figure 11–14 shows the formation just below the casing seat will hold a 13.8 ppg mud. Again this does not insure that all the open formations below 3,000 feet will hold 13.9 ppg mud, because 600 psi imposed, say at 6,000 feet would represent only a 1.9 ppg increase in mud weight. Thus the 6,000 foot formation has been tested to only 11.9 ppg in the test shown in Figure 11–14. However, in young sediments normally associated with most offshore and coastal area formations, the leak-off test results taken just below the casing shoe are generally indicative of the maximum mud weight that can be used.

Figure 11–15 is a leak-off test for the first sand below protective casing. It will be noted that the leak-off occurred when the surface pressure increase reached 1950 psi. This occurred with a 13.5 ppg mud in the hole and the surface pressure plus mud weight represents a formation fracture gradient equal to a mud weight of 17.25 ppg, calculated as follows:

$$13.5 + \frac{1950}{(10,000)(.052)} = 17.25 \text{ ppg}$$

After several days of drilling this formation resistance may increase; however field results show that in most cases subsequent tests will remain in the same approximate range as the initial test when testing below protective casing.

Special Considerations

Special considerations in running leak-off tests include: (1) the pumping rate, (2) the decision to test to a leak-off pressure, (3) which pressure to use if there is a difference in drill pipe and annulus pressure (4) changes in line slope during the test, (5) the frequency of testing and the effect on formation resistance, and (6) what is the maximum mud weight relative to that shown on a leak-off test.

The pumping rate should be kept at a low value, such as 0.25 to 0.5 Bpm. It will be noted that the tests in Figures 11–13, 11–14 and 11–15 were run at 0.3 Bpm. This means the normal rig pump should generally not be used. Exceptions would be with plunger type pumps where the suggested low volumes can be attained. A cementing unit, with pump and volume tank is generally preferred.

If pumping rates are too high, the leak-off test may follow the pattern shown in Figure 11–16. In Figure 11–16 there is no indicated leak-off pressure, the formation suddenly ruptured and whole mud was lost quickly. Even this type test will probably have no long range detrimental effects, the primary problem is that the objective of determining the leak-off pressure has not been attained.

Some operators are repelled by the concept of increasing surface pressure until some mud is lost to the formation. Others may feel that their drilling conditions do not justify such tests. If all the drilling is to be performed in formations with a normal pore pressure, leak-off tests would not be neces-

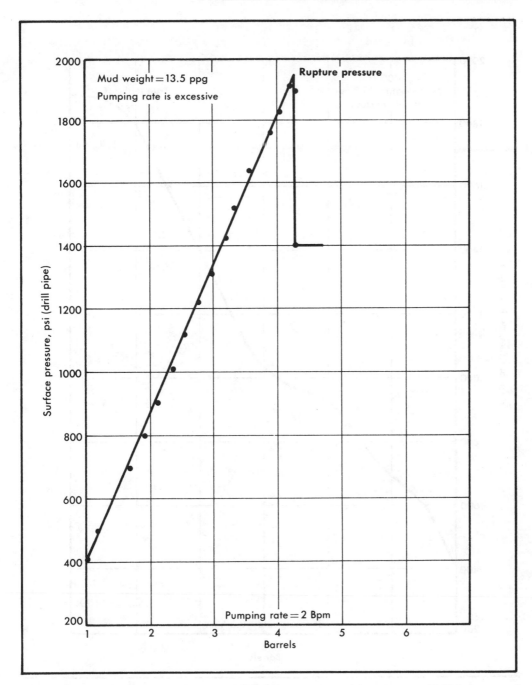

FIG. 11–16. *Leak-off test at first sand below protective casing seat at 10,000 feet mud weight = 13.5 ppg, pumping rate is excessive*

sary. The belief, commonly accepted in drilling, that once the formation is tested to leak-off that it will never again hold that much pressure is an outdated concept.

Several hundred leak-off tests have been conducted during a three-year

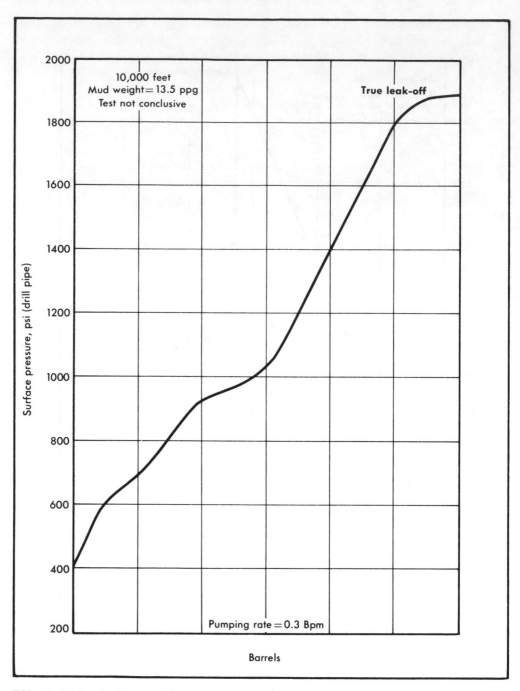

FIG. 11–17. *Leak-off test with 3,000 feet of open sand below protective casing seat at 10,000 feet, mud weight = 13.5 ppg, test not conclusive*

period in coastal areas of the United States and the tests have not been detrimental to subsequent formation resistance. There may, however be valid reasons for not testing to a leak-off pressure. At times surface equipment may not permit the surface pressure necessary to reach the leak-off point.

.ne maximum leak-off pressure is desirable under these conditions a re-test may be performed later after the mud weight has been increased.

Another reason for not testing to leak-off is that the operator knows based on offset well data or geologic information that future mud weights will not be high enough to justify a test to leak-off.

It is not uncommon to encounter differences between the drill pipe and annular pressure during leak-off tests. If the operator is pumping only through the drill string then it is suggested that he use the drill pipe pressure. One method to overcome the problem is to pump slowly into both the drill string and annulus and this has been done on a limited basis.

In some leak-off tests the pressure may not increase in a perfectly straight line as shown in Figures 11–13, 11–14, and 11–15. Figure 11–17 shows what has been observed in some past tests. It is noted that the pressure leaked-off at 600 psi and then continued to increase to 900 psi where it leaked-off again and then continued to increase to the true leak-off. The question that arises immediately is, "how would the operator know the first leveling of pressure build-up was not the true leak-off?"

Actually it may be the true leak-off; however, if the pressure at which this occurs is substantially below that anticipated, pumping should be con-tinued to the rupture pressure because in all probability some type remedial action will be necessary. The cause of this type test is unknown. There may be gas or air pockets in the annulus or some mud may be lost, where there is a substantial amount of open hole as shown in Figure 11–17, and the pressure build-up may continue.

The frequency of testing is normally not a problem because retests are not generally run immediately. However, a limited number of field tests have shown that a leak-off test may be repeated immediately when the pressure is bled-off. Thus retesting will generally not be a problem from the standpoint of hole conditions.

To determine the maximum permissable mud weight after a definitive leak-off test requires a knowledge of mud properties, circulating pressure losses in the annulus and an operator's requirements for safety factors. It is not uncommon to specify a mud weight of 0.5 ppg below the equivalent mud weight at leak-off below surface casing and 1.0 ppg below the equiva-lent mud weight at leak-off below protective casing.

These are general numbers. Circumstances may demand more precise limits. Circulating pressures may be calculated or estimated, then con-verted to mud weight and added to the actual mud weight. After this the operator may wish to add 100 or 200 psi as a safety factor. Actually the risk involved may determine the operator's decision.

If there is a risk of an underground blowout, then the operator will probably take less of a chance on losing circulation. One fact is obvious if the operator exceeds the leak-off pressure from his last test, he is inviting problems and in all probability he will not be disappointed.

Summary

Pressure testing below the casing seat has application in all areas of the world. An operator flirts with disaster if he uses a mud weight in excess of what test pressure shows to be safe. Testing to leak-off has primarily been applied in coastal areas where sediments are young geologically and the open formations are primarily shales and sands. Limestone and dolomite type formations introduce a different type problem.

If these formations are ruptured, the fracture may or may not heal. At this point there is not enough data to justify any routine practice of testing to leak-off in limestone or dolomite. The same may be true for geologically old shale zones, although recent data indicate compaction trends will occur in these shales. If this is generally true, then leak-off tests in these geologically old shales should have no permanent detrimental effect.

At this time, data are too limited to recommend leak-off tests in the so-called hard rock drilling. As information is developed more definitive suggestions can be made. In any event, pressure testing is still recommended where the operator is unsure of fracture gradients and where mud weights must be increased to control high pore pressure zones.

It is better to pressure test and fracture a zone than to raise the mud weight and fracture the zone with a permeable high pore pressure zone open that requires the higher mud weight. The fracture during the pressure test may result in costly remedial action. A fracture with the high mud weight required to contain the fluid in a high pore pressure zone may result in a disastrous underground blowout and eventually an uncontrolled surface blowout.

REFERENCES

1. Pennebaker, E. S., "Detection of Abnormal Pressure Formations From Seismic Field Data," API paper No. 926-13-C, March 6–8, 1968, San Antonio, Texas.
2. Jorden, J. R., and Shirley, O. J., "Application of Drilling Performance Data to Overpressure Detection," Journal of Petroleum Technology, November 1966.
3. Combs, George D., "Prediction of Pore Pressure from Penetration Rate," SPE of AIME paper No. 2162, September 1968.
4. Matthews, W. R., and Kelly, John, "How to Predict Formation Pressure and Fracture Gradient," The Oil and Gas Journal, February 20, 1967.
5. Hubbert, M. King and Willis, D. G., "Mechanics of Hydraulic Fracturing," SPE of AIME paper No. 210, October 1969.
6. Eaton, Ben A., "Fracture Gradient Prediction and It's Application in Oilfield Operations," SPE Journal, p. 1353, October, 1969.

NOMENCLATURE

c = percent of total fluid at the surface that is gas
d = "d" exponent, dimensionless
D = well depth, feet
D_b = bit diameter, inches
D_i = well depths at which rock matrix strength is normal, feet
F = fracture gradient, psi/ft
K = drillability constant
K_i = fraction, less than 1.0, in fracture gradient equation, dimensionless
N = rotary speed, Rpm
P = hydrostatic pressure exerted by a column of mud and gas, psi

P_f = pore pressure, psi
P_s = surface pressure, psi
S = pressure exerted by weight of the overburden rock, psi
T_a = average temperature, degrees Rankine
T_s = temperature at surface, degrees Rankine
W = Bit weight, pounds
W_s = weight of shale plus weight of water, ppg
Z_a = average gas compressibility factor, dimensionless
Z_s = gas compressibility factor at surface, dimensionless
ρ = mud gradient, psi/ft
σ = matrix rock strength, psi
γ = Poisson's ratio, dimensionless

12 Pressure Control

$\mathbb{D}$URING the 1960's, pressure control received more attention than any other phase of drilling practices. One reason for this emphasis was an increasing awareness of the problems of deep drilling and the extremely high drilling costs associated with abnormally pressured formations. In addition, ecology became an important part of the American scene and several blowouts with the associated problems of pollution gained international attention. Politicians, quick to seize an opportunity to condemn which seems to have become a stature building technique in some areas, made public statements and seemed to thrive on any problem that could be related to the industry.

In the meantime , engineers and scientists within the industry were working hard to further develop and put into practice technology that would prevent a blowout. They also were working on methods to bring the blowout under control if it occurred. In addition methods were being developed to handle oil spills on open waters. No expense has been spared to develop methods of prevention and cure.

One of the most important parts of any operation is the planning stage. With geological information and experience in the same or similar areas, wells are planned to reach total depth with no problems. Casing programs are designed to maximize safety, mud programs are selected carefully and control equipment is designed to handle potential problems.

Even with this planning exploratory or new areas present problems. Specific knowledge of the new area is not available; however, geophysical methods of exploration can now be used to estimate formation pressures and give an insight into formation characteristics. Continued development of these tools will be a substantial aid in future exploratory drilling.

298

In development or exploratory areas that can be related to known development areas, the operator can be more precise. He can predict formation pressures with a fair degree of accuracy. He can predict formation fracture gradients closely. He can test predictions while drilling.

This means that in the planning stage of a well he can design the best casing program and prepare for the problems that may or may not occur. In the actual drilling he can test using imposed well head pressures before raising mud weights to levels that might fracture some of the formations open to the well bore. The operator should never fight the so-called lost causes. This is a situation where all the available technology indicates practically no chance for success.

The objectives of this discussion on pressure control will be to: (1) show the pressure relationships that exist in a bore-hole and (2) illustrate the methods used to control well kicks and blowouts.

BOREHOLE PRESSURE RELATIONSHIPS

The operator in planning a well needs to have some knowledge of the overburden and formation fluid pressures in order to select the necessary hydrostatic or drilling fluid pressure. Industry has generally accepted a theoretical overburden pressure of 1.0 psi/ft which is calculated as follows:

Assume rock specific gravity = 2.5
Average void or pore space = 10%
Overburden Gradient = (.433) psi/ft (2.5)(.90) + .433 psi/ft
    (.10) = 1.018 psi/ft

In actual fact, some rocks such as dense limestone will have a specific gravity of 3.0+ and very little void space. Other rocks such as surface shales in young formations may have a pore space of more than 40 per cent of the bulk rock volume. This simply means that the pressure exerted by overburden rock may vary from 0.75 psi/ft to more than 1.3 psi/ft. The primary significance of using 1.0 psi/ft as the theoretical overburden weight is that it simply places an upper limit on well bore pressures.

Using this upper boundary, operators will always try to prevent any combination of surface plus hydrostatic pressure from exceeding 1.0 psi/ft. In actual operations formation fracture gradients are generally less than this upper boundry. However, with only surface casing set there is a danger of fluid penetrating through the earth to the surface if the 1.0 psi/ft is exceeded. This is obviously a much worse condition than simply fracturing an underground formation and losing circulation. With the communication of fluids to the surface, the operator has lost complete control of the well.

Formation fluid or pore pressures are frequently classified into three categories, subnormal, normal and abnormal. Subnormal pore pressures are those below what would be considered normal in a specific area. In the Gulf Coastal areas of the United States, the normal pore pressure gradient is 0.465 psi/ft.

In the North Texas and Mid-Continent areas, subnormal pore pressures result when formations have been produced for a number of years or when formation outcrops are below the surface location of a well and direct fluid communication is established with the outcrop. This latter condition is normal to several drilling locations in the Rocky Mountain areas of the United States.

Abnormal pore pressures are defined as those above normal for the specific area. Generally abnormal pore pressures are associated with fluids that have been trapped within the pore space of rocks by some permeability barrier. These permeability barriers may be faults, folds, salt domes, or permeability pinchouts.

A noted cause of permeability pinchouts in coastal areas is a reduction in the sand-shale ratio generally associated with the deep water deposition of sediments. This geologic feature may be associated with specific fossils of the time period and permit the early recognition of environmental conditions conducive to abnormal pore pressure.

In addition to the permeability barriers there needs to be some reason for the increasing fluid pressure within the pore spaces. In soft rocks, there is a normal compaction trend in shale zones with depth. Abnormal pore pressures result when this normal compaction trend is reversed by the partial support of the overburden rock by fluids within the pore spaces. In this type environment the upper limit on pore pressure is the pressure exerted by the overburden rock.

In hard high compressive strength rocks, there may be no evidence of compaction with depth. Yet in many of these areas, abnormal pressures are frequently encountered. Typical occurrences of this type are the deep high pore pressure zones found in the west Texas area and in shallower zones of this type in southern Iran. Because compaction is not normal with depth in these areas, it is generally believed that deposition occurred within the pore spaces and thus forced the same quantity of fluid to occupy a smaller volume.

The basic cause of abnormal pressure affects appreciably the detection methods and drilling programs. Compaction can be detected by a number of different logging methods. Where compaction is common, the abnormal pore pressures will generally follow an increasing trend with depth. In the hard rock areas, abnormal pressures may be detected: However, the logging methods are not generally definitive. For these areas, abnormal pressure zones frequently lie above normal pressure zones.

Compactible formations are typically found in coastal areas. The normal pressure gradient for these areas is 0.465 psi/ft. The top of the so-called pressure transition zone is the point at which the formation pore pressure begins to increase above normal levels. The bottom of this pressure transition zone is the point where protective casing is set. There is no specific level of pore pressure that signals the bottom of the pressure transition zone.

Generally the requirement to set protective casing is based on the length of the surface casing, the fracture gradients of normal pressure formations and the attitude and experience of the operator. Efforts are generally made to minimize casing requirements, by planned drilling programs.

One significant observation related to abnormal pore pressures is that many of the permeable zones are small in size.

For development drilling this means that high fluid pressures may be bled-off quickly or in fact may be non-existent in an off-set well. This means that well correlations need to be used carefully. Note this can work both ways, the operator may have encountered no abnormal pressure and suddenly find it unexpectedly in his next well. In coastal areas compaction may continue as hydrocarbons are removed from permeable zones and this may result in permeability shut-offs in the region of the well bore where pressure draw-downs are highest.

Another potential danger and a cause of abnormal pressure is thick gas sands. Consider Example 1.

■ EXAMPLE 1:

Assume: Gas sand top at 2000 feet
          Effective sand thickness 600 feet
          Normal pressure gradient for the area = 0.465 psi/ft
Solution: Bottom of the sand is at 2600 feet
          Fluid pressure at 2600 feet = (0.465)(2600) = 1210 psi
          Weight of the gas is negligible, thus pressure at 2000 feet is also about 1210 psi.
          At 2000 feet, 1210 psi gives a pressure
          gradient = 1210/2000 = .605 psi/ft
          Mud weight required to balance = 11.65 ppg

In new or exploratory areas this represents a potentially hazardous problem. Drilling with normal mud weights into a zone of this type may result in well kicks that are difficult to control properly. This becomes a significant problem for shallow gas zones, because by the time a pit level increase is noted the gas is at the surface. This type problem complicates the drilling and casing programs in new areas.

Hydrostatic pressure is the pressure imposed by a static column of drilling fluid. Most mud balances provide a direct reading of this pressure in psi/ft. If they give only the weight of the fluid in pounds per gallon or pounds per cubic foot, the pressure gradient can be found by simply converting the units as shown in Example 2.

■ EXAMPLE 2:

Mud weight ppg $\times$ .052 = psi/ft

Mud weight ppf^3 $\times \dfrac{1}{144}$ = psi/ft

The selection of mud weight is related to formation pressures, swabbing, formation fracture gradients, penetration rate and the general knowledge of an area. In Coastal areas of the United States, a mud weight of 9.0 ppg is required to balance most normal pressure formations: thus the selected mud weight would be at least 9.2 ppg. In the Rocky Mountain and Mid-Continent areas the required mud weight for normal pressure formations will be less than 9.0 ppg.

Because most pressure kicks in wells occur when pulling the drill string, the operator needs to also consider the fluid properties. If water is being used, the swabbing action will be minimal and the fluid weight can be kept at a minimum. When a viscous mud is being used, additional mud weight may be required because of swabbing. The effect of swabbing is discussed in the chapter on surge and swab pressures.

CAUSES AND INDICATION OF WELL KICKS AND BLOWOUTS

A pressure kick or a well blowout may occur for the following reasons:
1. Insufficient mud weight
2. Failure to keep hole full of fluid *– use a trip tank*
3. Swabbing
4. Lost circulation
5. Mud cut by gas or water

Early warning signals are enumerated as follows:
1. Increase in fluid volume at the surface, commonly termed a "pit level increase."
2. Sudden increase in drilling rate
3. Mud cut by gas or water
4. Reduction in pump pressure
5. Reduction in drill pipe weight.

Insufficient Mud Weight

There has been an emphasis on drilling with minimum mud weights in recent years in order to maximize drilling rates. This means that more operators are drilling closer to what is called balanced pressure conditions, where the hydrostatic pressure is barely sufficient to contain the formation pore pressure. In many cases, well kicks are taken to determine the specific pore pressure to help in the planning of future operations. In certain areas such as the Delaware Basin of West Texas, the operator may drill a section of the hole underbalanced, where the pore pressure is greater than the hydrostatic pressure. In these cases upper formations are stable and the productivities of permeable formations are low.

Mud weight requirements are not always known for certain areas. Operators may encounter shallow abnormal pressure zones. Shallow gas sands present a problem as noted in the discussion on abnormal pressures. Correlations around salt domes are always difficult. In addition any new or exploratory area presents a problem in selecting the correct mud weight.

Actually insufficient mud weight is not the primary cause of well kicks or blowouts. Probably the most common cause is the combination of not keeping the hole full and swabbing while pulling the drill string.

Failure To Keep Hole Full and Swabbing

The problem of keeping the hole full of fluid has been emphasized almost as long as fluid has been used to control subsurface pressures. Surveys made on the causes of well kicks and blowouts show that about 25 per cent of the time these problems occur when pulling the drill string. There should be no problems in keeping the hole full if the crews follow volume tables.

Data are abundant on pipe displacement volumes. Special precautions have to be followed with drill collars. A drill collar that weighs 100 ppf displaces five times as much fluid as 5-inch drill pipe. Thus one stand of this weight drill collars equals five stands of 5-inch drill pipe. This emphasizes the precautions that should be taken.

One problem in filling the hole is the measurement of the fluid that goes into the hole when pulling pipe. Flow-meters, pump strokes and various other measuring methods are employed to be certain that the hole is taking a volume of mud equal to the pipe removed. Accuracy of measurement is very important. It is desirable to have a small volume trip tank where actual mud volumes displaced can be determined.

In addition, the rig supervisor should be responsible for keeping a record of the mud required to fill the hole on each round trip with the drill string. Also a record of the water loss should be kept to explain variations in mud volumes required to fill the hole.

Lost Circulation

Lost circulation and its effects are discussed in some detail in Chapter III.

Mud Cut by Gas and Water

Gas cut mud has always been considered a warning signal, but not necessarily a serious problem. There are multiple sources of gas. Gas may enter the mud as a result of the following:

1. Gas in shales that form a base line for a continuous gas unit level.
2. Gas from sands that may cause sudden changes in the gas concentration level.
3. Connection gas associated with swabbing on connections.
4. Trip gas associated with swabbing following round trips with the drill string.
5. Gas that enters the mud because of insufficient mud weight to control formation fluids.

Because gas is a compressible material it often gives the appearance of being a more serious problem at the surface than actually exists. Many shales contain gas in the pore spaces and furnish a continuous level of gas in the mud. This continuous level of gas forms a base line reference and for given areas it is predictable. Very little attention is paid to this source of gas.

Some gas sands may increase the gas in the mud substantially and may result in severe reductions in surface mud weights. These effects were shown in the preceding chapter.

Connection and trip gas are both introduced into the mud by swabbing or just by the reduction in hydrostatic head when the pump is stopped. The plot of gas from this source should be maintained to determine trends. If there is a general increasing trend of gas in the mud attributable to this source, use it as a warning signal. Check other indicators closely, see if the mud weight needs to be increased. If other indicators are negative, watch the well closely and be prepared to take action if the well kicks.

Gas that enters when the mud weight is too low forms the basis for current well control technology. The next section involves the control of a well kick or blowout.

CONTROL OF WELL KICKS AND BLOWOUTS

It is difficult to start a discussion on controlling well kicks and blowouts, because everything should be said first. However, two specific points are essential in well control technology:

1. The surface equipment for proper control must be available.
2. The operator must understand the compression and expansion characteristics of gas.

From this beginning the specific control procedures are almost classical. The drill pipe pressure method of control has become standard practice. Refinements are made in application methods and efforts continue to make procedures more understandable. In addition specific control problems, casing setting depths and drilling practices are being examined constantly to ensure a minimum of trouble.

Surface Equipment

To be prepared to handle any type of well kick the following list of equipment is needed:

1. A hydraulic adjustable choke
2. Two hand adjustable chokes
3. Hydraulic control valve
4. A method for removing gas from the mud
5. A back-pressure valve
6. A full opening positive shut-off valve for the drill string
7. At least one bag type blowout preventer
8. At least two ram type preventers.
9. Accurate pressure gauges, mounted close to the drillers position.

One possible arrangement for the chokes, hydraulic control valve and degassing unit is shown in Figure 12–1. The flow lines need to be secured to handle high pressures and high flow rates and should be sized to prevent any excess back pressure on the annulus.

There are several types of hydraulically adjustable chokes on the market. Some of these chokes are limited to 3000 psi maximum pressure because of rubber closing mechanisms. If there is a doubt on maximum surface pressure requirements a choke with higher pressure ratings should be used and these are available. The hand adjustable chokes should be tungsten carbide trimmed to prevent excessive wear during use.

A hydraulic shut-off valve on the flow line upstream from the chokes

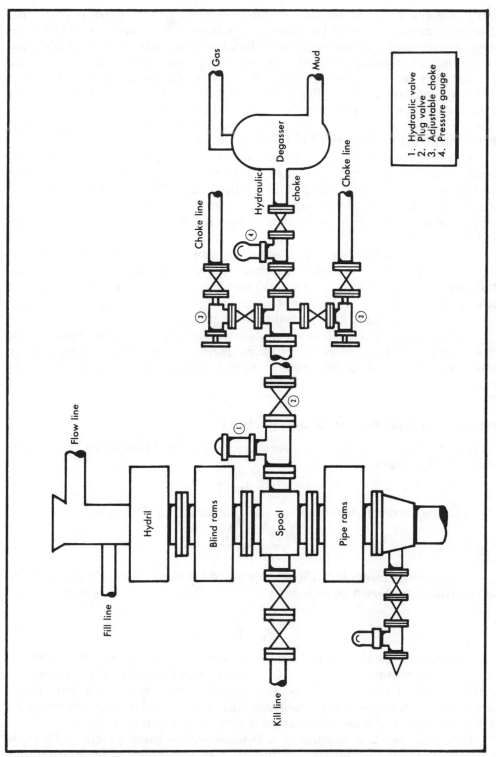

FIG. 12–1. Pressure control equipment – Low Risk Hook Up for low press.

Gas

Mud

Degasser

Hydraulic choke

Choke line

Choke line

Choke line

1. Hydraulic valve
2. Plug valve
3. Adjustable choke
4. Pressure gauge

Flow line

Hydril

Blind rams

Spool

Pipe rams

Fill line

Kill line

permits leaving one of the adjustable chokes open while drilling. A separator or degasser is needed to remove gas from the mud. A series of baffles may be sufficient for low risk wells, but for areas where trouble is a good possibility, arrangements need to be made for a degassing unit.

The full opening positive shut-off valve needs to be available for stabbing into the drill string if the well kicks while pulling pipe. It has already been mentioned that 25 per cent of the well kicks occur when pulling pipe. The regular back pressure valve offers a restriction to flow. If fluid flow is through the drill string the full opening valve is needed.

Blowout preventers are, of course, the key to successful control. For years, standard blowout preventer equipment on medium and high risk wells has included a so-called bag type preventer. In conjunction with this preventer, it is normal to include at least two ram type preventers. One ram preventer is equipped with pipe rams and the other is generally equipped with blind rams.

There are continuing discussions on blowout preventer and flow line choke arrangements. Without entering into this discussion, it can be said that the ram type preventers should be arranged to permit stripping pipe through them. This arrangement is needed as a back-up to the bag type preventer which is primarily used for this purpose.

Emphasis is placed on the availability of accurate pressure gauges. The spread between the control of entering formation fluids and losing circulation may be small. Quick responding gauges are also needed when controlling the well.

Compression and Expansion Characteristics of Gas

The relationship between pressure, volume and temperature on gas is shown in Equation 1.

$$PV = ZnRT \tag{1}$$

Another form of the same equation is shown in Equation 2.

$$\frac{P_1 V_1}{Z_1 T_1} = \frac{P_2 V_2}{Z_2 T_2} \tag{2}$$

Neglecting changes in the Z factors and temperature, Equation 2 can be rewritten as shown in Equation 3.

$$\frac{P_1}{P_2} = \frac{V_2}{V_1} \tag{3}$$

Equation 3 illustrates the danger of not permitting the gas to expand as it is pumped to the surface. If the volume is not allowed to change the pressure will change very little. Thus gas that enters the well bore from a formation with a given pore pressure will still be under the same pressure at the surface. A simple illustration is shown in Figure 12–2.

It is assumed that container A is filled with a mud weighing 12.5 ppg or 0.65 psi/ft. At the bottom of the 10 foot container, a pressure gauge should read 6.5 psi. In container B, one foot of gas is pumped into the con-

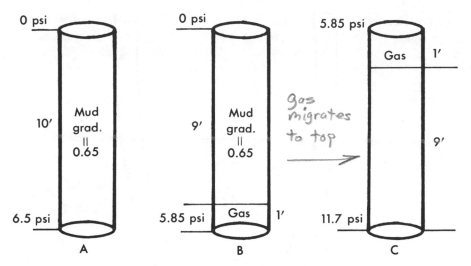

FIG. 12–2. *Inversion of pressure in closed container*

tainer and if no pressure is maintained at the top of the container, the pressure at the bottom of the container would be 5.85 psi. Container C shows what happens if container B is left closed and the gas migrates to the top by gravity segregation.

If the gas is not allowed to expand the pressure of the gas remains constant at 5.85 psi as shown in Equation 3. The pressure at the bottom of container C would increase to 11.7 psi.

Well Control Procedures

Two methods of controlling well kicks have been used, the annulus profile technique and the drill pipe pressure method. The only one in standard use is the drill pipe pressure method. To explain, note Figure 12–3, which is a U-tube diagram representing the inside of the drill string and the annulus.

If fluid of the same density is put into the U-tube, it will reach a constant level shown as A. This simply implies that what happens in one-half of the U-tube will be reflected in the other one-half. If the U-tube is filled with the same density fluid and pressure is applied, the gauge readings on both the drill string and annulus will be identical.

If a constant density fluid is put in the drill string portion of the U-tube, two or more fluids of different densities are allowed to enter the annulus, and pressure is applied at point B, then the pressure gauges on the annulus and drill string will show different values. However, if the entire U-tube is shut-in, the same total pressure will be exerted at point B by the combinations of pressure and fluid on each side of the U-tube.

On the drill string side there is a fluid of known density and a pressure which can be read at the surface. The addition of the drill string pressure reading and the pressure exerted by the known density fluid in the drill string will give the pressure being exerted at point B.

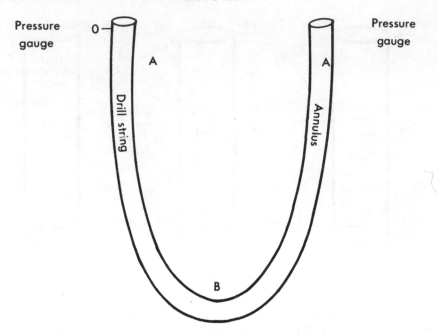

FIG. 12–3. *U-tube illustrating drill string and annulus*

In like manner the addition of the annulus pressure to the pressure exerted by the fluids in the annulus will give the pressure being exerted at point B. Using the annulus side for this purpose is complicated by: (1) the operator does not know accurately the total fluid that has entered the annulus; (2) fluid type is difficult to recognize and the formation fluid might be a combination of gas, oil and water; and (3) the hole may not be to gauge and the height of the formation fluid could not be determined even if an accurate measurement of fluid influx were possible. This explains why the drill string side of the U-tube is used.

Equation 3 shows how the pressure in each side of the U-tube can be equated:

$$P_d + P_{md} = P_a + P_{ma} + P_f \tag{4}$$

If the well kicks, while drilling, the formation fluid will follow the direction of fluid movement which is up the annulus. If the well kicks when pulling the drill string, the formation fluid will enter the entire well bore below the drill string. When stripping pipe back into the hole the drill string must be equipped with a back pressure valve which prevents entry of the fluid into the drill string. One exception might occur if when pulling pipe with no back pressure valve, flow continued long enough for formation fluids to enter both the drill string and annulus.

This exception would be recognized and it would place the operator in a position of guessing at the type fluid that entered, and formation pressure. Common differentiations have been introduced on control methods. The more simple method has been called the drillers method. Several names have been applied to other methods and for reference purposes it will be called the engineers method in this discussion.

Driller's Method

In this method, the driller has decided before any trouble occurs, the circulation rate he will use in killing the well and he has determined the pressure required to circulate the fluid at the kill rate. The use of safety factors and problems associated with casing setting depths and pressure ratings of surface equipment will be discussed in a later section.

When the well kicks, the driller follows a series of preplanned operations listed as follows:

Kick Occurs While Drilling

1. Shut-off the pump
2. Pull Kelly out of blowout preventers
3. Make certain the choke lines are open
4. Close the blowout preventers
5. Close choke lines if surface pressure conditions permit
6. Record the pressure on the drill pipe and annulus
7. Record the surface volume increase
8. Prepare to displace formation fluid

Kick Occurs While Pulling Pipe

1. Install full opening valve in drill string, close the valve
2. Install back pressure valve
3. Open full opening valve
4. Make certain the choke lines are open
5. Close the blowout preventers
6. Close choke line if surface pressure conditions permit
7. Record surface pressure and pit volume increase
8. Decide on the next course of action.
   This will be discussed in a later section.
   The use of the driller's method is illustrated by Example 2.

■ EXAMPLE 2:

Assume: Well depth = 10,000 feet
Hole size = $7^7/_8$ inches
Drill pipe size = $4^1/_2$ inches
Surface casing depth = 2,000 feet
Fracture gradient at 2,000 feet = 0.76 psi/ft
Mud weight = 9.6 ppg = 0.50 psi/ft
With drill string on bottom:
   Annulus pressure, P_a = 300 psi
   Drill pipe pressure, P_d = 200 psi
   Pit level increase = 10 barrels

1300 psi @ Csg Seat

Pore Press = 5200 psi

Friction Press = 500 (next page)

Normal circulation rate = 6 BPM at 60 SPM
Kill rate = 3 BPM at 30 SPM
Circulation pressure, P_c, at kill rate = 500 psi

Solution:

Formation pressure = $P_d + P_{md}$ = 200 + 5000 = 5200 psi

Pumping pressure = $P_c + P_d$ = 500 + 200 = 700 psi

Static pressure gradient at casing seat =

$$\frac{P_a + P_m}{2000} = \frac{300 + 1000}{2000} = 0.65 \text{ psi/ft}$$

Note: The fracture gradient is 0.76 psi/ft another 220 psi in the annulus above 2000 feet may result in lost circulation and no control.

The pumping pressure, will be 700 psi. The circulation rate must be kept constant. The pumping pressure will be controlled by adjusting the annulus choke. Annulus pressure will generally increase. Precautions will have to be taken to prevent fracturing during the displacement process. If the well is deeper with protective casing set, the displacement problem is minimized.

Engineer's Method Wait & Weight

In this method the initial procedures for controlling the well are the same as shown for the drillers method. It differs only because while circulating to remove formation fluids from the mud weight is being increased to control formation pressures for future drilling. In the drillers method mud weight would be increased after the formation fluids were displaced from the annulus.

The increase in mud weight may be done by the batch method, or continuously while circulating. These two methods will be discussed separately using the conditions shown in Example 3.

■ EXAMPLE 3:

Assume: Well depth = 15,000 feet
Hole size = $7\frac{7}{8}$ inches
Drill pipe size = $4\frac{1}{2}$ inches
Mud weight = 15 ppg
P_m at 15,000 feet = (15)(.052)(15,000) = 11,700 psi
Protective casing seat = 13,000 feet
Fracture gradient at 13,000 feet = 0.91 psi/ft
With drill string on bottom:
Annulus pressure = 1,000 psi
Drill pipe pressure = 700 psi
Pit level increase = 20 barrels
Normal circulation rate = 6 Bpm at 60 SPM
Kill rate = 3 Bpm at 30 SPM
Circulation pressure, P_c, at kill rate = 750 psi

Estimate: Bottom-hole temperature, T_b, = 250°F
Z_b = 1.4
Annulus temperature, T_s, = 120°F
Z_s = 1.1

Internal volume of drill pipe = 0.014 bpf = 210 bbls at 15,000 ft
Annulus volume = 042 bpf
Hole volume = .062 bpf

Batch Method

Pore pressure = $700 + 11,700 = 12,400$ psi

Initial pumping pressure = $750 + 700 = 1,450$ psi

Mud weight required = $\dfrac{12,400}{(15000)(0.052)} = 15.9$ ppg

Final circulating pressure = $\dfrac{(750)(15.9)}{15} = 795$ psi

Time required for weighted mud to reach bit

$$\frac{210 \text{ bbls. Min.}}{3 \text{ bbls.}} = 70 \text{ min. or 2,100 pump strokes}$$

Reduction in pump pressure as weighted mud is displaced through drill pipe to bit.

$$\frac{1450 - 795}{70} = \frac{9.4 \text{ psi}}{\text{min.}}$$

$$\frac{1450 - 795}{2100} = \frac{0.31 \text{ psi}}{\text{stroke}} = \frac{10 \text{ psi}}{32 \text{ strokes}}$$

Table 12–1 gives a pumping schedule for Example 3.

TABLE 12–1
Pumping Schedule

Time	Pump Strokes	Pumping Pressure, PSI
3:15	0	1450
3:20	150	1403
3:25	300	1356
3:30	450	1309
3:35	600	1262
3:40	750	1215
3:45	900	1168
3:50	1050	1121
3:55	1200	1074
4:00	1350	1027
4:05	1500	980
4:10	1650	933
4:15	1800	886
4:20	1950	839
4:25	2100	795

The actual mechanics of implementing the batch method would be as follows:

1. Increase the mud weight in the suction tank to 15.9 ppg and commence displacement.

2. Add barite to maintain mud weight at 15.9 ppg. Estimated requirements are 65 pounds of barite per barrel of mud. At a pumping rate of 3 bpm, the required barite additons will be 195 pounds or about 2 sacks of barite per minute.

3. After mud weighing 15.9 ppg reaches the bit, hold the drill pipe pressure constant at 795 psi until formation fluid that entered well bore has been displaced. This will generally require more than one cycle of the mud.

4. Additions of barite will have to continue until all the mud weighs 15.9 ppg.

5. If drilling is to continue the mud weight will have to be increased to allow for swabbing when pulling pipe or making connections. Swabbing effects can be calculated or a good estimate will be in the range of 150 psi of additional hydrostatic pressure per 5,000 feet of hole. For example 3, the mud weight would need to be increased to about 16.5 ppg if drilling is to continue.

Continuous increases in Mud Weight

The mud weight required remains the same as noted in the batch method. The decrease in pumping pressure is based on each point increase in mud weight. Pumping pressure reduction per point increase in mud weight.

$$\frac{1450 - 795}{9} = 72.8 \text{ psi/point}$$

Note: One point is equated to 0.1 ppg.
The pumping schedule with this procedure is shown in Table 12–2.

TABLE 12–2
Pumping Schedule

Mud wt.	Time in	Time out	Strokes in	Strokes out	Pumping Pressure, psi
15.0	2:00	3:10	0	2100	1450
15.2	2:15	3:25	450	2550	1304
15.4	2:30	3:40	900	3000	1158
15.6	2:45	3:55	1350	3450	1012
15.8	3:00	4:10	1800	3900	866
15.9	3:15	4:25	2250	4350	795

The actual mechanics of implementing the method of increasing mud weight continuously would be as follows:

1. Start pumping mud immediately and add barite at a maximum safe rate.
2. Measure mud weight at pump suction continuously.
3. Continue pumping schedule in a manner shown in table 2.
4. After 15.9 ppg mud reaches bit hold the pumping pressure constant at 795 psi until all of the mud has been increased to 15.9 ppg.
5. If drilling is to continue then follow the procedure shown in the batch method.

It will be noted that the pumping pressure schedule in Table 12–2 will impose additional pressure in the annulus because the downward adjustments in pumping pressure do not account for all of the mud weight increases. Actually the additional pressure in the annulus would be about 200 psi more than that shown for the batch method. For Example 3 this would impose no problems and the procedure would decrease the time required to displace formation fluids. In a critical situation more adjustments in pumping pressure could be made and this would minimize the excess annulus pressure.

In addition to the routine control of a well kick, many special problems may arise in the specific mechanics of application. The following section lists some of these special problems and methods that may be used to combat the problems.

Special Problems Associated with Pressure Control

1. Determination of circulating pressure at kill rate
2. Determination of shut-in drill pipe pressure
3. Type of formation fluid that entered the well bore
4. Annulus pressure profile
5. Gas volume when gas first reaches the surface
6. Well kick occurs when pulling the drill string
7. Lost circulation when a well kick occurs.

Determination of Circulating Pressure at Kill Rate

The circulating pressure at the kill rate may be determined before or after a well kick occurs. It is a common practice to change pump speed while drilling and record pump pressures. This pre-recording of pump pressure at different rates of circulation has many other useful functions which are discussed in other chapters.

The best method to determine circulating pressure at the kill rate is to measure this when pumping is initiated after the well kicks. The suggested procedure is as follows:

1. Open choke slightly and start pump slowly holding annulus pressure at the same level as the shut-in annulus pressure.
2. Bring speed of pump up to a pre-selected kill rate, holding annulus pressure constant by adjusting annulus choke size.
3. Record the drill pipe pressure and subtract the shut-in drill pipe pressure, the difference is the circulating pressure required to overcome friction.
4. The recorded pump pressure at the kill rate will be the initial pumping pressure.

The same procedure can be used if it is necessary to stop circulation and restart during the time formation fluids are being displaced. The only danger in using this procedure is when formation gas is very close to the surface and annulus pressures are changing rapidly.

Determination of Shut-in Drill Pipe Pressure

The shut-in drill pipe pressure is easy to obtain if there is no back-pressure valve in the drill string. After closing the blowout preventers and choke lines, the pressure may be read directly from a drill pipe pressure gauge. In some cases the drill pipe pressure may increase slowly and if gas begins migrating up the annulus it might be difficult to determine accurately the drill pipe pressure. Figure 12–4 shows a plot of drill pipe pressure versus time. The shut-in drill pipe pressure would be about the level shown as point A on Figure 12–4.

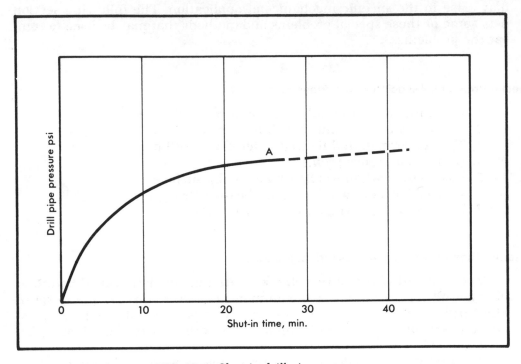

FIG. 12–4. *Shut-in drill pipe pressure*

A back-pressure valve in the drill string makes the determination of drill pipe pressure substantially more difficult. Where potential well control problems exist and a back pressure valve is considered necessary, it is possible to obtain a commercially available back-pressure valve with a small opening in the valve. Very little flow is permitted and there is a large enough opening to permit the communication of pressure to the surface. If a conventional back-pressure valve is being used, the approximate shut-in drill pipe pressure may be determined as follows:

Method 1:
1. Start pump slowly with the annulus choke closed and continue pumping until there is a sudden increase in pump pressure.
2. Watch annulus pressure, stop pump when annulus pressure starts to increase.
3. Read the drill pipe pressure at the point annulus pressure begins to

increase. If the annulus pressure is increased 50 psi above the shut-in annulus pressure, subtract the 50 psi from the pumping pressure to obtain the shut-in drill pipe pressure.

Method 2:
1. Predetermine the friction pressure at the circulation rate which will be used to kill the well.
2. Open choke and start pump slowly holding annulus pressure at the same level as the shut-in annulus pressure.
3. Bring speed of pump up to predetermined kill rate, keeping annulus pressure constant by adjusting annulus choke.
4. Read pumping pressure, subtract previously determined friction pressure from pumping pressure, the remainder will be the shut-in drill pipe pressure. For example assume the friction pressure was measured as 500 psi at 3 barrels per minute or 30 strokes per minute. When the well kicks and the pump rate of 30 SPM is achieved, while maintaining the annulus pressure constant, the pumping pressure is 1000 psi. Thus the shut-in drill pipe pressure is $1000 - 500 = 500$ psi.

There is no claim that these methods of determining the shut-in drill pipe pressure will provide results that are as accurate as those obtained when no back pressure valve is used. However, they will provide results that give a good approximation of the shut-in drill pipe pressure.

Type of Fluid That Entered The Well Bore

Of primary interest in the determination of fluid types is whether gas has entered the annulus. Actually it will be possible to have water, gas and oil enter when a well kicks. If only liquid is present the problems of control are simplified. If there is some doubt the operator should take no chances. To determine the type of formation fluid that has entered the well bore, the operator will need to have an accurate measurement of the quantity of formation fluid that has entered.

This estimate of the quantity of formation fluid entering the well bore is made from the increase in pit level at the surface. From this measurement the length of the annulus occupied by the formation fluid is estimated. To show this calculation use the data assumed in Example 3 and for calculation purposes use Equation 4.

$$P_d + P_{md} = P_a + P_{ma} + P_f$$

Solution:

$$700 + 11{,}700 = 1000 + (15000 - h_b)\rho_m + h_b\rho_f$$

$$h_b = \frac{20 \text{ bbl ft}}{.042 \text{ bbl}} = 476 \text{ feet}$$

$$12{,}400 = 1000 + 11329 + 476 \, (\rho_f)$$

$$476 \, \rho_f = 71$$

$$\rho_f = 0.149 \text{ psi/ft}$$

The gradient of 0.149 psi/ft indicates that the entering fluid is gas. A small error in the total fluid entering the well bore or an enlarged hole would make a substantial difference in the results. Thus, accuracy is emphasized; however, also emphasized is the philosophy of assuming gas is present unless the operator is positive there is no gas.

Annulus Pressure Profile and Gas Volume Changes

A calculation of the annulus pressure profile does not provide a means to control the displacement of formation fluids. It does provide the operator with a means to estimate the behavior of the annulus pressure during displacement and to be prepared for the pit volume increases that accompany the expansion of gas.

A prior knowledge of annulus pressure behavior may dictate the specific method to be used in displacing formation fluids. For example if there is a danger that the annulus pressure would exceed the safety limitations of surface equipment, it might be necessary to attempt to pump the formation fluid back into the formation. Also it may be desirable in some cases to increase the mud weight before displacement to insure a minimum surface pressure increase.

The annulus pressure profile for gas may be calculated using Equations 5, 6, 7, 8 and 9.

$$\rho_f = \frac{SP}{53.3 Z_b T_b} \tag{5}$$

$$P_{mg} = P_b - P_{ma_1} - 0.5\, P_f \tag{6}$$

$$h = \frac{P_b ZT}{PZ_b T_b}\, h_b \text{ (for constant annulus size only)} \tag{7}$$

$$P_a = P_b - (D - h)\, \rho_m - P_f \text{ (No change in mud weight)} \tag{8}$$

$$P_a = P_b - (D - h - D_a)\rho_m - D_a\rho_{m_1} - P_f \tag{9}$$

The annulus pressure, P_a, can be determined using these equations. The solution becomes trial and error when it is desired to determine the maximum pressure when the displaced gas first reaches a specific point in the annulus.

Equations 6, 7, 8, and 9 may be solved simultaneously to give Equations 10 and 11. Equation 10 can be used to calculate the pressure at the top of the gas at any point in the annulus, assuming no change in mud weight. Equation 11 can be used to calculate the pressure at the top of the gas at any point in the annulus, when the mud weight is increased by the batch method before displacement.

$$P = \frac{A}{2} + \left[\frac{A^2}{4} + \frac{P_b \rho_m ZTh_b}{Z_b T_b}\right]^{1/2} \quad \text{unweighted mud} \tag{10}$$

Let $A = P_b - (D - X)\rho_m - P_f$

$$P = \frac{A_1}{2} + \left[\frac{A_1^2}{4} + \frac{P_b \rho_{m_1} ZTh_b}{Z_b T_b}\right]^{1/2} \quad \text{weighted mud} \tag{11}$$

Let $A_1 = P_b - (D - X)\rho_{m_1} - P_f + D_a^*(\rho_{m_1} - \rho_m)$

The use of equations 10 and 11 will be illustrated in Examples 4, 5 and 6 using the basic well data from Examples 2 and 3.

■ EXAMPLE 4

Assume:
(1) Same well data as shown in example 2.
(2) Temperature at surface $= 70F$
(3) Temperature gradient $= 1.2F$ per 100 feet
Determine:
(1) The pressure exerted at 2,000 feet when the gas first reaches that point
(2) The maximum pressure exerted at 2,000 feet when the gas first reaches the surface.
Solution: Use equation 10
(1) $A = 5,200 - (10,000 - 2,000)(0.5) - 19$
    $A = 1,181$

$$P_{2000} = \frac{1181}{2} + \left[\frac{1181^2}{4} + \frac{(5200)(.5)(554)(238)}{650}\right]^{1/2}$$

$P_{2000} = 571 + 936 = 1527$ psi
(2) $A = 5200 - 5000 - 19 = 181$ psi

$$P_{surf} = \frac{181}{2} + \left[\frac{181^2}{4} + \frac{(5200)(.5)(554)(238)}{650}\right]^{1/2}$$

$P_{surf} = 91 + 732 = 823$ psi

$$h_{surf} = \frac{(5200)(530)(238)}{(823)(650)} = 1226 \text{ ft}$$

Mud column above casing seat $= 2000 - 1226 = 774$ ft
Pressure at 2000 ft $= 387 + 19 + 823 = 1229$ psi

It is noted that the maximum pressure at the casing seat, when the gas first reaches that point is 1527 psi and the pressure at the casing seat is reduced to only 1229 psi when the gas first reaches the surface. In this particular example the operator would have lost circulation just before the gas reached the surface casing seat. Consider next what happens if the mud weight is increased enough to control the well. This is shown in Example 5.

■ EXAMPLE 5

Assume:
(1) Same data as shown in Example 2 and Example 4.
(2) Mud weight is increased to control well before displacement is initiated.
Determine:
(1) The pressure exerted at 2000 feet when the gas first reaches that point.
(2) The pressure exerted at 2000 feet when the gas first reaches the surface.
Solution: Use Equation 11

(1) Mud weight increase $= \dfrac{200}{(10,000)(0.052)} \approx 0.4$ ppg

$$D_1 = \frac{141 \text{ bbls ft}}{.042 \text{ bbl}} = 3357 \text{ ft.}$$

$A_1 = 5200 - (10,000 - 2,000)(0.52) + 3357 (0.52 - 0.50) - 19$
$A_1 = 5200 - 4160 + 67 - 19 = 1088$

$$P_{2000} = \frac{1088}{2} + [295,936 + 548.504]^{1/2}$$

$P_{2000} = 544 + 919 = 1463$ psi
(2) $A_1 = 5200 - 5200 + 67 - 19 = 48$
$P_{surf} = 24 + 727 = 751$ psi

$$h_{surf} = \frac{(5200)(530)(238)}{(751)(650)} = 1344 \text{ ft.}$$

Mud column above casing seat $= 2000 - 1344 = 656$ ft.
Pressure at 2000 ft. $= 328 + 19 + 751 = 1098$ psi

By increasing the mud weight before displacing the formation gas, the operator may prevent the loss of circulation at the casing seat. The pressure gradient, not including the circulating pressure over 2000 feet, is 0.732 psi/ft. This compares with a predicted pressure gradient, not including the circulating pressure above 2000 feet, of 0.764 psi/ft. if the mud weight is not increased. The fracture gradient is 0.76 psi/ft.

It should be remembered that these calculations are based on a total influx of 10 barrels of formation gas. The complete annulus profile for Examples 4 and 5 is shown in Figure 12–5.

The displacement of formation fluid in Example 3 introduces a differ-

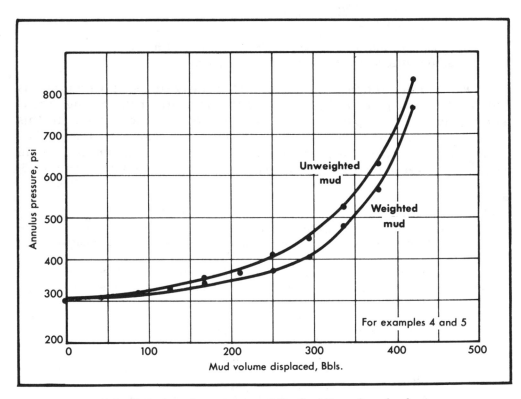

FIG. 12-5. *Annulus pressure of the function of mud volume*

ent problem because under normal conditions the gas has expanded very little by the time it reaches the casing seat at 13000 feet. This means the pressure at the casing seat will increase only a small amount as the gas is displaced. A primary concern is the maximum surface pressure. The maximum pressure at the casing seat and at the surface for Example 3 will be determined in Example 6.

■ EXAMPLE 6

Assume:
(1) Same well data as shown in Example 3
(2) Temperature at the surface = 70F
(3) Temperature gradient = 1.2F per 100 feet
Determine:
(1) The pressure exerted at 13000 feet when the gas first reaches that point
(2) The maximum surface pressure and the pressure at 13000 feet when the gas first reaches the surface.
Solution:
(1) Because no weighted mud has entered the annulus and because the pumping pressure is being adjusted downward to account for the increase in hydrostatic pressure, Equation 10 must be used to determine the maximum pressure at the casing seat at 13000 ft.

$A = 12,400 - (15000 - 13000)(0.78) - 71 = 10,769$

$P_{13000} = 5385 + [28,992,840 + 4,461,217]^{1/2}$

$P_{13000} = 5385 + 5784 = 11,169$ psi

(2) Use Equation 11 for maximum surface pressure

$X = 0$

$A_1 = 12,400 - 12,405 - 71 + 235$

$A_1 = 159$

$P = 80 + [1000 + 3,133,058]^{1/2}$

$P = 80 + 1770 = 1850$ (see Figure 12–6).

$$h = \frac{(12,400)(1.1)(580)(476)}{(1850)(1.4)(710)} = 2048 \text{ ft}$$

$P_{13000} = 1850 + (0.78)(5000) + (0.827)(5952) + 71 = 10743$ psi

It is noted that when the gas reaches the surface that the pressure of 10,743 psi at the casing seat at 13000 feet is less than the pressure of 11,140 psi at this point when the well was first shut-in.

The gas volume increase, when the gas first reaches the surface, is shown in the solution to Example 6. The height of the gas column is 2048 feet and this is equal to 86 barrels of gas. Thus the volume of gas increased from 20 to 86 barrels. If the gas were in bubble form, it would take 29 minutes for the gas to be released. In the actual case, the gas is commingled in the mud and the time required to remove all the gas will probably be much longer than 29 minutes. The operator should be prepared for a pit level increase of 66 barrels above the initial feed-in of 20 barrels.

The total increase of 86 barrels in fluid volume may result in an overflow of mud in surface pits before the gas is removed. If the drilling supervisor is not mentally prepared for this increase in volume, he might make a mistake and close the well-in at a crucial time. After the gas is released and if no loss of mud has occurred, the mud volume in the pit should return to the pre-kick level.

Questions often arise concerning: (1) the effect the pressure differential into the well bore has on control capabilities; and (2) the effect of an increased volume of feed-in.

These effects can be calculated directly using Equations 10 and 11.

Well Kicks While Pulling Pipe

It would be impossible to list all the specific problems that might be encountered if the well kicks while pulling pipe. However to illustrate some of the potential problems consider Example 7.

■ EXAMPLE 7:

Assume:
Well Depth = 15,000 feet
Hole size = $7\frac{7}{8}$ inches
Drill pipe size = $4\frac{1}{2}$ inches
Mud weight = 15 ppg
Protective casing seat = 13,000 feet
Fracture gradient at 13,000 feet = 0.91 psi/ft
Well kicks with bit at 13,000 feet while pulling pipe
P_a = 200 psi, P_d = 200 psi
Pit level increase = 20 bbls
Kill rate = 3 Bpm
T_b = 250°F
Z_b = 1.4
T_s = 120°F
Z_s = 1.1
Determine:
A. The rate of rise for the gas bubble.
B. Procedure to be used to force the gas bubble back into the formation.
C. Procedure for controlling the well if the pipe is not stripped back to bottom and if there is no circulation of fluid.
D. Procedure for controlling the well if fluid circulation is commenced with an increase in surface pressure.
E. Procedure for controlling the well if the mud weight is increased enough to control the potential rise of the gas bubble to the casing seat.
F. Procedure for stripping the drill string back to bottom.

Solution:

A. *Rate of rise of the gas bubble:* The rate of rise of the gas bubble can be determined from the rate of increase of the shut-in annulus pressure. Assume the following:

Initial shut-in annulus pressure = 200 psi.

Shut-in annulus pressure after one hour = 300 psi.

The increase in annulus pressure is 100 psi in one hour which means the hydrostatic head above the gas has been reduced by 100 psi.

Thus:

$$\frac{100 \text{ psi ft}}{0.78 \text{ psi}} = 128 \text{ feet of mud}$$

This shows that the gas has risen above 128 feet of mud in one hour for a rate of rise of 128 fph.

B. *Pushing the gas bubble back into the formation:* Pushing the gas bubble back into the formation should be a lubrication process. The annulus back pressure should be increased by pumping mud into the annulus. In the general case the increase in annulus pressure permitted can be determined by equipment limitations and the fracture gradient below the last string of casing.

For Example 7, the fracture gradient is 0.91 psi/ft at 13,000 feet, which is equivalent to a total pressure of 11,830 psi. The hydrostatic pressure is 10,140 psi. Thus circulation should not be lost until the shut-in annulus plus any circulating pressure reaches 1,690 psi (11,830 − 10,140). It is suggested that the annulus pressure in Example 7 be increased to 500 psi or to a sustained level that forces the gas back into the formation.

After the annulus pressure reaches 500 psi stop pumping to determine if the gas bubble will re-enter the formation. A gradual reduction in this annulus pressure shows the gas bubble is going back into the formation. A total of 20 barrels of mud will be required and this total process may take

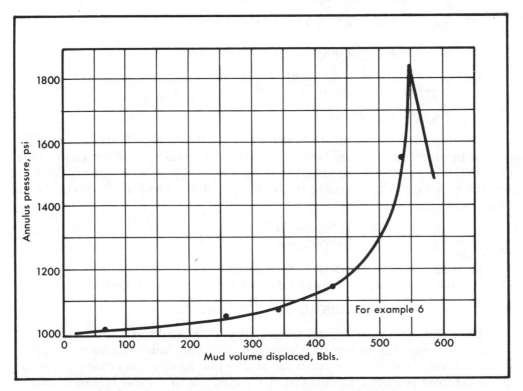

FIG. 12–6. *Annulus pressure of the function of mud volume*

several hours. When the gas bubble has been lubricated back into the formation the surface pressure should have returned to zero. At this point the drill string should be returned to bottom and the mud circulated while increasing the mud weight enough to prevent swabbing when pulling pipe.

C. *Procedure for controlling well if pipe is not stripped to Bottom:* First determine the maximum increase in surface pressure if the gas bubble is allowed to migrate to the casing seat while holding the total back pressure equal to the pore pressure.

$$\text{Let } A = P_b - (D - X)_m - P_f$$

$$P_{13,000} = \frac{A}{2} + \left[\frac{A^2}{4} + \frac{P_b \rho_m Th_b}{T_b}\right]^{1/2}$$

$$h_b = \frac{20 \text{ bbls ft}}{.062 \text{ bbl}} = 323 \text{ feet}$$

$$\text{Pore pressure} = 11{,}448 + 200 + P_f$$

$$\rho_f = \frac{SP}{53.3 \text{ ZT}}$$

Assume: gas specific gravity $= 0.6$

$$\rho_f = \frac{(0.6)(11{,}700)}{(53.3)(1.4)(710)} = 0.1327 \frac{\text{psi}}{\text{ft}}$$

$$P_b = 11{,}448 + 200 + 43 = 11{,}691 \text{ psi}$$

$$A = 11{,}691 + (2{,}000)(0.78) - 43 = 10{,}088$$

$$P_{13,000} = \frac{10{,}088}{2} + \frac{(10{,}088)^2}{4} + \left[\frac{(11{,}691)(.78)(686)(323)}{710}\right]^{1/2}$$

$$P_{13,000} = 10{,}362 \text{ psi}$$

The initial pressure at 13,000 feet was 10,340 psi. Thus the total increase in annulus pressure would be from 200 to 222 psi if the gas bubble is allowed to expand as required to keep the total pressure against the formation constant. The increase in the length of the gas bubble would be as follows:

$$\frac{22 \text{ psi ft}}{0.78 \text{ psi}} = 28.2 \text{ feet}$$

Thus the total equivalent length of the gas bubble would be 351 feet. (323 + 28). The amount of mud released would be as follows:

$$(28)(.062) = 1.74 \text{ barrels.}$$

Thus the procedure for controlling the well would be as follows:

Let the annulus pressure increase to 250 psi. Release enough mud to keep the annulus pressure constant at 250 psi. Keep an accurate record of the total mud released. If more than 1.74 barrels of mud is released, this is an indication the gas is above the casing seat. Also remember that with an annulus pressure of 250 psi, the total back pressure is still 28 psi above the pore pressure. However, if more mud is released it will be necessary to let the annulus pressure increase enough to compensate for the reduction in hydrostatic head.

D. *Procedure for increasing mud weight and circulating at the casing seat:* The required increase in mud weight will be as follows:

$$\frac{\bar{}200}{(13,000)(0.052)} \approx 0.3 \text{ ppg}$$

The procedure would be as follows:

1. Observe the annulus pressure while increasing the mud weight in the suction tank to 15.3 ppg. If there is no increase in annulus pressure, commence displacement.
2. Until the 15.3 ppg mud reaches the bit during displacement, the annulus pressure should remain constant with no adjustments in choke size and the drill pipe pumping pressure should be reduced by just under 200 psi. If the annulus pressure starts to increase before the weighted mud reaches the bit, the gas bubble is rising.
3. Assuming no change in annulus pressure until the 15.3 ppg mud reaches the bit, the annulus pressure will start down while holding a constant choke size after the 15.3 ppg mud enters the annulus.
4. The calculated annulus volume from 13,000 feet to the surface is 676 barrels. At a rate of 3 bpm the time required for the 15.3 ppg mud to fill the annulus would be 229 minutes or 3 hours and 49 minutes. Thus the annulus pressure should decline at a rate of about 0.87 psi/min. with no change in choke size. If this does not occur the gas bubble may be rising. If there is a large difference pumping should be stopped and the rate of rise of the gas bubble determined.

E. *Procedure for increasing the mud weight more than enough to control the 200 psi and circulating at the casing seat:* The primary problem with increasing the mud weight more than that required to control the 200 psi is the possible U-tube effect which would impose additional pressure on the formation if circulation is stopped. Assume in this case the mud weight is increased enough to impose an additional hydrostatic pressure of 300 psi at 13,000 feet.

$$\text{The required mud weight increase} = \frac{300}{(13,000)(0.052)} = 0.44 \text{ ppg}$$

The procedure would be as follows:

1. Observe the annulus pressure while increasing mud weight in the suction tank to 15.44 ppg. If there is no increase in annulus pressure, the gas bubble is not rising. In this case commence pumping. If the gas bubble is rising determine the rate of rise as shown. Determine the time required to increase mud weight and displace weighted mud. If the gas bubble is rising at the rate which will make it impossible to control the well with the mud weight being used, consider (1) a further increase in mud weight or (2) letting the gas bubble

rise to the casing seat and above before initiating any type of displacement.

2. If the gas bubble is not rising, the annulus pressure should remain constant at a constant choke size until the mud reaches the bit. Pumping pressure during this time will reduce by just under 300 psi. Time required for the weighted mud to reach the bit is 61 minutes. $(182 \div 3)$.

3. When the weighted mud enters the annulus the casing pressure will begin dropping. If there is no rise of the gas bubble, the annulus pressure should be reduced to zero after $2\frac{1}{2}$ hours of continued displacement. This would be when the weighted mud occupies two-thirds of the annulus.

F. *Procedure for stripping the drill string back to bottom:* First—assume no gas migration:

1. Start stripping pipe into the hole.

2. Release mud volume equal to pipe volume. This will be about 0.02 barrels of mud per foot of hole. Thus 100 feet of pipe in the hole would require the release of 2.0 barrels of mud.

3. Continue stripping pipe and releasing mud equal to the total pipe volume. Based on surface measurements the formation fluid will be encountered 323 feet from bottom or after stripping 1677 feet of pipe into the hole. At this point if no fluid migration has occurred about 33.5 barrels of mud will have been released and the annulus pressure is still about 200 psi.

4. When the pipe enters the formation fluid, the length of the annulus occupied by the formation fluid should be increased an amount equal to the total pipe displacement. For Example 7, 6.5 barrels of mud would need to be released to get the remaining 323 feet of pipe into the hole. During this time the annulus pressure would increase to about 320 psi and the drill pipe pressure should reduce to zero.

Second—assume gas migration:

1. Start stripping pipe into the hole.

2. Release a mud volume equal to the pipe volume. Watch annulus pressure, if gas is rising the annulus pressure will start to increase even though mud is released. Permit the annulus pressure to increase by about 50 psi to 250 psi.

3. Keep an accurate record of mud released. When 2 barrels of mud in excess of pipe volume has been released, this represents an expansion of 32.2 feet for the gas $\left[\dfrac{2\,\text{bbl ft}}{.062\,\text{bbl.}}\right]$. This removal of 32.2 feet of mud has reduced the hydrostatic pressure about 25.0 psi [32.2 × 0.78]. Now let the annulus pressure increase another 25 psi and continue this procedure until the top of the expanded gas column is reached.

4. Assume 4 barrels of mud have been released at the surface in excess of pipe volume when the top of the gas column is reached. The length of the gas column is now 387 feet and the annulus pressure is 300 psi, which gives a total back pressure on the formation which is 50.0 psi in excess of formation pressure $[100 - (2)(25.0)]$.

5. The pipe volume in the gas will expand the gas another 184 feet $\left[\dfrac{387 \times .02}{.042}\right]$. This represents a reduction in hydrostatic pressure of 144 psi $[184 \times .78]$. Thus the annulus pressure should be allowed to increase to 450 psi before bleeding off any more mud in excess of pipe volume.

6. At this point the shut-in pressure gradient is still only 0.815 psi/ft $\left[\dfrac{10,140+450}{13,000}\right]$ which is substantially below the fracture gradient of 0.91 psi/ft. Actually to be sure no additional gas entered the formation, another 50 psi increase in surface pressure would probably be desirable for this specific well.

Lost Circulation Associated with Well Kick

Warning signals that indicate lost circulation when a well has kicked are listed as follows:

1. Drill pipe pressure equal to or more than the annulus pressure.
2. No increase or a decrease in pumping pressure with increases in circulation rate.
3. An increase in annulus pressure when mud is released.

Precise surface control is not possible if circulation is lost and the operator's best hope is generally to spot a barite plug to seal-off the high pressure zone and leave his drill string in place.

Exceptions may be developed for all the suggested procedures for controlling wells when the noted problems occur. It will never be possible to anticipate all the problems and to develop a plan to handle all the contingencies. However, the anticipation of problems is important and the primary course of action should be planned before the problems occur. Operating decisions in precise detail will always require qualified people on location. The possibility always exists that preplanned action cannot be exercised because of limitations that were not previously considered.

NOMENCLATURE

A = defined in text, ~~dimensionless~~ psi

A_1 = defined in text, ~~dimensionless~~ psi

D = well depth, feet

D_e^* = height occupied by weighted mud in annulus, feet

D_i^* = height occupied in the annulus by unweighted mud inside drill string when displacing formation fluid with weighted mud, feet

h = height of gas column at any point, feet

h_b = height of gas column at bottom of hole, feet

n = pound moles of gas

p = pressure at any point, psi

P_a = pressure on annulus gage at the surface, psi

P_b = pore pressure, psi

P_d = shut-in drill pipe pressure, psi

P_{ma} = pressure exerted by mud in the annulus, psi

P_{ma_1} = pressure exerted by mud below gas in annulus, psi

P_{md} = pressure exerted by mud inside drill string, psi

P_f = pressure exerted by mass of formation fluid, psi

P_1 and P_2 = pressure on gas at specific points, psi

P_{mg} = pressure at midpoint of gas column, psi

R = units conversion constant

S = specific gravity of gas, dimensionless

T = temperature, degrees Rankine

T_b = temperature at bottom of hole, degrees Rankine

T_1 and T_2 = temperature of gas at specific points, degrees Rankine

V = volume of gas in cubic feet

V_b = volume of gas at bottom of hole, cubic feet

V_1 and V_2 = volume of gas specific points, cubic feet

X = distance from surface to the point of interest, feet

Z = gas compressibility factor, dimensionless

Z_b = gas compressibility factor at bottom of hole, dimensionless

Z_1 and Z_2 = gas compressibility at specific points, dimensionless

ρ_f = pressure gradient of formation fluid, psi/ft

ρ_m = pressure gradient of unweighted mud, psi/ft

ρ_{m_1} = pressure gradient of weighted mud, psi/ft

The Problem of Deviation
and Dog Legging
in Rotary Boreholes

ROBERT D. GRACE
Consulting Engineer
Amarillo, Texas

DEVIATION in drilling operations is not a new problem. The diamond core drill was invented in 1865 and widely used as a cable tool drill in mining operations. The first evidence of concern about hole deviation was the invention by Nolten in Germany in 1874 of the use of hydrofluoric acid to etch and predict hole deviation.

Later a South African miner named MacGeorge invented the clinostat to predict both deviation and direction. The clinostat consisting of a magnetic needle and a plumb immersed in gelatin was lowered into the hole and the gelatin was allowed to set. The instrument was then brought to the surface and deviation and direction were read directly. At a meeting of mining engineers in London in 1885 MacGeorge presented data illustrating deviations of 75 feet in 100 foot mine shafts.

The Petroleum Industry did not become aware of the problem until the Seminole, Oklahoma, boom of the middle 1920's. Town lot spacing was the primary factor contributing to the experiences of the industry. There are actual recorded incidents of offset wells drilling into each other, drilling wells drilling into producing wells, two rigs drilling the same hole, and wells in the geometric center of the structure coming in low or missing the field completely.

It was common drilling practice at that time to use only large drill pipe with no drill collars and all available weight since weight indicators were not available. Engineers and the industry in general made a concentrated effort to solve the crooked hole problem. As a result, most of the practices commonly used today in an effort to correct and control deviation were conceived, experimented with and adopted in the 1920's – 50 years ago.

The most effective practices adopted and still used today were the use of drill collars for weight and rigidity, the use of stabilizers at various points in the string to control deviation and provide rigidity, and the practice of fanning bottom to reduce hole angle. The first two have made the industry millions of dollars; the practice of fanning bottom has cost the industry millions of dollars.

For whatever reason early researchers were successful in their efforts. Wells surveyed in the greater Seminole, Oklahoma area with and without straight hole practices produced the following results:

	Without Straight-hole Practices	With Straight-hole Practices
Number of Wells	216	58
Total Feet Surveyed	910,232	233,341
Average Depth	4,214	4,023
Average Angle	13° 19'	5° 59'
Maximum Angle	46°-0'	19°-0'

These data would indicate that the engineers of the 1920's didn't solve the problem of deviation; but, the practices introduced are fundamental to the practices used today.

Very little research was performed in the area of deviation until Arthur Lubinski performed his work in the early 1950's, and real interest resulted from the advent and popularity of directional drilling. In the last few years considerable field experience has been reported which has contributed significantly to the total knowledge of this particular aspect of oil well drilling technology.

The economic incentive to develop techniques to cope with deviation is staggering. One contractor estimated that the additional surveys alone cost $0.50 per foot of hole drilled. The AAODC once estimated that a more realistic approach to deviation could save the industry $40 million per year, and these estimates did not extend to directional drilling. Field experience has proven that drilling time can be reduced by as much as 30–50 per cent of current practices with a realistic approach to deviation.

The sad thing about this deviation problem is that the available technology is not being applied. In many areas of the world drilling contracts are written in the same manner as they were 15 years or 50 years ago. That is, hole deviation is limited to one degree per thousand feet. This is truly an area where most operators and drilling people merely do what has been done for the last 15 years thereby impeding progress by not doing anything. The potential for advanced thinking in this area is unlimited. This problem has cost the drilling industry too much money for too many years.

With drilling costs what they are today, a lackadaisical attitude toward any phase of Drilling Technology, and particularly one so costly, should not be tolerated. In his early work Arthur Lubinski suggested it, recently in his series in *World Oil* Moak Rollins said it, and in this book I'm going to shout it around the world. Our industry can no longer afford to pay the prices that have been paid in the past for deviated holes. We can no longer afford to take twice as long to finish our work simply because we're afraid the hole might get crooked.

In this discussion we will analyze all available technology in deviation and dog-legging. Field experience will be evaluated in the light of available technology, and we will ultimately determine the most sophisticated approach to the problem in view of the total technology and available field experience.

THEORIES OF CAUSES OF DEVIATED HOLES

The anisotropic formation theory is widely accepted. Past theoretical studies have assumed that the bit drills in the direction of the resultant force on the bit in uniform or isotropic formations. This implies that the bit does not display a preferential direction of drilling. (Figure 13–1A) Stratified or anisotropic formations are assumed to possess different drillability parallel and normal to the bedding plane with the result that the bit does not drill in the direction of the resultant force.

Each formation is characterized by its anisotropic index and dip angle. The anisotropic index does not depend upon specific rock properties but is an empirical constant determined from drilling measurements. This theory has been applied to the computation of the equilibrium hole inclination angle for straight inclined holes.

The formation drillability theory seeks to explain hole angle change in terms of the difference in drilling rates in hard and soft dipping formations. (Figure 13–1B) Presumably angle in the hole changes because the bit drills slower in that portion of the hole in the hard formation. Inherent in this theory is the underlying assumption that the bit weight is distributed uniformly over the bottom of the hole. It predicts updip deviation when drilling into softer rock and downdip into harder rock.

The miniature whipstock theory is based on drilling experiments made by Hughes Tool Company (Figure 13–1C) in which an artificial formation composed of glass plates has been drilled with the hole inclined to the laminations. In these tests the plates fractured perpendicular to the bedding plane, creating miniature whipstocks. If such whipstocks are created when laminated rock fractures perpendicular to bedding planes, this could cause updip drilling. This theory offers a possible qualitative explanation to hole deviation in slightly dipping formations; however, it does not explain the downdip drilling which occurs in steeply dipping formations.

The drill collar moment theory suggests that (Figure 13–1D) when a bit drills from a soft to a hard formation, the weight on bit is not distributed evenly along the bottom of the hole. Since more of the weight on bit is taken by the hard formation, a moment is generated at the bit. Such a moment changes the pendulum length to the point of tangency as well as the side force at the bit. The variation of side force is not the same when drilling from soft to hard formations as when drilling from hard to soft and, therefore, can effect a change of hole inclination.

Raymond Knapp suggests that deviation results in dipping formations which vary in hardness and is directly related to the inability of the bit to drill a full gauge hole. (Figure 13–1E) All bits ream a small portion of the hole to gauge with the heal rows. Mr. Knapp contends that in going from a

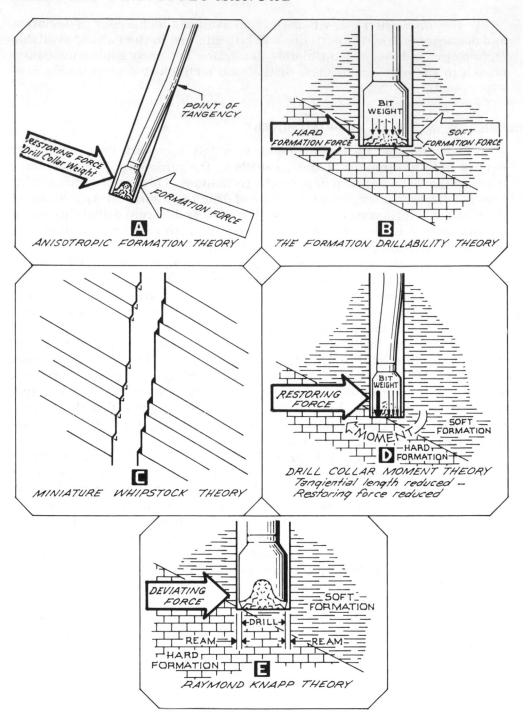

FIG. 13–1. *Theories of deviated holes*

soft to a hard formation the bit would be unable to ream the hard formation to gauge as fast as it could drill the soft formation; therefore, the bit would be deflected toward the softer formation. Random deviation would result.

Experience has shown that deviation occurs more often in laminated beds than thick, homogenous deposits. Deviation is almost always associated with areas of steeply dipping formations. Faulting or perhaps the stresses associated with faulting influence deviation.

In the final analysis there is no one satisfactory explanation for deviation. It appears to be related to geology. Deviation is never greater than bed dip. All theory and practice indicates that the maximum deviation is perpendicular to or parallel to the formation dip. In fact, Lubinski's model which is the most widely accepted suggests that total deviation will always be less than formation dip. This is an interesting and significant point. Why be afraid of deviation if bed dips are not severe?

CATEGORIZING CROOKED HOLES

If we are to fight a problem, we must decide when the problem is a problem and when the problem is only a potential problem. When we talk of hole deviation, we cover a multitude of evils; therefore, it becomes necessary to dissect the agglomeration into its component parts and evaluate the problems associated with each and the techniques normally associated with coping with those problems. First then, let's consider holes that deviate from vertical uniformly in one plane and then "Dog-Legs" or changes in direction will be considered.

Uniformly Deviating in One Plane

A rotary bore hole that is not vertical is deviated. Probably 95 per cent of the instruments commonly used in the industry today measure only deviation from vertical with no regard for direction or changes in direction. The assumption is made, then, that the borehole is planar as illustrated in Figure 13–2A.

With this type of deviation many reasons are given which brand this type hole as undesirable and necessitate the expenditure of multitudes of money to eliminate the problem. The more common anticipated problems are:

1. Inadequate and misleading subsurface information
2. Insignificance of surface location with respect to well spacing
3. Inadequate drainage of production sands
4. Crossing lease lines
5. Excessive production problems
6. Excessive drilling problems

With respect to the anticipated excessive drilling problems, directional drilling experience has proven that drilling problems are virtually non-existent in bore hole inclined in excess of 50°. As for production problems, many years experience at Long Beach, Huntington Beach, and other coastal fields has produced no severe production problems in holes deviated in

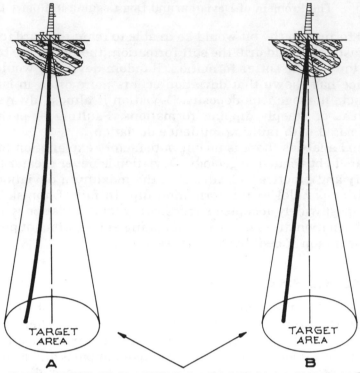

ROTARY BORE HOLES UNIFORMLY DEVIATING IN ONE PLANE

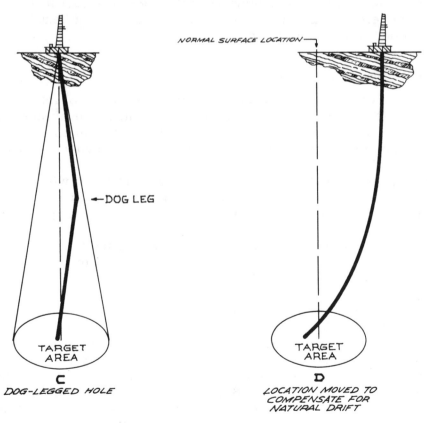

NORMAL SURFACE LOCATION

← DOG LEG

C
DOG-LEGGED HOLE

D
LOCATION MOVED TO
COMPENSATE FOR
NATURAL DRIFT

FIG. 13-2. *Reaching target area—deviated holes*

excess of 45°. The reason that very few problems result from deviation alone is best illustrated by Figure 13–3.

As illustrated, relatively insignificant forces result from deviation angle only. For comparative purposes a tool joint force less than 2,000 pounds is not considered detrimental in normal operations. Therefore, it must be concluded from the available experience and technology that deviation alone causes no severe drilling or production problems.

Lease line, spacing, subsurface information, or drainage cannot be considered a major problem since offset data has for years been used to predict natural drift. As illustrated in Figure 13–2D the surface location is simply moved to compensate for the anticipated drift and the hole is drilled in a normal fashion.

The only conceivable conclusion, then, from this discussion is that there is no good reason why deviation alone should be considered a major

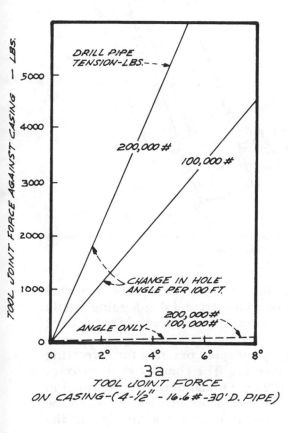

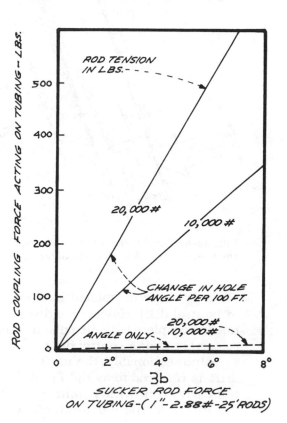

COMPARISON OF CASING AND TUBING WEAR FORCES

R.S. Hoch, World Oil 1962

FIG. 13–3. *Comparison of casing and tubing wear forces*

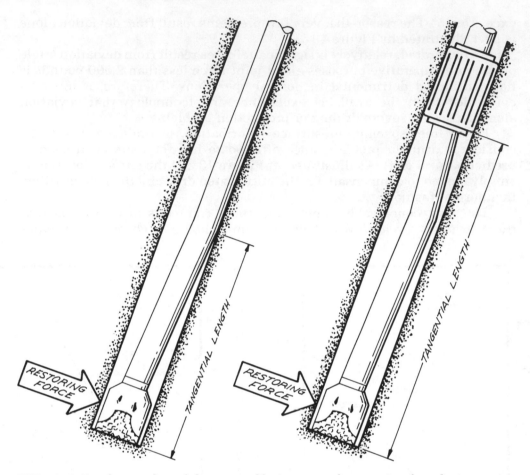

FIG. 13–4. *Fundamental pendulum assembly increases the restoring force by increasing the tangent length with a stabilizer in the proper position*

obstacle. Only in isolated instances should it become necessary to correct or control drift. However, a discussion of deviation and dog-legging would not be complete without a discussion of the techniques presently being used to correct and/or control drift.

Unquestionably, the most universally accepted practice for correcting drift is referred to as the Pendulum assembly. The concept evolved from the early work with stabilizers in the 1920's. Fundamentally illustrated in Figure 13–4, the Pendulum Assembly utilizes a stabilizer to increase the restoring force by increasing the drill collar length contributing to the restoring force.

Arthur Lubinski applied technology to the concept and explored the problem in his paper titled "Factors Affecting the Angle of Inclination and Dog Legging in Rotary Bore Holes" published in 1953. The purpose of the paper was to mathematically describe the effects of bit weight, formation characteristics, collar to hole clearance, and stabilizer placement on deviation. For mathematical simplicity it was assumed that:

1. The drill string always lays to the low side of the hole.
2. Dynamic forces caused by rotation are insignificant.
3. The drill string never helically buckles.
4. In an anisotropic formation as previously described, the bit travels in the direction of the resultant force including a formation force.
5. An equilibrium angle exists for a given set of drilling conditions and that equilibrium is necessary for approximately 100 feet if the data is to be properly applied.

These assumptions are basically sound and contribute simplicity to the mathematical model. However, the drill string does not always lay to the low side of the hole, the rotational forces may be significant and the drill string does helically buckle on occasion.

The conclusions reached by Lubinski were very significant to the industry and constituted a major contribution to drilling technology. Unfortunately, our industry does not now – 20 years later – utilize the technology developed. The most important, overlooked conclusion concerns dog-legs or changes in deviation which are to be discussed thoroughly later.

It was concluded that dog-legs normally result from a change in conditions such as a change in bit weight or a change in formation property. Dog-legs resulting from sudden changes in bit weight and sudden changes in formation dip were studied, and it was concluded that drilling with more bit weight and allowing the inclination to build did not result in sharper dog-legs!

The Pendulum Assembly resulted from the conclusion that the only force available to restore a deviated hole to vertical is the weight of the drill collars between the point of tangency and the bit (refer to Figure 13–4). Further the restoring force may be increased by increasing the length between the bit and the point of tangency by the proper use of stabilizers. The mathematical model described was used to determine the most effective stabilizer position under defined conditions.

The stabilizer position data were presented in many forms with the most effective being a set of tables available from Drilco, Readers Service Department, Box 3135, Midland, Texas. Complete instructions accompany these tables; however, a brief discussion of these tables is in order in this text. Let's assume that it has been determined that the formation dip is 20°, a 7⅞" hole is being drilled with 5½" drill collars, bit weight is 13,000 pounds, and deviation from vertical is 4°.

The tabular data developed by Lubinski, et. al. relating to their conditions appears as illustrated in Figure 13–5A. In the tables the hole class indicates the severity of the condition – i.e. "A" is most severe and "U" is least severe. Using the assumed conditions, it is then determined from Figure 13–5A that a class "Q" condition exists. This means that under the drilling conditions established, equilibrium has been established, the restoring force is equal to the deviating force, and the effects of increasing the restoring force or changing any of the drilling variables can be evaluated.

For the purpose of familiarizing ourselves with the material, let's look at some examples. Looking across the "Q" condition line, if a stabilizer is

A

7 7/8 INCH HOLE SIZE – 20 DEGREE FORMATION DIP

NOTE—Stab.=Stabilizer   DRILL COLLAR SIZES

HOLE ANGLE AND CLASS	5½" COLLARS BIT WEIGHT			6" COLLARS BIT WEIGHT			6½" COLLARS BIT WEIGHT			7" COLLARS BIT WEIGHT		
	No Stab.	Stab. @ Height		No Stab.	Stab. @ Height		No Stab.	Stab. @ Height		No Stab.	Stab. @ Height	
4° O	8217	10590	65-72	10038	12910	65-72	11899	15300	64-71	13514	17090	59-66
P	10429	13680	63-70	12912	16830	63-70	15548	20200	62-69	17987	22900	58-65
Q	13239	18110	60-67	16593	22200	61-68	20242	26800	60-67	23747	30600	57-63
R	16626	23540	56-62	21306	29540	58-64	26706	36400	58-64	32401	42800	55-61
S	20497	32600	48-53	27040	40200	50-56	35170	50200	53-59	44728	62200	52-58

B

SAME CONDITIONS AS **A** TABLE ~ Allowed to drift to 10 degrees

10° Q	29110	44800	40-45	38000	54800	44-48	48665			60000		

C

7 7/8 INCH HOLE SIZE – 45 DEGREE FORMATION DIP

4° R	11513	15280	62-69	14312	18900	62-69	17307	22500	61-68	20111	25780	58-64
S	15445	21550	58-65	19635	27000	59-66	24365	32800	58-65	29175	38300	56-62
10° R	21626	29800	48-53									
S	27776	41600	42-47									
15° S	35991	54600	38-42									

D

7 7/8 INCH HOLE SIZE – 20 DEGREE FORMATION DIP

15° N	32104	46000	40-44									

EXERT FROM DRILCO TABLES FOR EFFECTS OF COLLAR SIZE, BIT WEIGHT, AND COLLAR STABILIZERS ON DEVIATION

FIG. 13–5. *Stabilizer position data*

placed 60–67 feet above the bit, the bit weight can be increased to 18,110 pounds or 36.8 per cent. Since bit weight is directly proportional to penetration rate, penetration rate would be improved an impressive 36.8 per cent. Suppose we were brave and would run 7″ drill collars with a stabilizer 60 feet above the bit.

Looking under 7″ collars and the "Q" condition it is determined that bit weight could be increased to 30,600 pounds or 131 per cent which would improve penetration rate 131 per cent! Let's assume that we are near the bottom of the hole and we could live with a 10° deviation. Looking at Figure 13–5B for the "Q" condition and the 5½" collars with a stabilizer 40–45 feet from bottom, the bit weight could be increased to 44,800 pounds or 238 per cent which represents an increase of 238 per cent in penetration rate! Obviously, intelligently applied, this information can result in improved performance through faster penetration rates with a minimum risk.

What is the risk? Let's look at what would happen if the formation dip was 45° instead of the 20° assumed. Further, let's assume that we had to put the stabilizer at 40 feet on the 5½" collars, run the 44,800 pounds and let the hole drift to 10°. From Figure 13–5C, it is determined that the formation was probably as class "S". Under the conditions assumed Figure 13–5C indicates that the equilibrium hole angle would be slightly over 10°. Depending upon the circumstances, a 125 per cent error in formation dip may or may not constitute a risk at all. A radical change in formation classification would be critical.

Let's assume that the bit weight has been increased to 44,800 pounds with the stabilizer at 40–45 feet on the 5½ inch drill collars, and the formation suddenly changes from a type "Q" to a type "N", the deviation would drift to 15° from vertical before equilibrium was achieved.

Studying these charts reveals that the obviously best solution to the problem of drift is to pack the hole with the largest acceptable collars and stabilizers from the very start. This realization led to the development of the square drill collar which is, unquestionably, the best tool available for controlling direction or drift.

Tools other than conventional directional tools available and used in the field to correct or control drift are the Woodpecker Drill Collar, the Hammer Tool, Two Cone bit, and the DM bit.

The Woodpecker drill collar is an ordinary 30 foot drill collar with 1 inch holes drilled into it at 1 foot intervals. This causes the collar to rotate about a center other than its geometric center. Therefore in an inclined hole the restoring force would be increased by the centrifugal force of the collar mass rotating about its center of mass. However, the deviating force would be increased by the centrifugal force minus the collar weight. Figure 13–6 illustrates the principles involved. At equilibrium the Restoring Force is by definition equal to the Deviating Force. If the Woodpecker Drill Collar is introduced, both the Restoring Forces and the Deviating Forces are increased by some Mass multiplied by its acceleration (MA). The net effect is that the Woodpecker Drill Collar has no effect on the equilibrium conditions.

The Two Cone bit is widely used in attempting to reduce hole angle. The thought is that the bit weight per chisel is reduced (if each tooth can be called a chisel), permitting faster penetration rates at lower bit weights. In other words penetration rate can be maintained while lowering bit weight thereby reducing hole angle at no sacrifice of drilling time. Field data has never substantiated this position. In addition sufficient data has never been printed to permit the systematic application of the principle. It is my conclusion that the tool as it is now being used is worthless.

THE WOODPECKER DRILL COLLAR

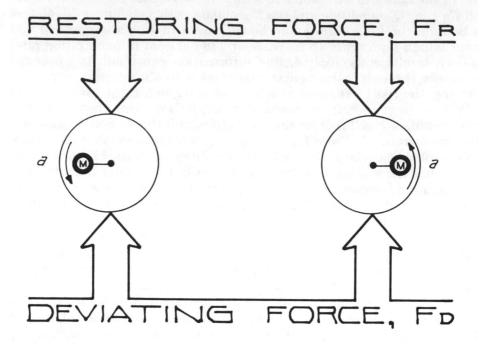

With the Woodpecker Drill Collar:
   RESTORING FORCE = DRILL COLLAR WEIGHT + Ma
   DEVIATING FORCE = DEVIATING FORCE + Ma

At equilibrium without the Woodpecker Drill Collar

$$F_R = F_D$$

Introducing the Woodpecker Drill Collar
   $F_R = F_R + Ma$
   $F_D = F_D + Ma$
Or the equilibrium is altered as follows
   $F_R + Ma = F_D + Ma$
   or $F_R = F_D$
   or the equilibrium condition is unaltered
   by the Woodpecker Drill Collar.

FIG. 13–6. *The woodpecker drill collar*

When air drilling was popular, the hammer tool was promoted as an aid in preventing deviation. The hammer tool adds the principle of cable tool drilling to rotary drilling. The thought was that penetration rates could be maintained at lower bit weights. The researchers of the 1920's found that cable tool holes were as crooked as rotary holes. Further, the position concerning the hammer tool has never been substantiated by field data. In

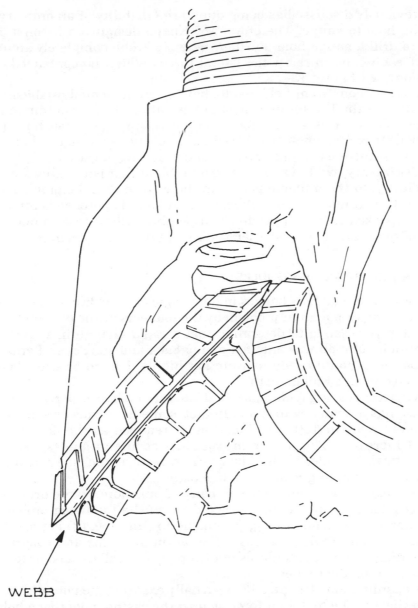

FIG. 13-7. *Webb bit*

fact, field data examined by this author found no improvement in crooked hole conditions could be attributed to the use of the hammer tool. In my opinion the hammer tool or percussion drill is worthless as a tool to control or correct deviation and that when sufficient collars are available for optimum bit weights, it is worthless as a tool to increase penetration rate.

The DM bit was designed by Mr. Raymond Knapp of Ardmore, Oklahoma. Mr. Knapp has many years experience in the oil field and is as fine a person as I have ever had the pleasure of meeting. His crooked hole theory was previously discussed as being due to the inability of an ordinary bit to drill the hole to gauge. Therefore, Mr. Knapp designed a bit on a 39° pin angle to drill a gauge hole. Further, he put a webb completely around the heel of each cone to resist the lateral forces which he contended caused deviation (See Figure 13–7).

Again no consistent field results have been presented establishing the usefulness of the DM bit in controlling deviation. To the contrary informal investigation by this author found no correlation between bit type and deviation. It is my conclusion based on extensive experience that the DM bit is not useful as a tool in combating or controlling deviation.

In summary, we have examined a problem that isn't really a problem. The solution to the problem is to plan the well so that the problem can be ignored. If drift must be controlled, the only sound approach is through the use of a packed hole or a pendulum assembly. All other techniques are a waste of time and money. Now let's talk about the real problem, dog-legging.

DOG-LEGGING IN ROTARY BOREHOLES

Dog-Legs or sudden changes in hole angle or hole direction were recognized as a major potential problem by the pioneers of the drilling business. When it was possible to determine that a rapid change in angle had occurred their solution was automatic-plug back and start over. Perhaps it is well that detection procedures were not highly defined or else a hole may never have reached total depth.

Modern surveying techniques indicate that no hole is perfectly vertical. Any hole has a tendency to spiral. In fact, some holes surveyed made three complete circles in 100 feet. Spiraling is reduced as the deviation from vertical increases. The maximum spiraling occurs at angles less than 3° from vertical. At angles greater than 5° from vertical, the hole may move in a wide arc, but spiraling is almost non-existent.

Dog-Legs are a major factor in many of our more severe drilling problems. Dog-legging should be suspected when the following problems are encountered: (1) unable to log, (2) unable to run pipe, (3) key seating, (4) excessive casing wear, (5) excessive wear on drill pipe and collars (6) excessive drag, (7) fatigue failures of drill pipe and collars, and/or (8) excessive wear on production equipment.

Re-examination of Figure 13–3 partially explains the potential hazards. With respect to the tool joint force against the casing, note that a hole with 8° deviation from vertical, the tool joint wear force is essentially negligible. However, under 100,000 pounds of tension a dog-leg of 8° in 100 feet produces a wear force exceeding 4,000 pounds. It is obvious that the dog-legging is the most severe problem, and it is primarily a drilling problem.

The major problem facing the industry was to define a severe dog-leg within the industry's ability to survey dog-legs. Author Lubinski made the first efforts to define a severe dog-leg in his paper entitled "Maximum Permissible Dog-Legs in Rotary Boreholes" published in 1961. Lubinski recognized that severe dog-legs created major drilling problems and pro-

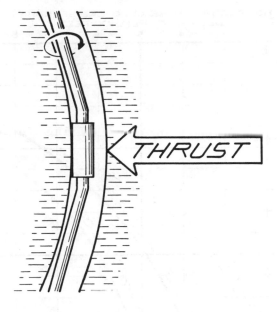

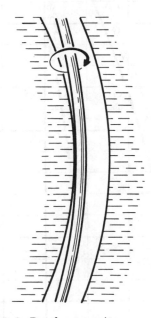

FIG. 13–8. *Dog leg severity*

posed that a dog-leg was too severe if any one of the following conditions existed:

(Refer to Figure 13–8)

1. The stress reversals when rotating in the dog-leg were sufficient to to fatigue the drill pipe.

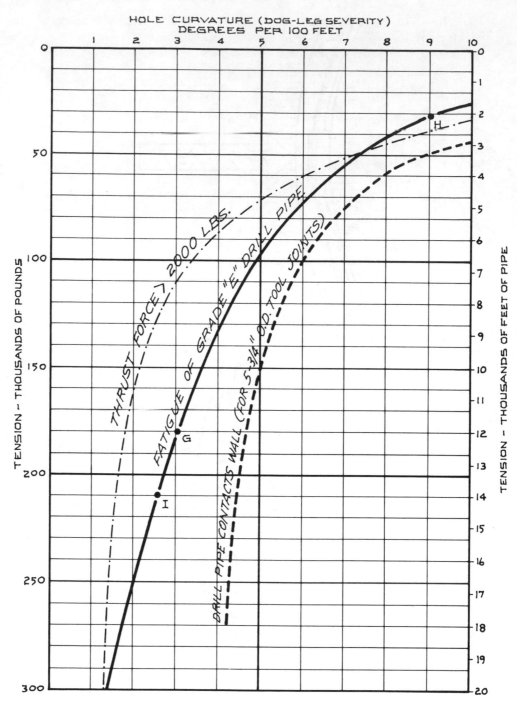

FIG. 13–9. *Gradual and long dog-leg hole curvature vs. tension for 4½ inch grade E drill pipe*

2. The thrust force on the drill pipe tool joint in the dog-leg was sufficient to cause the tool joint to dig into the formation and cause a key seat or produce casing wear.

3. The stress reversals when rotating in the dog-leg were sufficient to fatigue the drill collars.

Lubinski concluded that these conditions should be avoided and each section of the hole should be evaluated in view of the limiting conditions in order to determine the maximum permissible dog-leg at any given depth.

First, let's consider drill pipe fatigue. Figure 13–9 illustrates the maximum permissible dog-legs to avoid drill pipe fatigue as a function of tension on the drill pipe or depth for $4\frac{1}{2}$ inch Grade E drill pipe. From Figure 13–9, for example, with 300,000 pounds of tension on the drill string which is comparable to approximately 20,000 feet of pipe, a dog-leg in excess of $1\frac{1}{2}$ degrees in 100 feet at/or near the surface.

At a point 3,000 feet from the bottom of the 20,000 foot hole, the tension would be only 45,000 pounds, and a $7\frac{1}{2}$ degree change in 100 feet of hole would be required to produce drill pipe fatigue. Thus potential drill pipe fatigue constitutes a limiting condition in determining a maximum permissible dog-leg, and the tension or load on the drill pipe is the major factor affecting fatigue. Obviously, a much larger change in angle can be tolerated at total depth, whereas only very small changes can be tolerated at the surface in very deep holes.

Another consideration is the force caused by the drill string at a dog-leg in the hole or casing. In order to determine the maximum permissible thrust force, Lubinski assumed that a dog-leg of $1\frac{1}{2}$ degrees in 100 feet never caused any trouble. The deepest holes of that time were 16,000 to 18,000 feet in depth. At 17,000 feet, the drill pipe load in a $1\frac{1}{2}$ degree per 100 feet dog-leg results in a 2,000 pound thrust force on the formation or casing in the dog-leg. Based on this experience then, it was assumed that a 2,000 pound or less thrust force would never create a drilling problem. The dashed curve in Figure 13–9 represents the maximum permissible dog-leg to prevent excessive thrust forces as a function of tension on the drill string. Obviously, casing wear and key seats associated with excessive thrust forces are more critical near the surface in deep wells. Larger dog-legs can be tolerated nearer total depth without danger to hole and casing.

The final limiting conditions, according to Lubinski, is drill collar fatigue. Lubinski studied various conditions for different collar sizes, and calculations were made of the abrupt dog-leg angle for which the connections would be subjected to a bending moment sufficient to produce fatigue failure. It was concluded that the critical angle is a function of collar to hole clearance, the amount of tension or compression to which the collars are subjected in the dog-leg, hole inclination, and whether the inclination is increasing or decreasing.

In packed holes the critical angle is independent of all factors except clearance. For example, in Figure 13–10A, $6\frac{1}{4}$ inch collars in a 7 inch hole can tolerate a dog-leg of 1.2 degrees in 30 feet or 3.6 degrees in 100 feet with fatigue failure. On the other extreme the behavior of $6\frac{1}{4}$ inch collars in a $12\frac{1}{4}$ inch hole is erratic as illustrated in Figure 13–10F. The more

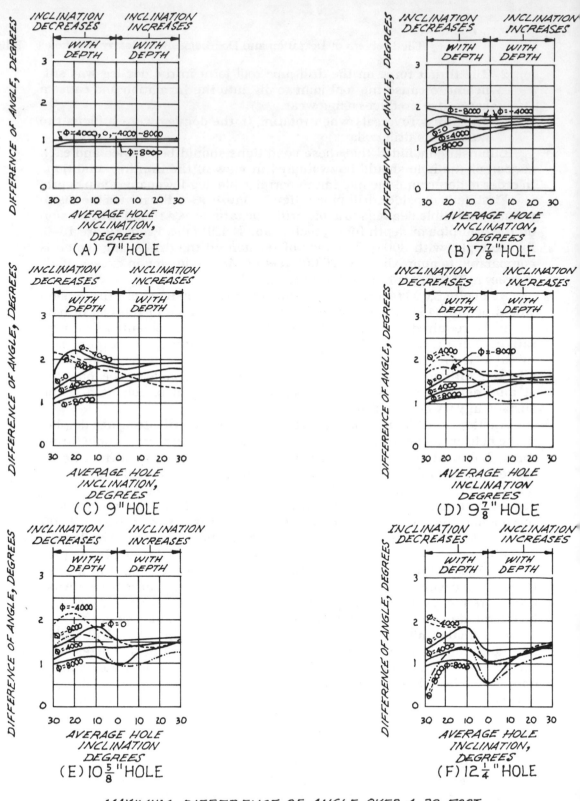

MAXIMUM DIFFERENCE OF ANGLE OVER A 30 FOOT
INTERVAL TO AVOID DRILL COLLAR CONNECTION
FAILURES — 6¼ INCH DRILL COLLARS.
POSITIVE φ = TENSION AND NEGATIVE φ = COMPRESSION

From Lubinski

FIG. 13–10. *Average hole inclination*

common combination of 6¼ inch collars in a 7⅞ inch hole (Figure 13–10B) will tolerate a more severe dog-leg without failure than will the packed hole combination under normal operating conditions and is essentially a function of clearance. Conversely it is obvious from Lubinski's work that reduced collar to wall clearance resists dog-legging.

Lubinski concluded that each section of the hole, every 1,000 feet, for example, should be evaluated with respect to these limiting conditions of pipe and collar fatigue, key seating, and casing wear. The condition permitting the smallest dog-leg would dictate the operating limitations imposed on that section of the hole. The normal result is that potential key seating and casing wear will dictate the maximum permissible dog-leg angle tolerable to moderate depths, and will be less then 1½ degrees per 100 feet only in ultra-deep holes.

Potential drill pipe fatigue will normally dictate the maximum permissible dog-leg in the middle of the string and to the top of the collars. Potential drill collar failure will dictate the maximum permissible dog-leg angle both at the bottom of the hole and from surface to total depth as well. Obviously, the collars must drill and pass through all of the hole. For example, for 6¼ inch collars in a 7⅞ inch hole, the maximum permissible change in angle is approximately 3½ degrees in 100 feet; therefore, nowhere in the hole should a greater change be tolerated.

In an effort to apply experience to Lubinski's work, the API Mid-Continent Study Committee on Straight-Hole Drilling published in 1963 a study of 1,094 dog-legs in the Gulf Coast, Mid-Continent, and West Coast areas. Elaborate efforts were made to associate problems with the dog-leg and its severity. Dog-legs up to 25 degrees per 100 feet were reported.

The committee's study substantiated the minimum limitation of 1½ degrees per 100 feet. This minimum limit should be re-evaluated when drilling ultra-deep holes. Further, the study substantiated the limitations with regard to drill collar fatigue. Essentially, the committee suggested that the bottom of the hole equal to the length of the collar string should be limited by collar fatigue (i.e. 3½ degrees per 100 feet for 6¼ inch collars) at the bottom and limited to 1½ degrees per 100 feet over the remainder of the hole.

The experience evaluated indicated that no problems would be encountered under this schedule regardless of the ultimate total deviation from vertical. However, it should be noted that no dog-leg was reported below 11,000 feet and that such a program should be re-evaluated when drilling deep tests.

Two other significant conclusions were reached in these papers. First, Lubinski concluded that dog-leg severity was independent of the weight on the bit. That is, drilling with more weight on the bit would not result in sharper dog-legs; conversely, drilling with reduced bit weights or "fanning bottom" would not reduce dog-leg severity. In fact, "fanning bottom" is detrimental to the drill string since reduced bit weights result in increased string tension which increases the potential of casing wear, pipe fatigue, collar fatigue, and key seating as previously explained.

Secondly, the API Study Committee divided the dog-legs into "soft" and "hard" formations. The previous discussions applied to "soft" formations. Of the 140 "hard" formation dog-legs reported, only one resulted in a serious problem; therefore, it was concluded that fewer problems could be expected in hard formations as a result of changes in angle.

Since dog-leg severity is primarily a function of clearance, the only practical method available to reduce dog-leg severity is to reduce the collar to wall clearance or pack the hole and to increase stiffness by increasing hole size (last resort). Packed hole assemblies include all forms of reamers, integral blade stabilizers, spiral collars, and rotating stabilizers at almost any spacing combination. However, the best tool available today is the square drill collar with corner to corner measuring 1/16th inch less than hole size.

Numerous papers have reported successful experiences all over the world including South America, Gulf Coast, Mid-Continent, Rocky Mountains, and Canada, and under all conceivable drilling conditions. In all the literature reviewed no dog-leg severity greater than $3\frac{1}{2}$ degrees per 100 feet was reported while using a square drill collar regardless of drilling conditions.

Other practices designed to merely cope with severe dog-legs are as follows:

1. Increase frequency of drill collar inspection.
2. Use non hard-banded drill pipe through the dog-leg to avoid excessive casing wear.
3. Reduce rotary speed while drilling through the dog-leg to reduce the number of stress reversals.
4. Minimize off-bottom rotation to reduce unnecessary stress reversals with maximum tensile stress.
5. Use packed holes to reduce dog-leg severity.
6. Keep the kick-off point in a directional well as deep as practical.
7. Use heavier casing through working dog-legs.
8. String reamers will often reduce dog-leg severity and prevent key seats.

In summary, dog-leg severity is a serious drilling problem. Dog-leg severity is a function of collar clearance. The best tool available to control dog-leg severity is the square drill collar. In reports from around the globe covering over one million feet of hole, the square drill collar has been credited with improving hole conditions, reducing fishing jobs, improving penetration rates, improving bit runs, decreasing survey frequency, and decreasing dog-leg severity. Fanning bottom is of no benefit in controlling dog-leg severity. In fact, it is detrimental to the drill string.

A UNIQUE APPROACH

The purpose of this section is to describe this author's experience in deviation and dog-legging. The fundamental premise was that the reduced penetration rates and high costs associated with severe deviation and dog-legging would not be accepted; an acceptable solution would be found.

Research was difficult. Data was confusing. Correlating and evaluating field data was extremely difficult. A particular bottom hole assembly would appear successful on one job and a dismal failure on the next. Controlled drilling practices might appear successful and fail on an offset. The problem seemed to be that the variables were not being adequately isolated and evaluated.

The approach adopted, then, was to include all conceivable variables over a broad application and find some variable or combination of variables common to all successful ventures. The area evaluated was the Arkoma Basin of Southeastern Oklahoma. The drilling practices common to the area comprised the variables and were as follows:

1. *Subsurface Geology*—Detailed regional geology was prepared. In addition, specifics studied at each location included faulting, bed dips, formation tops, and any other anomolies.

2. *Bottom Hole Assembly*—Surely every conceivable bottom hole assembly was in use in the Arkoma Basin. Common techniques included stiff stabilizer assemblies, square drill collars, woodpecker drill collars, the hammer drill, and the pendulum assembly.

3. *Bit Type*—All types of conventional bits were common including insert bits. Common "straight hole" bits included the DM Bit, the Two Cone Bit, and various other webbed bits (T and U Teeth).

4. *Drilling Fluid*—The Arkoma Basin offered a unique opportunity to study the variation in drilling fluids from dry air or gas to various forms of mist drilling to aerated mud drilling to mud drilling. All phases were common to the area.

5. *Hole Size and Collar Size*—Not much variation was noted in the area. Hole size was almost always $7\frac{7}{8}$ inches in diameter, and contractors generally furnished 6 inch collars.

6. *Surveys and Survey Frequency*—Only single shot data providing deviation without regard for direction was available. Surveys were frequently run at 15 foot intervals; the interval between surveys never exceeded 90–120 feet.

7. *Rotary RPM*—The rotary speeds studied varied 60–120 RPM with speeds closer to 60 RPM being more common. No significant variation was noted.

8. *Penetration Rate*—Penetration rates varied from as little as ten feet per day to 1,500 feet per day. A wide variation was available.

9. *Bit Weight*—A wide variation was available for study. In some air holes it was questionable whether or not the bit was on bottom; in some mud holes bit weight was as high as 60,000 pounds on a $7\frac{7}{8}$ inch bit.

As discussed, the approach was to compile this data on as many wells as possible and evaluate the success at each location in controlling deviation and dog-legging.

This study resulted in a very startling conclusion—within the realm of the research and on the limited number of wells analyzed, as long as the subsurface geology was consistent, no variable nor group of variables reviewed had a consistent and correlatible effect on the ultimate deviation. (Since only single shot data was available, dog-leg severity was not rigorously evaluated.) Now, that's a conclusion which challenges some time-honored concepts relating to deviation and dog-legging and is not readily acceptable. So, let's examine the variables for consistency within the conclusion.

It was conceded that the variation in rotary RPM and clearance was not sufficient to provide significant data. The data indicated no benefit in ultimate deviation resulted from Bit Type or Bottom Hole Assembly. (Again, it must be emphasized that dog-leg severity was not rigorously evaluated due to insufficient data. The literature is filled with competent field data stressing the relationship between clearance and dog-leg severity and is not denied or ignored.) There was no correlation between the type of drilling fluid and the ultimate deviation.

The conclusion that really sticks in the throat is that no correlation was found between bit weight and ultimate deviation. "Anyone who has been in the oil field overnight knows that a reduction in bit weight will reduce hole angle," asserted a very capable toolpusher and this statement adequately describes the position of the industry. But, the conclusion in this area was inescapable.

A comparable offset to an air hole was drilled with mud. Bit weights on the air hole were from essentially zero to a high of 8,000 to 10,000 pounds; the bit weights on the mud hole would range from 20,000 to 60,000 pounds; and no significant variation in ultimate deviation was recorded. Even more dramatic was the comparison before and after an air hole was mudded up. Bit weight while air drilling would typically be 8,000 to 10,000 pounds; immediately after mudding up, the bit weight would be routinely increased to 30,000 pounds. The results—you guessed it—no significant change in ultimate deviation. It's still hard to swallow.

The most interesting variable to evaluate was survey frequency. As mentioned, surveys were routinely run at 15 to 30 foot intervals which meant that routinely over half of any day was consumed running surveys describing only total deviation. (Since it was decided previously that dog-legging was the problem and deviation offered no problem, this routine practice might be difficult to defend.) It is logical that more surveys mean more data which would enable drilling personnel to make better decisions and ultimately drill straighter holes. But, the holes with high survey frequency were no straighter than those with fewer surveys.

Two questions immediately follow: (1) What is accomplished by running surveys every 15 to 30 feet? Is it exercise—something done routinely from habit? (2) If some formula, variable or combination of variables, consistently influenced deviation, then shouldn't that condition logically correspond with either increased survey frequency resulting in more data producing better decisions, or reduced survey frequency resulting from straighter holes? The fact that survey frequency could in no way be correlated with deviation further suggested that even though the rig personnel knew the condition, nothing done significantly effected the ultimate deviation. It's bitter, isn't it?

In summary, it was concluded that the degree of deviation is a function only of subsurface geology and can be correlated and predicted and that no field practice common to the Arkoma Basin had a significant, consistent effect on the total deviation.

The obvious recommendations which followed were:
1. Predict deviation from control wells chosen from subsurface geology.
2. Spot the surface location so that the bottom hole location would be acceptable.
3. Run a square drill collar to control dog-leg severity.
4. Run optimum bit weights and rotary speeds.
5. Pick the best bit for drilling.
6. Run surveys every 500 feet as required by law.

Nine wells were drilled under this philosophy. All were development locations and anticipated total deviation was predicted from offset information. Bit weights were increased to 20,000 to 30,000 pounds (all available weight) in air holes and held constant. In no instance were surveys run more often than every 500 feet. The results were impressive. The total deviation was as predicted from subsurface correlations and no problems associated with dog-legs were encountered.

Drilling time was reduced to $\frac{1}{3}$ to $\frac{1}{2}$ of that previously required and the number of bits required was reduced 50%. The nine wells were drilled for the price of three drilled conventionally. One well was drilled to approximately 7,000 feet and completed for less than $24,000. The drilling contractors total invoice on one well to 7,000 feet was less than $9,000.

A typical comparison is offered as Figure 13–11. Wells "A", "B", and "C" in Figure 13–11 are offsets and geologically similar. Well "A" represents one of the nine previously described. Well "B" represents an air hole drilled by the same operator using conventional practices, and well "C" was drilled with mud by a major operator. As illustrated, well "C" was drilled with mud in 30 days to 7,000 feet. Bit weights were as high as 50,000 pounds. The maximum measured deviation was $6\frac{1}{2}$ degrees and 33 total bits were required. Well "B" was drilled to 7,000 feet in 22 days. Bit weight was routinely 4,000 to 6,000 pounds.

The maximum measured deviation was $5\frac{1}{2}$ degrees and 9 bits were required. Well "A", representing one of the nine, was drilled to 7,000 feet in 10 days which represents a 70% improvement over the time required for well "C" and a 55% improvement in the time required for well "B". Total measured deviation at well "A" was 9 degrees. Well "A" required only five bits or 85% fewer than well "C" and 44% fewer than well "B".

The tenth well in the series offered some interesting observations and presented the unique opportunity to drill a well twice. The tenth was a step-out and control was poor. In addition, a known fault was not well defined at the location. However, one mile offsets to the north and west were chosen for control wells and are illustrated as Figure 13–12.

Deviation to 12 degrees was recorded in the north offset; however, 12 degrees is not critical in 640 acre spacing. The north offset in Figure 13–12 illustrates the inconsistent logic related to drilling fluid type. Note that to 200 feet the bit weight is held to 5,000 to 15,000 pounds while gas drilling. After mudding up, the weight was increased to 30,000 pounds and the total

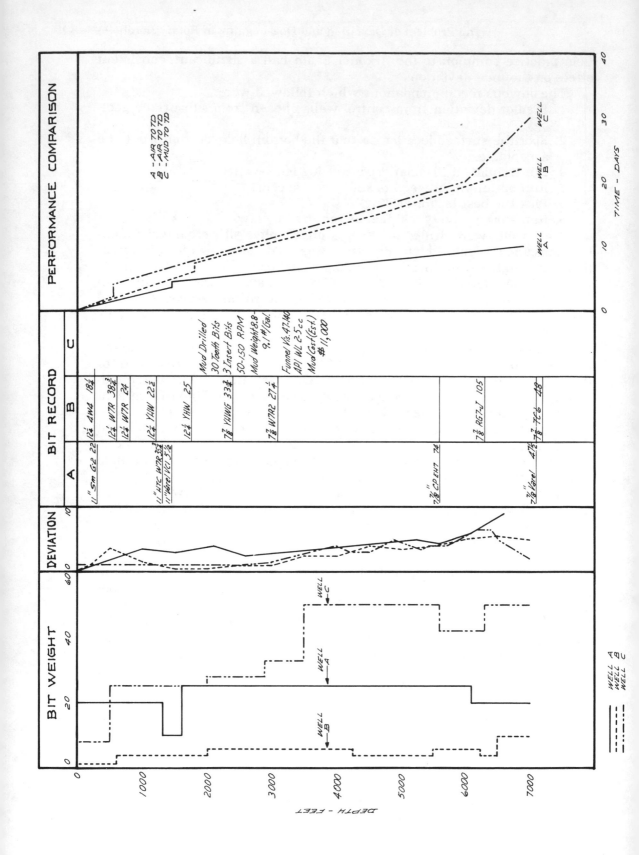

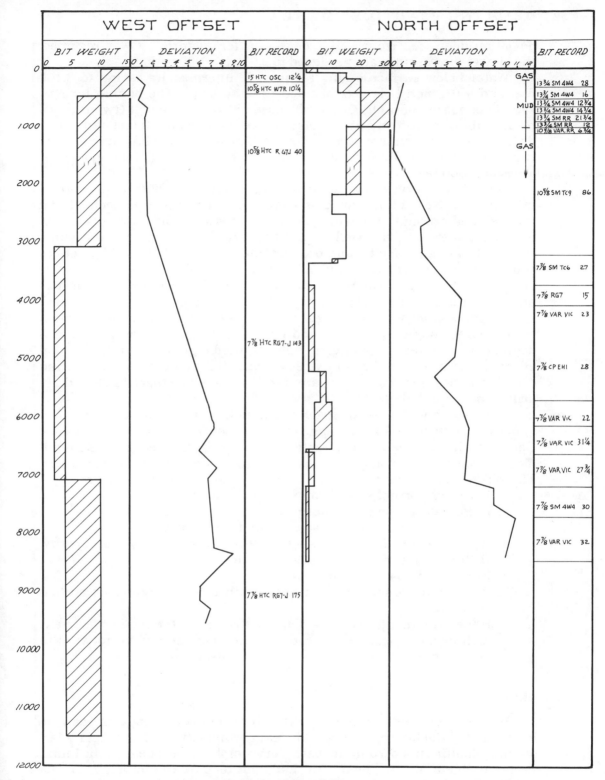

FIG. 13–12. *Offset drilling*

deviation decreased from 1 to ½ degree. After gas drilling was resumed, the bit weight was ultimately reduced to 2,000 to 4,000 pounds.

Water below surface casing required 7″ intermediate casing to 3,100 feet and a 6¼ inch hole was planned to total depth. The bottom hole assembly consisted of a bit, Circle "C" assembly above the bit (two integral blade stabilizers 10 feet apart), and a sleeve stabilizer at 30 feet and 5 inch drill collars. Bit weights were 10,000 to 12,000 pounds and surveys were run every 500 feet. Total deviation was 2 degrees at 3,000 feet and 41 degrees at 7,500 feet.

Directional surveys indicated dog-leg severity as high as 3 degrees per 100 feet. No drilling problems were encountered. It was decided that the well had drifted in the wrong direction, crossed a fault, and missed the pay zone; therefore, the well was plugged back to 4,900 feet and whipstocked to 1 degree. No bottom hole assembly was used. All field practices commonly associated with prudent practice in crooked hole country were employed. Bit weight was reduced to 1,000 to 2,000 pounds, straight hole bits were run, and surveys were run every 15 to 30 feet.

Deviation after whipstocking at 5,750 feet was 2 degrees. The deviation at 6,950 feet was 37½ degrees and dog-leg severity was measured at 5 degrees per 100 feet. The original hole required eight days while the final effort required sixteen days or twice as long. All the care, slow drilling, and "safe" practices common to the area used in the whipstocked hole did not drill a straighter hole.

Experience with this approach is limited; however, it may well have application in hard rock areas such as the Mid-Continent region and Canada. Certainly the experience of directional drillers in the coastal regions indicates that this approach does not have application in the coastal areas. The recommendations for drilling in crooked hole country are as follows:

1. Establish control wells from subsurface geology.
2. Predict deviation from control wells.
3. Locate surface to compensate for undesired drift.
4. Run a square drill collar above the bit to reduce dog-leg severity.
5. Drill with optimum weights and speeds.
6. Run surveys as required by law.
7. Run a directional survey at total depth if severe problems are encountered.
8. Have faith that all is being done that is within the realm of current technology. If excessive problems are encountered, go forward to new technology – not backward 50 years to fanning bottom.

SUMMARY

What's been said? It is submitted that the drilling industry can no longer afford the luxury of crooked holes. The industry will no longer tolerate the penalties imposed in the past. Very simply, a solution must be found to any problem in any given area. Let's not tolerate this problem. In the Arkoma Basin the problem was ignored. A pendulum assembly may be completely satisfactory elsewhere. Let's change our thinking. Dog-leg severity is the problem and square drill collars combat dog-leg severity to the best of our ability. Deviation is not the problem.

Experience off-shore has proven that operating problems are not severe in holes intentionally deviated to angles far in excess of naturally deviated angles anywhere in the world. Let's quit routinely worrying about deviation. Let's solve this problem or learn to live with it and quit paying an unnecessary premium for drilling in crooked hole country.

14 Rotary Drilling Bits

ROBERT D. GRACE
Consulting Engineer
Amarillo, Texas

THIS discussion is primarily concerned with tungsten carbide insert bits and includes conventional bits. This discussion includes advancements in inserts, seals and friction bearings; bit selections and limitations; and current operating practices and trends. Finally, the economics of a bit run will be considered along with a discussion of dull bit evaluations.

DEVELOPMENT OF INSERT BITS

By the late 1940's the industry was venturing into deep drilling. In most areas this meant harder rocks such as limestone and chert, slow penetration rates, and reduced bit life. Bit records of that time were filled with typical runs of only five to ten feet in four to five hours at depths below 10,000 feet. Conventional mill tooth bits were simply inadequate for the drilling environment encountered. Then in 1949 Hughes Tool Company introduced the first three cone bit using tungsten carbide inserts in the cutting structure and named it the "Chert Bit."

As illustrated in Figure 14–1, the "Chert Bit" was characterized by short and closely spaced inserts. This cutting structure was durable and the "Chert Bit" performed its task well. Typically, the five to ten foot runs were increased to fifty to one hundred foot runs; and the four to five hours became twenty-five to forty hours, and often as not failure was due to bearing failure and not structure failure.

From this point the development and use of insert bits lagged. Conventional mud systems were characterized by high viscosities, high weights, and a high total solids content. In this environment insert bits could not

354

FIG. 14-1

FIG. 14-2

compete with conventional mill tooth bits in the softer rocks found at shallower depths because bit life was primarily a function of bearing life.

The experience gained by the industry from the advent and popularity of air drilling in the 1950's provided insight into the potential application of insert bits. In fluid free air drilling operations the bearings were not subjected to the abrasive wear of drilling mud solids normally encountered in conventional operations. As a result, in hard rock areas such as the Arkoma Basin of Southern Oklahoma conventional mill tooth bit life was limited to the life of the cutting structure.

At that point, an indestructible cutting structure such as offered by the insert bits became desirable, and some perceptive drilling man initiated the use of insert bits in air operations. The results were beyond expectation. Insert bits were eventually used from top to bottom in air operations. Insert bit life routinely ran into days instead of hours as previously recorded with conventional mill tooth bits.

Seven thousand foot wells were routinely drilled from below surface casing at 500 feet to total depth with one (two at the most) insert bits. In addition, low solids, low weight muds were becoming popular and bit runs in this environment were significantly improved, and perceptive bit manufacturers recognized the potential of developing a carbide insert bit which would be capable of drilling the softer formations at penetration rates approaching those achieved by conventional mill tooth bits. In the late 1950's Smith Tool Company introduced the TC-6 which was the first insert bit to have long, shaped inserts and offset to the cones. Figure 14-2 illustrates the Smith TC-6.

Today, modern insert bits are routinely used in many areas from top to bottom in low solids drilling mud systems. Recently, in the Oklahoma Panhandle two insert bits were required to drill a 6,500 foot well; not so long ago twenty conventional bits would have been used in as many days in a comparable operations. Advancements in technology of tungsten carbide insert bits have been rapid in recent years, and have significantly improved drilling costs.

It is doubtful that operations such as the Lone Star Baden recently drilled to a record depth of 31,500 feet would have been feasible without the modern selection of insert bits. The industry is not through. Insert bits will one day be used from top to bottom routinely throughout the world.

FUNDAMENTALS OF BIT DESIGN

In order to understand the development and future of insert bits, it is necessary to review and understand the fundamentals of conventional bit design. In particular, it is necessary to understand the fundamental differences in design between the hard formation bit and the soft formation bit.

The first three cone bits and the Chert bits were hard formation bits. That is the primary drilling mechanism is intrusion. The chisels or teeth are driven into the rock by the weight on the bit and pulled from the rock by the rotary action. The cones rotate very nearly about a true center and have no cone profile.

Characteristically, hard formation bits have heavy cone shells, short teeth, and heavy journals in order to withstand the high bit weights necessary to fracture the harder formations. Traditionally, drilling operations with hard formation bits were characterized by high bit weights and low rotary speeds. (See Figure 14–3.)

At the time the roller bit was introduced, soft formations were drilled with drag bits or fish tail bits. Drilling was accomplished by the gouging and tearing action of the bit on bottom which was very efficient in the soft formations. However, the superiority in design of the three cone roller bit was evident, and the modern soft formation mill tooth bits were developed incorporating characteristics of both the hard formation three cone bit and the drag bit. The gouging, tearing action of the drag bit was accomplished by extending tooth length and offsetting the cones slightly so that the cones did not rotate about their true center. Offset is illustrated in Figures 14–4, 14–5, and 14–6. As illustrated, if the cones are forced to rotate about a center other than their true center, they must slide or drag on the bottom of the hole and produce the desired gouging action.

The drilling action is by both intrusion and drag. The cone offset has additional significance in design and practice. As illustrated in Figure 14–5, the offset cones do not drill a gauge hole; it is necessary that gauge be obtained from the reaming action of the teeth on the heel of each cone. In addition, the dragging action created by the cone offset produces tangential forces on the teeth which can cause breakage when loading is excessive. Further, the long, slender teeth dictate smaller bearings, thinner cone shells, and lighter journals. In normal drilling operations these features combined with the cone offset and drilling mechanism impose lighter bit weights and higher rotary speeds. (See Figure 14–3.)

Figures 14–7 and 14–8 illustrate typical bottom hole patterns of hard and soft formation mill tooth bits, respectively. Note that in the bottom hole pattern for the hard formation bit in Figure 14–7, each tooth row is clearly

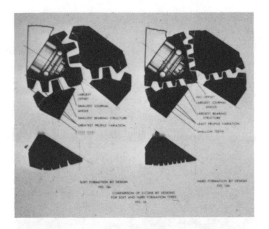

FIG. 14–3

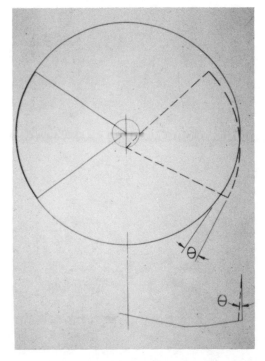

FIG. 14–5

FIG. 14–4

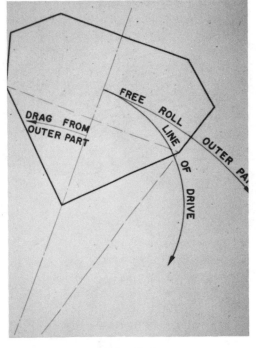

FIG. 14–6

FIG. 14-7

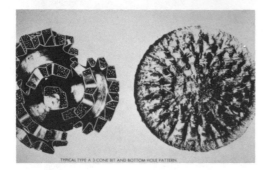

FIG. 14-8

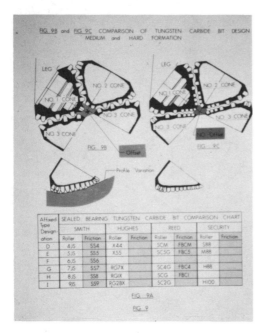

FIG. 14-9

FIG. 14-10

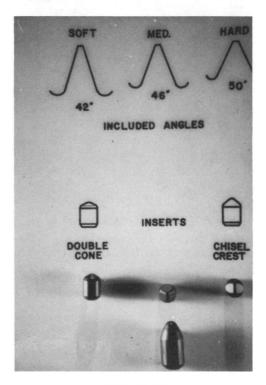

FIG. 14-11

defined which illustrates the intrusive drilling mechanism. By comparison, there is no clear tooth definition to the bottom hole pattern of the soft formation bit illustrating that the drilling mechanism is a combination of intrusion and drag in that the cones do slide across the bottom of the hole.

THE INSERT BIT

With the popularity of clean, low weight drilling fluids, bit life, or more particularly, bearing life was significantly improved, and cutting structure life became significant to bit life. The potential of the carbide insert structure had been demonstrated in the air operations; however, in mud operations the penetration rates achieved in the soft to medium formations with the hard formation insert bits was unacceptable.

In combating this problem bit manufacturers took advantage of their experience with mill tooth bits and designed insert bits with longer, shaped inserts, cone profile, and cone offset. (See Figure 14–9.) As mentioned, the first such bit was the Smith TC-6 illustrated in Figure 14–9. Since the TC-6, the industry has developed a completely new line of drilling tools typically illustrated in Figure 14–10.

A comparison between modern mill tooth and carbide insert bits is an indication of the future carbide insert bits. Figure 14–11 is a comparison of teeth and inserts. As illustrated, the included tooth angle varies from 42 degrees on a soft formation bit to 50 degrees on a hard formation bit while the smallest included angle on an insert is 70 to 75 degrees.

The brittleness of the tungsten carbide currently used dictates the high included angles. Obviously, if the use of insert bits is to be expanded to softer formations, the included angle must approach that of conventional mill tooth bits, and bit manufacturers are experimenting with tungsten carbide of varying hardness which might permit a reduction in included angle.

Table 14–1 is a further comparison between conventional mill tooth bits and carbide insert bits.

It is interesting to note that the longest insert extension (0.312 inches) isn't as long as the shortest tooth (0.421 inches) and isn't half as long as the

TABLE 14–1
Design Comparison

$7^7/_8''$ Bit
Mill Tooth (MT) versus Carbide Insert (CI)

Bit Type	Height inches		Offset inches		Bearing Angle degrees	
	MT	CI	MT	CI	MT	CI
S	0.730	0.312	$^3/_8$	$^1/_8$–$^1/_4$	33	36
M	0.531	0.250	$^1/_8$–$^1/_4$	0–$^1/_8$	36	36
H	0.421	0.163	0	0	36	36
XH		0.148		0		36

(Also, refer to Figure 14–14)

longest tooth (0.75 inches). Obviously, the future insert bits will have longer inserts. Recently, a major manufacturer experimented with extensions of 0.437 inches with only limited success due to excessive breakage of brittle inserts. In addition, note that the offset used in insert bits is considerably less than that used in mill tooth bits. A future reduction in insert hardness might reduce the brittleness and permit bits with longer, more slender inserts and increased cone offset capable of competing with mill tooth bits in soft formations.

Figure 14–12 illustrates the bottom hole pattern of a typical modern soft formation insert bit. The noticeable lack of insert definition indicates that some drag is being obtained on bottom; however, it does not compare with that pattern illustrated in Figure 14–8 which was obtained with a conventional mill tooth bit. Insert bits in future years must have longer inserts, more slender inserts, and more cone offset if they are to compete with mill tooth bits in soft formations.

The insert bits are not without their problems. When a bit is pulled with all the inserts missing, how many times has it been said that "shale sucked them out?" The shale didn't "suck them out." That's propaganda; the problem is cone hardness.

Cone steel has a Rockwell hardness of 35 to 40 when received. In the manufacture of conventional mill tooth bits the cone is case hardened by carbonizing at 1700 degrees Fahrenheit with natural gas for 24 hours and annealing at 400 degrees Fahrenheit. The cone surface then has a hardness of about 60 R_c. Since carbonizing effects only the surfaces, the cone is progressively softer with depth. As illustrated in Figure 14–13 the cone hardness decreases from 60 R_c at the surface to the original hardness of 35R_c 0.10 inches from the surface.

In contrast, the cone used for the insert bits is case hardened to a surface hardness of only 45R_c. A hardness profile for the insert bit cone is also presented in Figure 14–13. The surface hardness is reduced to permit the insert holes to be drilled.

The significance is that the cone used in the manufacture of insert bits has a relatively soft surface and will erode when exposed to circulating drilling mud. Compounding this problem is the fact that for any given bit size, mill teeth are at least twice as long as the longest insert which would at least double the fluid velocity across the face of the bit.

As illustrated by the hardness profiles in Figure 14–13, the cone becomes progressively softer and easier to erode as the erosive process continues. At some point the cone is no longer capable of supporting the inserts, and the obvious consequence is that the inserts fall out leaving a clean hole in the cone and causing bit failure. In cases of premature failure due to lost inserts the bit should be returned to the manufacturer and the cone hardness profile determined by destructive testing.

Another related problem is structural burial. If the inserts are excessively buried, the flow area is significantly reduced which accelerates erosion. Further, when burial is excessive, the bit load is transmitted to the relatively soft cone shell which can cause the cone to crack and break.

FIG. 14–12

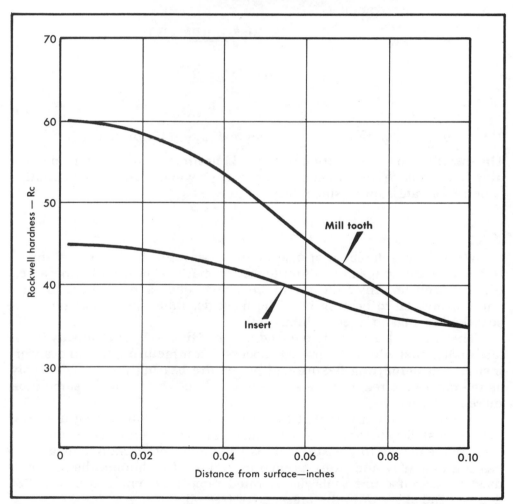

FIG. 14–13. *Variation of hardness with distance from the surface*

Manufacturers have suggested that structural burial should not exceed 80 percent. The limiting penetration rate may be determined as follows:

$$\text{Per cent structural Burial} = \frac{20 \text{ DR}}{\text{RPM } E_t}$$

Where: 20   = Units conversion constant
DR   = Penetration rate, ft./hr.
RPM = Rotary revolutions per minute
E_t = Tooth or insert extension

Example 1:
Insert extension = 0.312 inches (Table 14–1)
RPM   = 50
Maximum burial = 80%
Rearranging

$$\text{DR} = \frac{\% \text{ Burial (RPM)}(E_t)}{20}$$

$$= \frac{80 \ (50)(0.312)}{20}$$

$$= \underline{\underline{62.4 \text{ ft. 1 hour,}}}$$

or an instantaneous rate of approximately 1 minute per ft.

The instantaneous penetration rate should be limited to one minute per foot in this example to prevent cone damage. An obvious alternative to limiting penetration rate is increasing rotary speed.

Seals

With the development of insert cutting structures capable of competitive penetration rates in medium to soft formations, bearing life became a primary problem. Any mud system must have solids, and solids are abrasive and detrimental to bearing life. The most desirable solution would be to seal the bearings in an ideal lubricant.

The first seal bearing bits introduced by Hughes Tool Company in the late 1950's met with only limited success. Manufacturing tolerances were insufficient permitting flexure and premature bearing failure. Obviously, as tolerances are reduced the seal is more effective and, at the same time, more necessary.

Another problem was that the viscosity of lubricants being used was increased at elevated temperatures and pressures. As a result the lubricants were not sensitive to pressure surges which permitted mud to invade the bearing assembly and cause premature failure. In addition, the materials used to manufacture seals deteriorated rapidly at temperatures of 250 degrees and above or in the presence of free oil.

The first and still most popular seal is the Belleville Spring illustrated in Figures 14–15 and 14–16. As illustrated, the Belleville Spring is merely

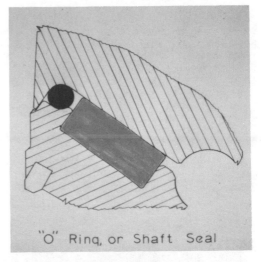

OFFSET COMPARISON								
STC	HTC	REED	SEC	APPROX OFFSET	STC	HTC	REED	SEC
DS	OSC3A	YT3A	S3S	3/8				
DT	OSC3	YT3	S3	1/4 – 3/8				
DG	OSCIG	YTIA	S4	1/4				
K2	OSC	YTL	SG	1/4	3		SCS5	S8BS
V2	OWV	YSI	M4N	1/8	4	J44	SCM5	S88
T2	OWC	YM	M4L	1/8	4-7		SCM	
L4	W7	YH	H7T	0	5	X55R	SCH5	M88
W4	W7R2	YHW2	H7U	0	6			
WC	WR	YBV	HC	0	5-7		SCH	
				0	7	RG7X	SC4G	H88
				0		RGIX	SCG	H99
				0	9	RG2B	SC2G	HIOO

FIG. 14–14

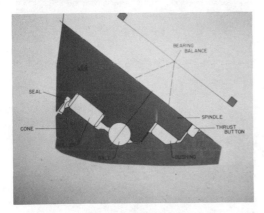

FIG. 14–15

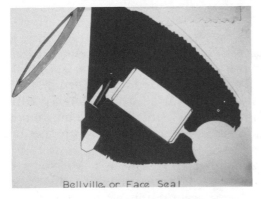

Bellville, or Face Seal

FIG. 14–16

"O" Ring, or Shaft Seal

FIG. 14–17

FIG. 14–18

FIG. 14–19

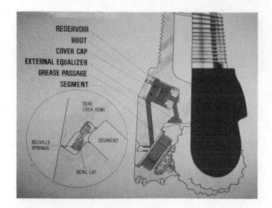

FIG. 14–20

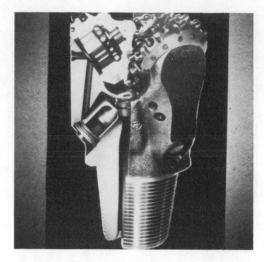

FIG. 14–22

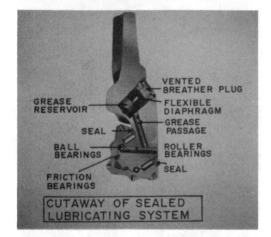

FIG. 14–21

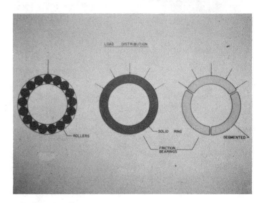

FIG. 14–23

circular spring steel encased in rubber which seals against the face of the shank and the face of the cone (sometimes referred to as a face seal). All major manufacturers except Hughes Tool use the Belleville Spring. The "O" Ring seal is the most effective seal, is being used by Hughes Tool in the "J" series and is illustrated in Figures 14–17 and 14–18. The major problem confronting the "O" Ring is tolerance which must be precise in order to maintain an effective seal.

An understanding of lubricants and lubricating systems is essential to successful insert bit operations. Figures 14–19 through 14–22 illustrate typical seal and lubricating systems. All are essentially the same and are composed of an external equalizer located either under the bit or on back of the shanks, a grease reservoir with some sort of expandable diaphragm to distribute the grease, and some sort of distribution system to the bearings. In addition, Hughes Tool Company offers a pressure relief valve to release any trapped pressure which might otherwise rupture the seals and ruin the bit.

Pressure surges are detrimental to seal systems. As pressure and temperature increase, the viscosity of the lubricant increases. As a result, the system cannot instantaneously compensate for abrupt changes in pressure due to surges going in the hole, making connections, etc., and small quantities of mud invade the system. With the close tolerances necessary for effective sealing, mud solids can be even more damaging.

Adequate cleaning is even more important with sealed bearing bits. If drilled cuttings are allowed to build up around the shirt tail, seal damage and premature bearing failure may result. By similar analogy, gauge protection is important to seal and bearing life. Seal damage could occur from shirt tail wear caused by inadequate gauge protection. Any time a sealed bearing bit is to be re-run, the seals and shirt tails should be carefully examined for excessive wear or grooving.

Friction Bearings

With improved seal assemblies to protect the bearings from abrasive solids, improvements in the bearings themselves became practical. In review, the conventional bearing structure is illustrated as Figure 14–15. The heel bearing is a roller bearing, carries most of the load, and receives most of the wear. The ball bearings in the middle hold the cone on the journal and resist thrust or longitudinal loads in either direction. The nose bearing consists of a special case-hardened bushing pressed into the nose of the cone and a male piece hardfaced with a special material which displays a low coefficient of friction and resists seizure and wear.

This configuration is termed roller-ball-friction or "RBF" and is standard on all bits between six inches and nine inches in diameter. Bits larger than nine inches in diameter have a roller-ball-roller or 'RBR" arrangement or some combination thereof, while bits smaller than six inches have a friction-ball-friction or "FBF" arrangement because there is not enough room for a roller in the smaller sizes.

The industry has long realized that replacing the heel rollers with a bushing bearing such as illustrated in Figure 14–20 would offer many advantages toward prolonging bearing life. As can be seen from Figure 14–23 in a $7\frac{7}{8}$ inch conventional bit there are approximately 28 roller heel bearings which represent 28 points of contact or 28 points of loading. It should be obvious that a primary factor in roller bearing failure is the flexure caused by loading and unloading as the cone turns.

In contrast, a bushing would have an infinite number of contact points and, consequently, an infinite number of loading points. The thought of expanding from a limit of 28 loading points to an infinite number is exciting.

Figure 14–24 is a summary of 144 controlled laboratory tests comparing $7\frac{7}{8}$ inch rollers and bushings. Note that with 40,000 pounds the roller bearings spalled in 60 hours while the segmented bearing had no wear in 200 hours. With 100,000 pounds of bit weight at 60 RPM the roller bearings spalled in 7 hours while the bushing had only 0.006 inch wear in 10 hours. The potential of a bushing bearing is clear.

Actual performance has not been as dramatic. The bushing bearings

WEIGHT LBS.	RPM	HOURS	
		ROLLER BEARING	SEGMENT BEARING
40,000	60	60 HRS TO SPALL	200 HRS NO WEAR
50,000	60	35 HRS TO SPALL	200 HRS NO WEAR
75,000	60	16 HRS TO SPALL	30 HRS .004 SEGMENT WEAR
100,000	60	7 HRS TO SPALL	10 HRS .006 SEGMENT WEAR

$7\frac{7}{8}$ 3-CONE BEARING TEST, ROLLER VS SEGMENT BEARING. COMPILED FROM 144 CONTROLLED LABORATORY TESTS.

FIG. 14-24

FIG. 14-26

FIG. 14-25

must fit to very close tolerances. Consequently, the bit weight and rotary speed generate extreme heat which is detrimental to bearing life. Bit weight and rotary speed must be limited. In addition, the bushing is even more sensitive to foreign materials such as mud solids due to the necessary tolerance; therefore, an adequate seal assembly is even more critical. If the seal fails, bushing life is significantly reduced. Due to these problems field performance has not been consistent.

Another persistent complaint is that the bushing bearing does not generate sufficient torque on failure to prevent leaving cones in the hole. Some operators are hesitant to use the bushing for this reason. Others have limited bit runs to a predetermined number of total revolutions. Only extensive field experience will establish a satisfactory solution to this problem.

Figure 14-23 illustrates the two bushing bearings currently available. Murphy and Security offer the bushing while Smith Tool offers the segmented bushing. Murphy, Hughes Tool, and Security offer a Journal Bearing which is merely a solid journal. Figures 14-18 and 14-22 illustrate the

Hughes Journal Bearing which is the most sophisticated bushing currently available. As illustrated, the journal wear surface is made from sintered tungsten carbide to resist journal wear. In addition, the cone area is lined with strips of a silver alloy designed to lubricate and dissipate the extreme heat (Figure 14–25).

Regardless of the problems, insert bits with sealed bushing bearings represent a significant contribution to oil well drilling technology. These bits are the bits of the future and only with extensive field experience will their potential be realized. Some suggestions for better bit performance from world wide field experience are as follows:

1. Stabilize the bit—This will improve any bit performance.

2. Maintain minimum mud weight, sand, and solids. This is very important. The best performance is obtained in air where no solids are present. Performance in clear water is second. Any solids are detrimental. High solids in conventional clay-water systems almost precludes the successful use of insert bits. It is extremely important to take advantage of all aspects in order to optimize performance and minimize total cost. For example, only minimum mud properties are necessary due to the long hours on bottom. Some operators drill with a low carrying capacity system and slug the hole prior to making a trip.

3. Adequate bottom hole cleaning must be maintained in order to protect the seals.

4. Avoid pressure surges while making trips and connections. Do not spud on the bit. Protect those seals.

5. Do not routinely regrease a bit or return them to the manufacturer for repair. It can readily be determined in the field if a bit should be rerun. Examine the seals for leakage or damage. Examine the cones for cracks or excessive erosion. Remove the cap and (Figure 14–26) diaphragm from the grease reservoir. If the diaphragm is ruptured, the bit should be discarded; if the reservoir is empty but the diaphragm is not ruptured, fill the reservoir with lubricant and rerun; if the reservoir still has grease in it, re-run the bit.

6. Keep oil from the mud. Oil is detrimental to seal life and reduces penetration rate.

7. Do not store a sealed bearing bit in diesel oil or any other oil. It is not necessary to store it in anything; but, if you must store it in water.

8. Current operating trend is 5500 to 6000 pounds bit weight per inch bit diameter at relatively low rotary speeds of 40 to 50 revolutions per minute. Lower bit weights are common in field operations; however, this practice is of dubious value and should be carefully evaluated. A West German operator reported an excellent run in Southern Arabia. An $8\frac{1}{2}$ inch Hughes Tool Company J-44 made 900 feet in 80 hours with 80,000 pounds bit weight at 40 to 60 RPM. Conven-

tional hard rock mill tooth bits would routinely cut 100 feet in 14 hours. Interestingly, the bit weight in this example is over 9400 pounds per inch which is a 57 per cent increase in the operating trend. In the future the industry will routinely run more weight on these bits.

INSERT BIT SELECTION AND EVALUATION

"So we have a new line of tools, so what?" "I don't know when to run an insert bit and whether or not I have made or lost money." That remark is typical of competent drilling personnel world wide. Unfortunately, the computers cannot tell us when to run an insert bit. No MCD program currently available adequately relates to insert bits, sealed bearings, or bushing bearings.

Today, the decision to run an insert bit can only be based on experience and judgement. Conscientious drilling engineers will plot offset bit records on electric logs and calculate the cost per foot for each bit run and the cumulative cost per foot for each bit run. (These calculations are presented later in this text. Also, see the section on well planning.) Each bit run on the well being drilled is plotted, cost calculations are made, and the data is compared with that from the offsets; and then, the drilling engineer knows the competence of his decisions.

Only in this way can field personnel gain the experience required for successful and optimum application of this advance in drilling technology.

Example 2 illustrates the application of cost per foot data in evaluating the economics of insert bits.

■ EXAMPLE 2:

Applicable costs are as follows:

Rig, per day	$1,500.00
Bits, mill tooth, each	260.00
insert	$1,250.00
Mud, per day	500.00
Water, per day	200.00
Desilter, per day	150.00
Supervision, per day	250.00
Total daily cost	2,600.00
Hourly rig cost	$ 108.00

Trip time is 0.7 hour per 1000 feet
The cost equation is:

$$C_T = \frac{B}{F} + \frac{C_r T_d}{F} + \frac{C_d T_t}{F}$$

Where:
C_T = drilling cost, $/ft.
B = bit cost, $/bit
F = footage drilled, ft.
C_r = rig cost drilling, $/hr.
T_d = time drilling, hrs.
T_t = time tripping, hrs.

C_d = rig cost on trip, \$/hr.

Assume:

1) Comparable lithology
2) Cr = Cd = \$108 per hour
3) The well in question is to be deepened from 6000 to 7650 feet

Determine:

The economics of insert bits

Solution:

The bit record from the offset control well is presented as Table 14–2.
The cost per foot for each bit run is calculated as follows:

Run No. 17

Drilling hours = 10.5

Trip hours = 0.7 hr/1000 ft. 5.958(1000 ft.)

        = 4.1 hrs.

Total hours = 14.6 hours

Total footage = 160 feet

Therefore

$C_T = 1/F \ [B + C_R(T_d + T_t)]$

   = 1/160 feet [\$260 + \$108/hr.(10.5 + 4.1) hrs.]

   = \$11.51 per foot

Similar calculations are made for each bit run and recorded on the bit record.
The bit record for the deepening job is presented as Table 14–3. Inserts were run
  below 6500 feet.

Cost per foot data was calculated for each bit and is presented.

    From Table 14–3, inserts drilled 1132 feet at a cost of: \$17,205

    From Table 14–2, conventional bits from the offset cost: 27,803.

       Savings with inserts: \$10,597.

How do we decide when to pull a bit? A bit should be pulled for one of
two reasons – the bearings are worn out or the cutting structure has worn to
the point that it is no longer economical to drill with that bit. That sounds
simple enough, but the decision is not always simple in field operations.
Conventional roller bearings will spall and lock the cone.

This condition is clearly visible at the surface in shallow wells and is
detected from abnormal torque which may be obtained from several sources.
The number of torque turns in the rotary table is a good indicator as is the
conventional torque gauge. On older rigs experienced personnel use the
compound engine vacuum gauge as a torque indicator; on electric rigs the
rotary ampmeter is used. In deeper wells the problem becomes more complex.

High angle drilling adds a new dimension as do the bushing bearing
bits. As previously mentioned, some operators have resorted to running a
bit for some predetermined number of revolutions. Certainly this approach
relieves everyone of the responsibility of making a decision; otherwise, it
is of dubious merit.

In the deep Anadarko Basin where production is below 20,000 feet,
contractors are using the finest torque gauges available to accurately predict bearing wear and maximize bit life. Let's get good torque, gauges on
the rigs, train our personnel to use and understand them, and get the best
from our equipment.

TABLE 14-2
Cost of Insert Bits

COUNTY	PARK
FIELD	HOODOO RCH. W/C
STATE	WYO.
SECTION	7
TOWNSHIP	50N
RANGE	101W
OPERATOR	A.G. HILL
RIG NO.	3
CONTRACTOR	SIGNAL DRLG. & EXPLOR. INC.
LOCATION	GOVT. NO.1
SPUD	11-23-64
US	11-25-64
INTER	
TOTAL DEPTH DATE	12-22-64
TOOL PUSHER	BILL SUMMERS
TOOL JOINTS / GIVE SIZE & TYPE	1. 4 1/2XH-HTC-FWT 2.
DRILL COLLARS	1
STOCKPOINT NO.	POWELL 17353
DRAWWORKS & POWER	UNIT 36 A 2-21000 AC
FUEL	DIESEL
WATER	HAUL
SALESMAN	DON DAUGHERITY
DIVISION	CASPER, WYO.

TOTAL COST	COST $/FT	NO.	SIZE	MAKE	TYPE	REG	JET 32nd IN	SERIAL	DEPTH OUT	FEET	HOURS	WT 1000 LBS	RPM	DEV VERT	PUMP PRESS	NO.1 SPM	NO.1 LINER	NO.2 SPM	NO.2 LINER	MUD WT	VIS	W.L.	T B	OTHER	DULLS
1842	11.51	17	8 3/4	HTC	OWVJ	✓	3-13	39081	5958	160	101 1/2	60	80	4	900	68	5½			9.1	40	7.4	3 5		✓
2784	9.22	18	-	-	-		-	93147	6260	302	19	-	-	4	-	-	-			9.3	-	6.0	2 5		✓
1116	16.17	19	-	REED	YTTAJ		2-13 / 1-14	442322	6329	69	3 1/2	-	75	NS	850	-	-			-	-	-	5 3		
2427	17.34	20	-	HTC	OWCJ		3-13	69454	6469	140	15 1/2	-	80	3½	-	-	-			-	-	-			✓
1787	18.61	21	-	REED	YTLJ		1-14	D44348	6565	96	9 1/2	-	-	4	-	-	-			-	-	-	3 6		
1733	14.44	22	8 3/4	SMITH	K2PJ		2-14	98420	6692	127	9	60	80	4	850	68	5½			9.3	44	7.0	5 7		✓
2498	13.52	23	-	HTC	OSCJ		2-13 / 1-14	2966	6873	181	15 1/2	-	90	NS	800 / 850	-	-			9.4	42	9.0	4 7		✓
2524	15.97	24	-	-	-		-	6783	7031	158	16	-	-	3/4	-	-	-			-	-	-	4 7		✓
2806	18.83	25	-	-	-		-	1346	7180	149	18 1/2	-	-	NS	850	-	-			-	-	11.0	4 7		✓
1993	31.63	26	-	-	OWVJ		-	66410	7243	63	11	-	90 / 80	4¼ / 750	850	-	-			-	45	6.0	4 7		✓
1993	38.33	27	8 11/16		◇CC		-	H1038	7295	52	11	24	70	NS	550 / 650	64	-			-	60	-			
2112	39.85	28	8 3/4	SMITH	L4H	✓	2-13 / 1-14	12957	7358	53	12 1/2	260	80 / 60	-	600	68	-			9.5	53	7.0			
2177	32.49	28	-	REED	YH-GJ		2-13 / 1-14	044583	7425	67	12 1/2	-	50 / 65	-	750	-	-			-	-	6.8			
1907	54.49	29	-	SMITH	L4H	✓		12931	7460	35	10	55	70	-	400	-	-			-	46	-			
2080	31.04	30	-	HTC	W7R-2J		2-13 / 1-14	52116	7527	67	11 1/2	-	-	-	700	-	-			-	58	5.0			✓

TABLE 14-3
Cost of Conventional Bits

COUNTY	PARK	OPERATOR	EARL T. SMITH & ASSOC.
FIELD	WILDCAT	SECTION	7
STATE	WYO.	TOWNSHIP	50N
CONTRACTOR	TRUE DRILLING COMPANY	RANGE	101W
LOCATION	FROST RIDGE	TOOL PUSHER	BILL BAKER
RIG NO.	9	SALESMAN	TOMMY EVERHART

TOTAL COST	COST $/FT.	NO.	SIZE	MAKE	TYPE	REG	JET 32nd IN.	SERIAL	DEPTH OUT	FEET	HOURS	WT 1000 LBS.	RPM	DEV	PUMP PRESS.	NO.1 SPM	LIN-ER	NO.2 SPM	LIN-ER	MUD WT.	VIS.	W.L.	T	B	OTHER
2773	15.94	2	8 3/4	SMITH				K-2	6008	174	19	40	100		1400					9.6		6.4	6	5	
2838	10.92	3	✓	SMITH				D-G	6268	260	19 1/2	50	110		1400					9.7		5.6	6	8	
3456	13.82	4	✓	SEC				M4N	6518	250	25	55	72		1400					9.4		8.2	8	8	
12,619	13.62	5	✓	SEC				S88	7444	926	993 1/2	35/40	46		1600					9.0		20	8	8	
4586	22.26	6	✓	REED				SC5G	7650	206	25 1/2	35/40	46		1600					8.9		9	1	1	

If a bit is not worn out, the only other reason to pull it is because it has quit drilling. Translated, that statement means that the cutting structure has deteriorated to the point that it is no longer economical to drill. In the past the decision that the bit has quit has too often been arbitrary or merely a judgement based on experience. Although the experienced judgement is often fairly accurate, it is no longer acceptable.

Drilling costs are too high for the industry to tolerate anything but the finest efforts from drilling personnel. If a bit has not worn out and evaluation of data indicates no unexpected changes in lithology, then the decision to pull the bit should be based on simple, fundamental cost per foot calculations. Example 3 illustrates the application.

■ EXAMPLE 3:

The calculations of Example 2 are made throughout the bit run. The cost per foot is then plotted versus total footage. When the cost per foot begins to increase, the bit should be pulled.

Rig Cost is $70 per hour

Trip Time is 1 hour per 1000 feet

Bit Cost is $300 per bit

From Example 2

Cost $/ft. = 1/Footage [Bit Cost + Rig Cost (Rotating Hours) + Rig Cost Trip hours)]

Bit #6 in @ 2817 feet at 11:25 A.M.

Time	Depth	Calculation	Cost per foot
2:25	2942	$1/125[300 + 70(3.0 + 3)]$	$5.28/ft.
3:45	2974	$1/157[300 + 70(4.5 + 3)]$	5.25
4:55	3004		4.86
6:12	3036		4.56
6:39	3046		4.50
6:57	3056		4.40
7:26	3066		4.36
7:57	3076		4.34
8:22	3086		4.28
8:58	3096		4.30

The data are graphically illustrated as Figure 14–27.

The bit should be pulled at 280 feet after 9.76 hours. This is a simple calculation. One of the world's largest drilling companies has successfully used a variation of this approach for many years to pull bits. Nomographs such as Figure 14–28 speed the calculations and are available from most bit manufacturers.

Now the only remaining question is which bit to run back in. This decision must reflect the bit just pulled and any anticipated changes in lithology. Assuming no changes in lithology, the bit to choose need only compensate for whatever adverse conditions experienced by the bit just pulled.

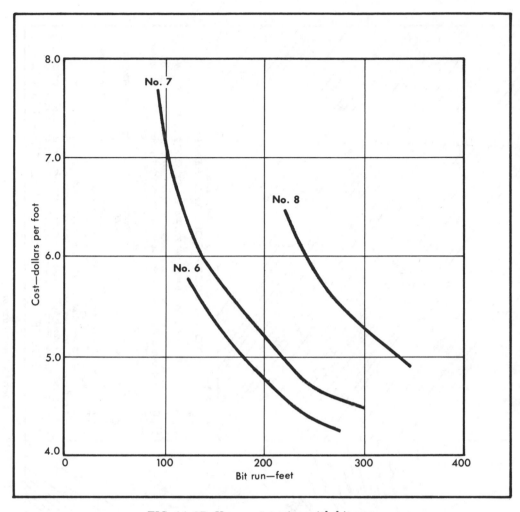

FIG. 14–27. *How cost varies with bit run*

If the old bit looks like Figure 14–29, the selection was at least adequate since there is no structural damage. A bit designed for softer drilling might improve penetration rate and should be considered.

If only heel row inserts are broken as in Figure 14–30, the energy levels, particularly rotary speed, are excessive. This is impact breakage and occurs on the outside or drive rows of the cones since this is the area subjected to the highest loading and greatest impact forces. The solution is to reduce the rotary speed and/or weight.

If all of the inserts are broken off, as in Figure 14–31, several things might have happened. The abuse could be a severe case of excessive impact loading, junk, or formation too hard. Since some of the holes appear to be completely empty, erosion could have caused some inserts to fall out and break the others. This type of occurrence is common. If the bit made a good run, evaluate solids and pump volume. Consider cone hardness and alternate bit selections.

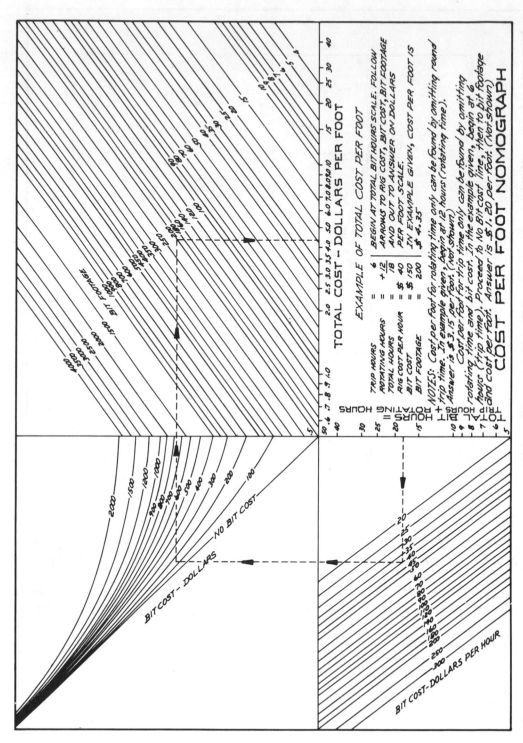

FIG. 14-28. *Cost-per-foot nomograph*

FIG. 14–29

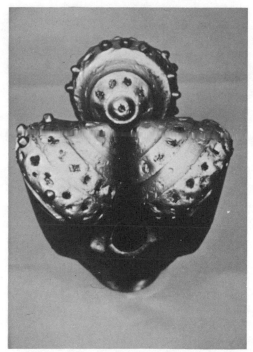

FIG. 14–31

FIG. 14–30

FIG. 14–32

FIG. 14–33

FIG. 14–35

FIG. 14–34

FIG. 14–36

FIG. 14–37

FIG. 14–39

FIG. 14–38

TABLE 14–4
Proposed Classification of Rock Bits
(Courtesy of The Petroleum Engineer, March, 1972)

These classifications are general and are to be used only as simple guides. All bit types will drill effectively in formations other than those specified. These charts show relationships between specific bit types.

G. W. MURPHY INDUSTRIES

	Series	Formations	Types	1 Standard	2 "T" Gage	3 Gage Inserts	4 Sealed Roller Bearings	5 Seal. Brgs and Gage Insert	6 Sealed Friction Bearing	7 Sealed Friction Brg. & Gage	8 Other	9 Other
MILLED-TOOTH BITS	1.	Soft formations having low compressive strength and high drillability.	1.	YT3A			ST3A					
			2.	YT3	YT3T		ST3					
			3.	YT1A	YT1T			ST1AG				
			4.									
	2.	Medium to medium hard formations with high compressive strength.	1.	YS1/YS4		YS1G/YS4G	SS1	SS1G				
			2.									
			3.	YM		YMG	SM	SMG				
			4.									
	3.	Hard, semi-abrasive or abrasive formations.	1.	YH		YHG	SH	SHG		FHG		
			2.	YHW2		YHWG						
			3.									
			4.	YBV		YBVG	SBV	SBVG	FV			
INSERT BITS	5.	Soft to medium formations with low compressive strength.	1.									
			2.							FCT/FBCT		
			3.					SCS5		FCS5/FBCS5		
			4.									
	6.	Medium hard formations of high compressive strength.	1.									
			2.					SCM5		FCM5/FBCM5		
			3.					SCM		FCM/FBCM		
			4.									
	7.	Hard, semi-abrasive and abrasive formations.	1.									
			2.			YC5G		SCH5		FCH5/FBCH5		
			3.					SCH		FCH/FBCH		
			4.									
	8.	Extremely hard and abrasive formations.	1.									
			2.			YC4G		SCH4		FCH4/FBCH4		
			3.			YC2G		SCH2		FCH2/FBCH2		
			4.									

DRESSER OILFIELD PRODUCTS DIV. (SECURITY ROCK BITS)

	Series	Formations	Types	1	2	3	4	5	6	7	Dev/Dir. Control	Center Circ.
MILLED-TOOTH BITS	1.	Soft formations having low compressive strength and high drillability.	1.	S3S		S3SG	S33S				S3SJD	S3SJ4
			2.	S3	S3T	S3TG	S33				S3JD	S3J4
			3.	S4	S4T	S4TG	S44					
			4.	S6	S6T	S6TG	S66				DS-DSS	
	2.	Medium to medium hard formations with high compressive strength.	1.	M4N		M4NG	M44N					
			2.	M4		M4G						
			3.	M4L		M4LG	M44				DM	
			4.	M5								
	3.	Hard, semi-abrasive or abrasive formations.	1.	H7	H7T	H7TG	H77					
			2.	H7U		H7UG	H77U					
			3.			H7SG		H77S				
			4.	HC		HCG	H77C		H77C			
INSERT BITS	5.	Soft to medium formations with low compressive strength.	1.									
			2.					S84				
			3.					S86				
			4.	S8				S88		S88		
	6.	Medium hard formations of high compressive strength.	1.									
			2.									
			3.									
			4.	M8				M88		M88		
	7.	Hard, semi-abrasive and abrasive formations.	1.									
			2.									
			3.									
			4.	H8				H88		H88		
	8.	Extremely hard and abrasive formations.	1.									
			2.	H9				H99		H99		
			3.									
			4.	H10				H100				

TABLE 14–4 (Cont'd.)
Proposed Classification of Rock Bits

These classifications are general and are to be used only as simple guides. All bit types will drill effectively in formations other than those specified. These charts show relationships between specific bit types.

HUGHES TOOL COMPANY

	Series	Formations	Types	1 Standard	2 "T" Gage	3 Gage Inserts	4 Sealed Roller Bearings	5 Seal. Brgs and Gage Insert	6 Sealed Friction Bearing	7 Sealed Friction Brg. & Gage	8 Other	9 Other
MILLED-TOOTH BITS	1.	Soft formations having low compressive strength and high drillability.	1.	OSC-3A			X3A					
			2.	OSC-3			X3					
			3.	OSC-1G	C1C	ODG	X1G	XDG				
			4.									
	2.	Medium to medium hard formations with high compressive strength.	1.	CWV/OW4		ODV/OD4	XV	XDV				
			2.	WO								
			3.	CWC			XC					
			4.									
	3.	Hard, semi-abrasive or abrasive formations.	1.	W7		WD7	X7	XD7				
			2.	W7R-2								
			3.									
			4.	WR		WDR	XWR	XDR	J8	JD8		
INSERT BITS	5.	Soft to medium formations with low compressive strength.	1.									
			2.							J33		
			3.									
			4.									
	6.	Medium hard formations of high compressive strength.	1.					X44		J44		
			2.					X55R		J55R		
			3.					X55		J55		
			4.									
	7.	Hard, semi-abrasive and abrasive formations.	1.									
			2.			RG7J		RG7XJ		J88	RG7AJ	
			3.									
			4.	R1				RG1XJ				
	8.	Extremely hard and abrasive formations.	1.									
			2.									
			3.					RG2BXJ				
			4.									

SMITH TOOL COMPANY

	Series	Formations	Types	1	2	3	4	5	6	7	8	9
MILLED-TOOTH BITS	1.	Soft formations having low compressive strength and high drillability.	1.	DS			SDS				DJ	BHDJ
			2.	DT	DTT		SDT					
			3.	DG	DGT	DGH	SDG	SDGH				
			4.	K2		K2H						
	2.	Medium to medium hard formations with high compressive strength.	1.	V1		V1H						
			2.	V2		V2H	SV	SVH				
			3.	T2		T2H	ST2					
			4.									
	3.	Hard, semi-abrasive or abrasive formations.	1.	L4		L4H	SL4	SL4H				
			2.	W4		W4H						
			3.									
			4.	WC		WCH	SWC	SWCH				
INSERT BITS	5.	Soft to medium formations with low compressive strength.	1.									
			2.									
			3.					3JS		SS3		
			4.									
	6.	Medium hard formations of high compressive strength.	1.									
			2.					4JS		SS4		
			3.					47JS				
			4.									
	7.	Hard, semi-abrasive and abrasive formations.	1.					5JS				
			2.					6JS		SS6		
			3.					7JS				
			4.									
	8.	Extremely hard and abrasive formations.	1.					8JS				
			2.									
			3.					9JS				
			4.									

Figures 14–32 and 14–33 illustrate severe erosion. Note that the cone face is void of definition. This condition is often accompanied by a circumferential crack at the roller where the cone is thinnest. The solution is to check metallurgy, reduce pump rate, and reduce mud weight and solids.

If only the inner rows of inserts are broken off as in Figure 14–34, the formation is too hard for the bit. The tangential loads are excessive. The solution is to run a harder formation bit with less offset for a long interval or reduce bit weight over a short interval.

Figures 14–35 and 14–36 illustrate bearing failure. Note that the cones have been running together to the extent that definite ridges have been cut in each cone by the inserts from adjacent cones. Also, note that in Figure 14–35 the interference is so severe that one cone has been dragging on bottom.

The random breakage illustrated in Figure 14–37 probably resulted from junk. Figure 14–38 illustrates a broken spear point on the No. 1 cone, and Figure 14–39 is the result of continued drilling once the spear point has been broken off.

Proposed classification of rock bits is shown in Table 14–4.

Corrosion Control in Drilling Operations

CHARLES C. PATTON
PetroTech, Ltd.
London, England

T H E cost of drill pipe failure has been estimated to be approximately $1 per foot of hole drilled.[1] This is a significant fraction of drilling costs and is a point of extreme concern to those in the drilling business.

Since some form of corrosion is the cause of the majority of drill string failures an understanding of the fundamentals of corrosion control is mandatory for minimum cost drilling.

THEORY OF CORROSION

Most metals are found in nature as metallic oxides or salts. Refining to produce pure metal requires a large energy input. This energy is "stored" and is available to supply the necessary driving force to return the metal to its original state – an oxide or salt. This means that metals are unstable with respect to most environments and have a natural tendency to return to their original lower energy state, or "corrode."

Corrosion is an electrochemical process. This means that electrical current flows during the corrosion process. In order for current to flow, there must be a driving force, or a voltage source, and a complete electrical circuit.

Voltage Source

The source of voltage in the corrosion process is the energy stored in the metal by the refining process. Different metals require different amounts of energy for refining and therefore have different tendencies to corrode. This is illustrated in the following table:[2]

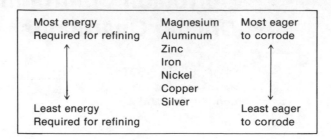

| Most energy Required for refining ↑ ↓ Least energy Required for refining | Magnesium Aluminum Zinc Iron Nickel Copper Silver | Most eager to corrode ↑ ↓ Least eager to corrode |

The Electrical Circuit

The electrical circuit of the corrosion process consists of four parts.

1. Anode

This is the point where metal dissolves or goes into solution. The chemical reaction for iron is:

$$Fe \rightarrow Fe^{++} + 2e$$

2. Cathode

Metal does not dissolve at the cathode although a chemical reaction does occur at this point.
If oxygen is absent it is often:

$$2H^+ + 2e \rightarrow H_2 \uparrow,$$

or if oxygen is present,

$$O_2 + 2H_2O + 4e \rightarrow 4OH^-$$

in neutral or basic solutions.

3. Electrolyte

A solution capable of conducting electricity is called an "electrolyte." Water is an electrolyte which increases in conductivity as the amount of dissolved salts or ions increase. The electrolyte conducts current from the anode to the cathode.

4. Electronic Conductor

The anode and cathode must be connected by something which will conduct electrons (electrical current) in order to complete the circuit and provide a path for current to flow from the cathode back to the anode.

A very simple example of a corrosion cell (that is useful) is a typical flashlight battery sketched in cross-section below:

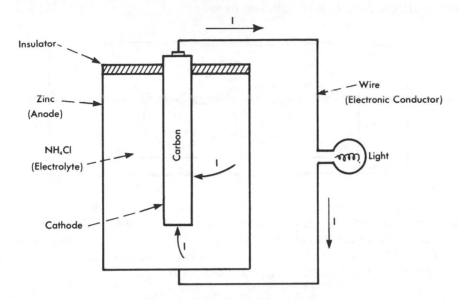

The zinc case acts as the anode (corrodes), the ammonium chloride is the electrolyte, and the carbon rod acts as the cathode (does not corrode). The zinc case and the carbon rod are prevented from touching each other by an insulator. However, once we connect the carbon rod (cathode) and the zinc case (anode) with a piece of wire, the circuit is completed and current begins to flow from the anode through the electrolyte to the cathode, and then through the wire back to the anode to complete the circuit. Current will continue to flow until the zinc corrodes away, or until we disconnect the wire and break the circuit.

Now the question arises as to how all of this applies when a single piece of metal, such as steel, corrodes.

The Nature of Steel

Steel is primarily an alloy of iron and carbon. Pure iron is a relatively weak, ductile material. When it is alloyed with small amounts of carbon (usually 0.2 to 1.0 per cent), a much stronger material is created. However, as a result of reacting part of the iron with carbon, we now have a metal composed of two materials: pure iron and iron carbide (Fe_3C), the product of the iron-carbon reaction.

The iron carbide is distributed within the iron as tiny microscopic islands, and these islands of iron carbide have a lower tendency to corrode than does the pure iron. The Fe_3C and the pure iron are in intimate contact (allowing electron flow), so when the steel is placed in water (an electrolyte), the electrical circuit is complete and current flows through thousands of tiny microcells on the steel surface. The pure iron acts as the anode and corrodes, while the Fe_3C acts as the cathode.

If we were to look through a microscope and concentrate on two adjacent grains of Fe and Fe_3C on the surface of a piece of steel immersed in water, it might look like the sketch below:

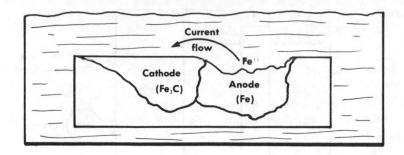

The important thing to realize is that current flows during the corrosion process, and the amount of current flowing is a measure of the severity of the corrosion. It is dependent on both the nature of the metal and the type of environment, or electrolyte. One ampere of current flowing for one year represents the loss of twenty pounds of iron.

THE EFFECT OF METAL COMPOSITION

As previously stated, different metals have different tendencies to corrode. However, a discussion of a wide range of metals is not really warranted here. Steel is the primary metal used in the oilfield and so we will stick pretty closely to the subject of how and why water corrodes steel.

There are, of course, many "steels." Simple low carbon steel is universally used in production operations for lines, tubing, tanks and treaters. However, some steels containing chromium and nickel (which add corrosion resistance) are used for a few items, and some non-ferrous alloys (contain no iron) are used, particularly in pumps.

THE EFFECT OF ENVIRONMENT COMPOSITION

Conductivity

As previously stated, the corrosivity of an electrolyte increases as the conductivity increases. Distilled water is not very conductive and is not very corrosive. Salt water is quite conductive and is corrosive. The corrosivity increases as the water gets saltier.

pH of Electrolyte

The corrosivity of the electrolyte usually increases as the pH decreases (becomes more acidic). At higher pH values protective scales (iron hydroxide, or carbonate scale) may form on the steel surface and prevent or slow down further corrosion.

Dissolved Gases

Oxygen, carbon dioxide or hydrogen sulfide dissolved in water drastically increases its corrosivity. In fact, dissolved gases are the primary cause of most oilfield corrosion problems. If they could be excluded and the electrolyte maintained at a neutral pH or higher, most systems would have very few corrosion problems.

Dissolved Oxygen

Of the three dissolved gases mentioned, oxygen is by far the worst of the group and is the major cause of the corrosion of drill pipe.[3] It can cause severe corrosion at very low concentrations (<1 ppm), and if either or both of the other two gases are dissolved in the water it drastically increases their corrosivity.

The solubility of oxygen in water is a function of pressure, temperature, and the chloride content. Oxygen is less soluble in salt water than in fresh water.

> (a) Mechanism of Oxygen Corrosion
> Anode Reaction $Fe \rightarrow Fe^{++} + 2e$
> Cathode Reaction $O_2 + 2H_2O + 4e \rightarrow 4OH^-$
>   or, combining the two
> $4Fe + 6H_2O + 3O_2 \rightarrow 4Fe(OH)_3 \downarrow$

Providing the pH is above 4, ferric hydroxide is insoluble and precipitates as shown.

Oxygen accelerates corrosion drastically under most circumstances. It does this in two ways. First, it acts as a "depolarizer." This means that it will consume electrons at the cathode and allow the corrosion reaction to proceed at a rate limited primarily by the rate oxygen can diffuse to the cathode. Without oxygen, the energy it takes to evolve hydrogen gas from the cathode is a major bottleneck in the corrosion reaction and keeps it slowed down. When oxygen is present, it acts as an additional acceptor of electrons at the cathode surface and allows the reaction to speed up.

Secondly, the oxidation of ferrous ions (Fe^{++}) to ferric ions (Fe^{+++}) causes the reaction to speed up if the pH is above 4. This happens because ferric hydroxide is insoluble and precipitates from solution. If this occurs away from the metal surface, then the corrosion reaction speeds up as it attempts to maintain equilibrium by supplying more Fe^{++} ions to the water solution. If the Fe^{++} are rapidly oxidized to Fe^{+++}, it is a hard race to win and the corrosion reaction runs rampant. So:

$$\text{Normally, } Fe \rightleftharpoons Fe^{++} + 2e,$$
$$\text{but if} \quad Fe^{++} \rightleftharpoons F^{+++} + e$$
$$\text{then} \quad Fe \rightleftharpoons Fe^{++} + 2e$$

Oxygen attack is often pitting in nature.

It is important to realize that the corrosivity of "pure" water increases as the dissolved oxygen content increases—up to a point. If there is sufficient oxygen in the water, the $Fe^{++} \rightarrow Fe^{+++}$ oxidation may occur very rapidly, before the Fe^{++} ions have a chance to diffuse away from the metal surface. In this case the $Fe(OH)_3$ can form on the metal surface and become protective. However, if sufficient chloride ions are present, they interfere with the formation of a protective film and corrosion rates continue to increase with oxygen concentration.

Very small oxygen concentrations (< 1 ppm) can be very damaging. Also, because of its role as a depolarizer, it will drastically increase the corrosivity, resulting from other dissolved gases such as H_2S or CO_2.

Concentration cells, or differential aeration cells, can cause preferential attack or pitting. Any time there is a difference in the oxygen content of the electrolyte in two areas of a system, attack will take place preferentially in the area exposed to the lowest oxygen concentration. Typical examples are water-air interfaces, crevices and "oxygen tubercles" in water systems.

Dissolved Carbon Dioxide

When carbon dioxide dissolves in water it forms carbonic acid, decreases the pH of the water and increases its corrosivity. It is not as corrosive as oxygen, but usually results in pitting.

The solubility of CO_2 in water, like all gases, is a function of the partial pressure of CO_2 in the atmosphere above the water. The greater the partial pressure, the greater the solubility. Therefore, corrosion rates go up as the partial pressure of CO_2 increases.

In waters containing bicarbonate alkalinity, the amount of CO_2 present in solution to cause corrosion is a function of pH due to the CO_2-bicarbonate-carbonate equilibria.

$$CO_2 + H_2O \rightleftharpoons H_2CO_3$$
$$H_2CO_3 \rightleftharpoons H^+ + HCO_3^-$$
$$HCO_3^- \rightleftharpoons H^+ + CO_3^=$$

As the pH decreases (the number of hydrogen ions increases), bicarbonate ions convert to carbonic acid, more CO_2 results and corrosion increases. Conversely, as the pH rises, the water will tend to become scale forming and the corrosivity will decrease. It should be mentioned that a reduction in system pressure which allows CO_2 to come out of solution can accomplish this rise in pH.

As previously mentioned, the presence of any oxygen increases the corrosivity of CO_2.

Dissolved Hydrogen Sulfide

Hydrogen sulfide is very soluble in water and, when dissolved, behaves as a weak acid. It usually causes pitting. The combination of H_2S and CO_2 is more aggressive than H_2S alone and is frequently found in oilfield environments. Once again, the presence of even minute quantities of oxygen can be disastrous.

Another problem with H_2S is that some of the hydrogen ions at the cathodic areas will enter the steel instead of evolving from the surface as a gas. This can result in hydrogen blistering in low strength steels or hydrogen embrittlement in high strength steels.

Hydrogen sulfide can also be generated by sulfate reducing bacteria and by thermal degradation of organic additives.

Physical Variables

Water Temperature

Corrosion rates usually increase as temperature increases because all of the reactions involved speed up.

In a system open to the atmosphere, corrosion rates may increase at first and then with a further temperature increase, the corrosion rate may drop off due to dissolved gases coming out of solution. If the system is closed, the corrosion rate will continue to increase with temperature, because the dissolved gases have no place to go.

When the water contains bicarbonates, increased temperature will promote scale formation which may slow down the corrosion reaction. However, it can also lead to decomposition of bicarbonates and produce additional CO_2.

System Pressure

Pressure also has an effect on chemical reactions. The primary importance of pressure is its effect on the solubility of dissolved gases. More gas goes into solution as the pressure is increased.

Velocity

(a) Stagnant or low velocity water usually gives a low general corrosion rate, but pitting is more likely.

(b) Corrosion rates usually increase with velocity, but there are exceptions.

(c) High velocities and/or the presence of suspended solids or gas bubbles can lead to erosion-corrosion. Any protective film is constantly removed or eroded away, leaving a bare metal surface which eagerly corrodes.

(d) High-Low velocity areas

Systems may contain adjacent areas which are exposed to different water velocities. If no oxygen is present the high velocity area is anodic to the low velocity area and corrodes. If oxygen is present, the low velocity area receives less oxygen, acts as the anode, and corrodes.

(e) Increased velocity in an oxygenated system can cause an initial increase in corrosion rate (supplies more oxygen), then actually slow the reaction as the velocity is further increased due to formation of $Fe(OH)_3$ on the metal surface. A further increase may rip the protective film off of the surface.

(f) Cavitation

Cavitation can result from high velocity where vapor bubbles can form for a split second (due to a drop in pressure), then collapse (due to a rise in pressure). The collapse of the bubbles physically rips tiny pieces of metal from the surface. This often happens in pumps.

FORMS OF CORROSION

Thus far we have been primarily concerned with the corrosion of a piece of steel immersed in an electrolyte such as water. The corrosion may be uniform in nature, resulting in uniform thinning, or it may be localized in the form of discreet pits or larger localized areas. Localized corrosion is usually the most disastrous and the toughest to detect.

Localized attack takes several different forms and is caused by a variety of different situations.

Bimetallic Corrosion

When two different metals are placed in contact in an electrolyte, the more reactive one will corrode and the other one will not. Thus, if copper and steel are connected together and placed in water, the steel will corrode. This principle is utilized in a beneficial way in cathodic protection. Steel is connected to a more reactive metal, such as zinc, aluminum or magnesium, and is "protected" (does not corrode). The steel has become a cathode and the more reactive metal the anode.

Mill scale is less reactive than steel, so if pipe is covered with mill scale, which is subsequently partially removed or knocked off, the mill scale will act as the cathode and the bare steel will act as the anode and corrode.

A similar phenomenon often occurs when new pipe is connected to old pipe. The new pipe acts as the anode and corrodes preferentially.

Localized corrosion can result from heat treatment gradients, and is really a form of bimetallic corrosion. When pipe is welded, the welding process results in a localized heat treatment, creating a steel microstructure near the weld which differs from the microstructure of the parent steel. The two areas have different tendencies to corrode and "weld line" corrosion can occur. This is avoided by proper welding practices.

Concentration Cells

Localized corrosion can also be created by differences in electrolyte composition at two points on the metal surface.

Crevice Corrosion

Crevices promote the formation of concentration cells. This is especially serious in oxygenated systems where the oxygen in the crevice may be consumed more rapidly than fresh oxygen can diffuse into the crevice.

This causes the pH in the crevice to decrease, resulting in a more acidic environment, which accelerates corrosion.

Air-Water Interface

The water at the surface contains more oxygen than the water slightly below the surface. The difference in concentrations will cause preferential attack at the water line.

Oxygen Tubercles

This is a form of pitting which results from the same type of mechanism as crevice corrosion. It is encouraged by the formation of a porous layer of iron oxide or hydroxide which partially shields the steel surface.

Scales and Sludges

The deposition of any solid on a metal surface which is not sufficiently tight and non-porous to completely protect the metal surface can cause increased corrosion under the deposit. Even tight, adherent scales can create problems if they form only in spots rather uniformly over the metal surface. Sulfate reducing bacteria thrive under scales and sludges, creating H_2S and causing localized pitting.

THE EFFECT OF STRESS: FATIGUE AND CORROSION FATIGUE OF STEEL

Corrosion fatigue has been recognized as the predominant cause of drill string failures for many years. As such, it deserves special attention.

Fatigue of Metals in Air

Before discussing corrosion fatigue, it seems appropriate to first give a very brief review of metal fatigue concepts and terminology.

When metals are repeatedly stressed in a cyclic manner in air they will fail at stresses far below the yield or tensile strength. There is however, a limiting stress below which the material may be cyclically stressed indefinitely without failure. This stress is called the endurance limit and is always lower than the yield and tensile strengths. The performance of materials subjected to cyclic stressing is normally described by plotting the stress at failure against the logarithm of the number of cycles to failure for a series of stress levels. This type of plot is known as an S-N curve (Figure 15-1).

The endurance limit for ferrous metals is usually 40% to 60% of the tensile strength, depending upon the microstructure and heat treatment. (Figure 15-2) Rapid quenched and tempered steels normally have better fatigue properties than slow cooled carbon steels. Since the tensile strength of ferrous metals is roughly proportional to hardness over a wide range, the endurance limit is also proportional to hardness over this range.

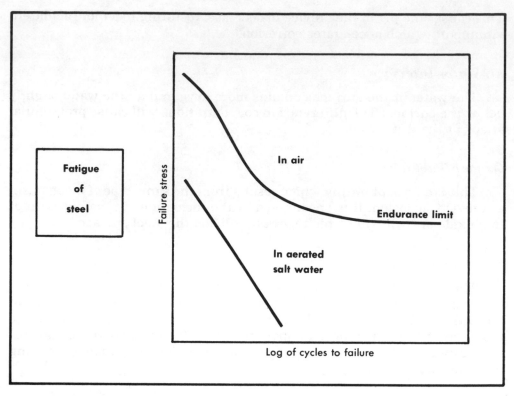

FIG. 15–1 (Ref. 4). *Typical S-N curves for steel in air and in aerated salt water*

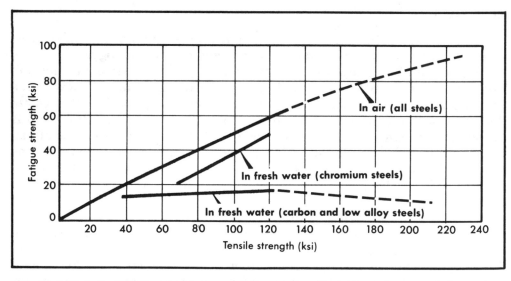

FIG. 15–2 (Ref. 4). *Relationship between fatigue strength and tensile strength for carbon, low-alloy, and chromium steels in air and in fresh water*

Since fatigue cracks usually start at the metal surface the fatigue performance of any item is drastically effected by surface conditions. Notches or metal inhomogenities such as inclusions or porosity act as stress risers and the actual stress at the root of a notch may be many times the nominal applied stress. Heat treating processes can decarburize the surface of a metal part and the fatigue strength is dependent on the strength of this lower carbon content surface layer.

Fatigue performance is also influenced by the stress history to which the part in question has been subjected. Prior cyclic stressing at a high stress reduces the fatigue life of a part subsequently cycled at a lower stress, when both stresses are above the endurance limit. Little or no fatigue damage may result from cyclic pre-stressing at low stresses. In fact, the fatigue performance is sometimes increased by a procedure known as coaxing, where the specimen is subjected to a series of stress cycles at successively increasing stress levels starting at some stress below the endurance limit.

Corrosion Fatigue—The Effect of a Corrosive Environment On Fatigue Life

The fatigue life of a metal is substantially reduced when the metal is cyclically stressed in a corrosive environment. The simultaneous occur-

Corrosion fatigue of steel in brine	
Dissolved gas	**% Decrease from air endurance limit**
H_2S	20%
CO_2	41%
$CO_2 + air$	41%
$H_2S + air$	48%
$H_2S + CO_2$	62%
Air	65%

FIG. 15–3 (Ref. 4). *Effect of dissolved-gas composition on corrosion fatigue performance of steel in salt water*

rence of cyclic stress and corrosion is called corrosion fatigue. Hence, the presence of the corrosive media augments the fatigue mechanism and hastens failure as a result of corrosion. Hydrogen embrittlement can also contribute to fatigue failure in high strength steels. The critical characteristic of corrosion fatigue which must be recognized is that the metal no longer exhibits an endurance limit as it did in air. (Figure 15–1) Corrosion fatigue performance is normally characterized by the "fatigue limit" which is an arbitrarily defined quantity. The corrosion fatigue limit is commonly defined as the maximum value of stress at which failure no longer occurs after 10^7 cycles.

In corrosion fatigue the corrosivity of the environment becomes extremely important. The presence of dissolved gases such as H_2S, CO_2 or air causes a pronounced increase in corrosivity resulting in decreased corrosion fatigue life (Figure 15–3). A pitting or localized attack is most damaging from the standpoint of corrosion fatigue, but even slight general corrosion will substantially reduce fatigue life.

Other variables which affect corrosion rates are also important. The fatigue life of carbon steel in salt water or drilling mud has been shown to increase markedly above pH 11 (Figure 15–4).

The corrosion fatigue performance of carbon and low alloy steels is in-

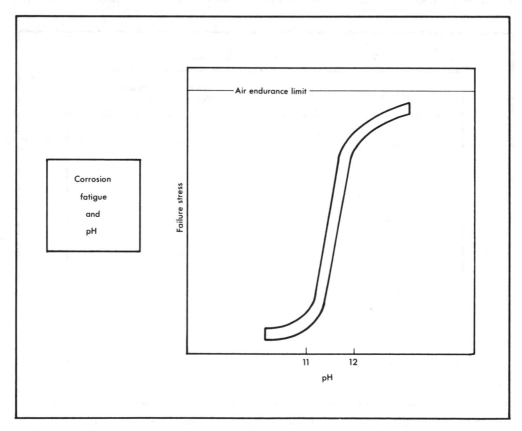

FIG. 15–4 (Ref. 4). *Influence of pH on the corrosion fatigue behavior of carbon steel*

dependent of strength.[4] This has been shown both for aerated fresh water (Figure 15-2) and for brine containing H_2S, CO_2, air, or some combination. Hence, heat treatment and alloying are most important as related to corrosion resistance rather than to physical properties.

Stress history and surface conditions are just as important in corrosion fatigue as in air fatigue performance. Notches also remain critical.

It must be realized that the frequency of the applied stress and therefore the time available for corrosion damage is extremely important in the corrosion fatigue behavior of materials. Tests which are run at high cycle speeds may show little deviation from air fatigue performance over a wide portion of the S-N curve. In contrast the same tests repeated at a very slow cycle speed will show a significantly smaller number of cycles to failure for any given stress. Therefore cycle speed must be considered as a major variable in corrosion fatigue performance.

THE EFFECT OF STRESS: SULFIDE CRACKING (HYDROGEN EMBRITTLEMENT)

When hydrogen atoms are formed on a metal surface by a corrosion reaction, some of them combine to form gaseous molecular hydrogen (H_2) and are released to the environment. However, a portion of the atoms are absorbed by the metal. When hydrogen sulfide is present, the sulfide ion reduces the rate at which the atoms combine, creating a larger concentration of atomic hydrogen on the metal surface. Thus, a greater portion of the hydrogen atoms enter the metal.

The presence of atomic hydrogen in steel reduces its ductility and it tends to break in a brittle fashion rather than bend. This phenomenon is known as embrittlement and affects only high strength steels, generally those having yield strengths of 90,000 psi or higher. If the metal is under a high tensile stress, brittle failure can occur.

Failures due to hydrogen embrittlement often do not occur immediately after application of the load or exposure to the hydrogen producing environment. Usually there is a time period during which no damage is observed, followed by a sudden, catastrophic failure. This phenomenon is referred to as delayed failure. The time period prior to failure is referred to as the incubation period, during which hydrogen is diffusing to points of high triaxial stress.

The time to failure decreases as the amount of hydrogen absorbed, applied stress, and strength level of the material increases.

Until a steel containing hydrogen actually cracks, there is no permanent damage. In many cases, the hydrogen can be baked out by suitable heat treatments and the original properties of the steel restored.

Spontaneous brittle failure that occurs in steels and other high strength alloys when exposed to moist hydrogen sulfide and other sulfidic environments is frequently referred to as sulfide cracking. It is generally thought to be a form of hydrogen embrittlement.

Although the mechanism of sulfide cracking is not completely understood, it is generally accepted that the following conditions must be present before cracking can occur:

1. Hydrogen sulfide
2. Water—even a trace amount of moisture is sufficient
3. A "high strength" steel.* The exact strength level varies with the composition and microstructure of the steel.
4. The steel must be under tensile stress or loading. The stress may be residual or applied.

If all of these conditions are present, sulfide cracking may occur after some period of time. It is important to realize that sulfide cracking usually

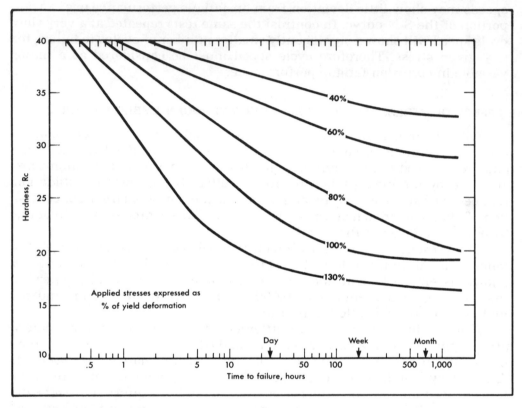

FIG. 15-5 (Ref. 9). *Approximate failure time vs. hardness and applied stress for carbon steel, 3,000 ppm H_2S in a 5% solution of NaCl*

does not occur immediately after exposure to the sour environment, but may take place after hours, days, or years of service. The susceptibility of a material to failure by this mechanism is primarily determined by the following variables.

1. Yield strength or hardness

"Plain" carbon steels with yield strengths below 90,000—100,000 psi are generally considered to be immune to sulfide cracking.[5] This corresponds to a hardness of R_c22. Strengths above this level are susceptible to cracking. The higher the strength, the shorter the time to failure (Figure 15-5).

---

* Other materials are also susceptible to sulfide cracking.

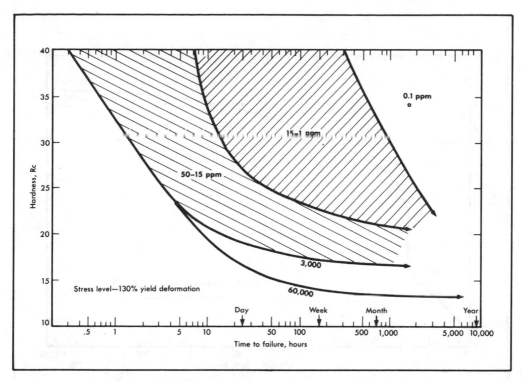

FIG. 15–6 (Ref. 9). *Approximate failure time of carbon steel in 5% NaCl and various ppm of H₂S*

If steel is alloyed with other materials, such as nickel, failure can occur at hardness levels less than R_c22. Conversely, certain heat treatments can raise the maximum permissible hardness level above this value.

2. Stress level (either residual or applied)

The time to failure decreases as the stress level increases. (Figure 15–5) In most cases the stress results from a tensile load or from the application of pressure, or both. However, residual stresses and hard spots can be created by welding or by cold-working the material. (Cold bending, wrench marks, etc.)

3. Hydrogen sulfide concentration

The time to failure increases as the H_2S concentration decreases.[5] Delayed failures can occur at very low concentrations of H_2S in water (0.1 ppm)[5] and at partial pressures as low as 0.001 atmosphere,[6] although the time to failure becomes very long (Figure 15–6).

4. pH of Solution

The cracking tendency increases as the pH decreases.[5] Failures can be drastically reduced if the pH of the solution is maintained above 9.0 (Figure 15–7).

5. Temperature

There is a considerable amount of data indicating that cracking susceptibility decreases above approximately 150°F. This phenomenon is not well defined at this time.

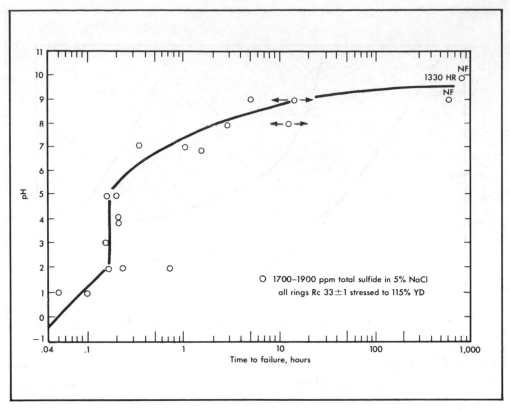

FIG. 15–7 (Ref. 9). *Steel fails much less rapidly when pH is above seven or eight*

CORROSION PREVENTION

There are a number of possible approaches to corrosion control. In most cases the primary objective in drilling operations is to extend corrosion fatigue life rather than to simply prevent metal loss.

Conventional corrosion control procedures have proven to be quite effective in extending corrosion fatigue life. However, both field experience and published data indicates that the requirements are often much more stringent in the case of corrosion fatigue control than in the case of normal weight loss corrosion control. It seems logical to conclude that this is due to the fact that a single corrosion fatigue crack can result in catastrophic failure. Thus, even though crack initiation may have been prevented over 99% of the metal surface, cracks initiating in the remaining unprotected 1% can be fatal.

When drilling in sour areas, there is the additional concern of preventing sulfide cracking.

Regardless of the specific problem, the basic philosophy behind failure prevention is the same: the corrosion reaction must be slowed to an acceptable rate using conventional corrosion control procedures.

There are several methods of corrosion control which can be considered in drilling operations.

Alter Metal or Design

As previously shown, the use of higher strength carbon steel or low alloy steels will result in little if any improvement in corrosion fatigue life. The cost of high alloy steels is prohibitive in most situations and it would be extremely difficult to economically justify high alloy drill pipe.

In short, the use of materials which are immune to attack by oil field production environments is normally prohibited by cost except in extremely severe situations. This simply means that other alternatives are usually much more economically attractive.

It is always advisable to examine the situation and see if the cyclic stress can be reduced or removed.

Low strength steels should be used when operating in a sour environment to minimize the possibility of sulfide cracking.

The use of heavy wall pipe will reduce the stress and help to reduce both fatigue and cracking problems. Also, good handling practices should be followed so that notches due to rough handling are avoided or minimized.

Alter Environment

Another approach to corrosion control is to remove or neutralize the corrodents.

As previously mentioned, maintenance of a high pH mud system can be extremely beneficial from the standpoint of both corrosion fatigue and sulfide cracking.

Since dissolved oxygen is so damaging, one approach is to remove the oxygen from the mud using a chemical scavenger such as sodium sulfite. This chemical reacts with oxygen to produce sodium sulfate.

Scavengers can also be added to a drilling fluid to remove small amounts of hydrogen sulfide. Copper carbonate has been used in the past, but is not recommended because of its adverse effect on corrosion rates and corrosion fatigue life.[7] This results from the fact that the copper plates out on the steel creating many tiny copper-steel limetallic corrosion cells.

Another form of electrolyte modification is the use of oil base muds in place of water base fluids. The low conductivity of an oil base mud makes it considerably less corrosive than most water base fluids.

Coatings

Coatings are another common approach to corrosion control utilized in petroleum production. Plastic coatings have been widely accepted for various applications. Both laboratory data and field experience show that plastic coatings do work; however, some coatings seem to work better than others. One of the primary problems is that it is almost impossible to obtain holiday-free coatings. Holidays or mechanical damage which results in the removal of the coating from part of the surface can completely nullify the beneficial effect of the coating. Coating inspection immediately after application and at regular intervals during service is essential.

There is general agreement that plastic coating of drill pipe has resulted in drastically increased service life for drill strings. This is probably the best data point we have on the effectiveness of plastic coatings for a corrosion fatigue application in the petroleum production industry.

Corrosion Inhibitors

Corrosion inhibitors have been shown to be effective against weight loss corrosion and can also be helpful for corrosion fatigue. It is difficult to completely prevent sulfide cracking with a corrosion inhibitor.

Inhibitors are normally wiped onto the drill pipe while going into the hole or batched down the pipe.

In conclusion, successful control of corrosion in drilling operations normally entails a combination of control techniques. Attention to drill pipe materials, handling procedures, control of mud properties and the use of coatings and inhibitors are all important and must often be used in concert to achieve acceptable drill pipe life. The application of a single control technique seldom gives the desired results.

RECOMMENDED PRACTICES

Sour Service

1. Use drill pipe corresponding to C-75 tubing or casing specifications. Some Grade E meets this requirement. Grade D drill pipe not usually susceptible to sulfide cracking. Should hardness screen pipe and throw out hard joints.
*2. Use controlled makeup on tool joints — stay at the low end of recommended torque range.
3. Avoid use of low-friction lubricants on small joints. Allows high stresses to be "locked in" with low makeup torque.
*4. When temperatures above 300°F are anticipated, sulfur containing drilling fluid additives and thread compounds should be avoided.[8]
5. Train crews to avoid impact loads in breaking out joints.
*6. Try to avoid creating notches. Common causes are tongs and slips.
7. Limit exposure of drill string to H_2S on drill stem tests. A maximum exposure time of 1 hour has been recommended. Can run inhibited cushion. Can circulate string through pumpout sub after tool is closed.
*8. Internally coat drill pipe and keep it coated. Inspection during coating strongly recommended. Inspection at regular intervals during usage necessary to determine coating condition.
*9. Inhibitors can also be used. Organic inhibitors are available for continuous addition to mud system or batch application. Can be batched down pipe, sprayed on pipe or wiped on pipe.
*10. Use of copper carbonate to reduce H_2S content of mud not recom-

mended. Copper will "plate out" on steel and set up bimetallic corrosion cell. Steel will corrode.

*11. Maintain a mud pH above 11.0. This will decrease both cracking and fatigue problems.

*12. Primary problem in drill strings in sour service is tool joints which normally have yield strengths of 120,000 – 135,000 psi and hardnesses of Rc 30 – Rc 37. Desperately need lower strength materials, even if they are heavier wall.

*13. Heavy wall pipe will reduce the stress level and extend life. Recommended for severe service applications.

*14. Throw out badly damaged joints. Set up procedure to spot them and get them out of the string.

*15. Perform regular electronic and visual inspection of drill pipe. Look for pitting, fatigue cracks, coating damage. Don't hesitate to downgrade if damage is indicated.

*16. Consider using an oil-base drilling fluid.

Normal Service (No H₂S)

When H_2S is not present or anticipated, the list of recommendations can be shortened. Here, only the recommendations marked with an asterisk need to be considered.

General

A major amount of drill pipe corrosion occurs while the pipe is out of the hole. This form of corrosion can be minimized by the following practices:

1. Wash down pipe in derrick with fresh water.
2. Spray pipe with an atmospheric corrosion inhibitor for storage or transit.

REFERENCES

1. Shaffer, Paul D.: "Malfunctions Push Well Costs Out of Proportion," *World Oil*, Vol. 173, No. 2, p. 57 (Aug. 1) 1971.

2. Patton, C. C.: *Oilfield Water Systems*, John M. Campbell & Co., Norman, Oklahoma (1974).

3. Bradley, B. W.: "Oxygen Cause of Drill Pipe Corrosion," *Petroleum Engineer*, Vol. 42, No. 13, p. 50 (Dec.) 1970.

4. Patton, C. C.: "Corrosion Fatigue Problems in Petroleum Production," presented at Corrosion/71 National Conference of NACE, Chicago, Ill. March, 1971.

5. Hudgins, C. M., McGlasson, R. L., Mehdizadeh, P., and Rosborough, W. M., "Hydrogen Sulfide Cracking of Carbon and Low Alloy Steels," *Corrosion*, Vol. 22, No. 8, p. 238 (Aug.) 1966.

6. Treseder, R. S. and Swanson, T. M., "Factors in Sulfide Corrosion Cracking of High Strength Steels," *Corrosion*, Vol. 24, No. 2, p. 31 (Feb.) 1968.

7. Perricone, A. C. and Chesser, B. G.: "Corrosive Aspects of Copper Carbonate Drilling Fluids," *Oil & Gas Journal*, Vol. 68, No. 37, p. 82 (Sept. 14) 1970.

8. Mauzy, H. L.: "Minimize Drillstem Failures Caused by Hydrogen Sulfide," *World Oil*, p. 65 (Nov.) 1973.

9. Hudgins, C. M., Jr. "Hydrogen Sulfide Corrosion Can be Controlled," *Petroleum Engineer*, Vol. 42, No. 13, p. 33 (Dec.) 1970.

16 Cements and Cementing

DWIGHT K. SMITH
Technical Supervisor
Halliburton Research

OIL-WELL cementing is the process of mixing and displacing a cement slurry down the casing and up the annular space behind the pipe where it is allowed to set, thus bonding the pipe to the formation. No other operation in the drilling or completion process plays as important a role in the producing life of the well as does a successful primary cementing job (Figure 16–1).

The first verified use of portland cement in an oil well, for shutting off water that could not be held with a casing shoe, was in 1903.[1] After placing the cement, the operator normally waited 28 days before drilling the cement and testing. Improvements in cements, understanding WOC times, and the use of admixes have reduced WOC time to a few hours under present-day practices.

Cementing procedures may be classified into primary and secondary phases. Primary cementing is performed immediately after the casing is run in the hole. Its objective is to obtain an effective zonal separation and help protect the pipe itself. Cementing also helps in:

1. Bonding the pipe to the formation.
2. Protecting producing strata.
3. Minimizing the danger of blowouts from high pressure zones.
4. Sealing off "lost-circulating zones" or other troublesome formations as a prelude to deeper drilling.

Secondary cementing, or squeeze cementing, can be described as the process of forcing a cement slurry into holes in the casing and cavities behind the casing. These operations are usually performed for repairing or altering a well at some later date or it may be used during the initial drilling process. Squeeze cementing is necessary for many reasons, but probably

400

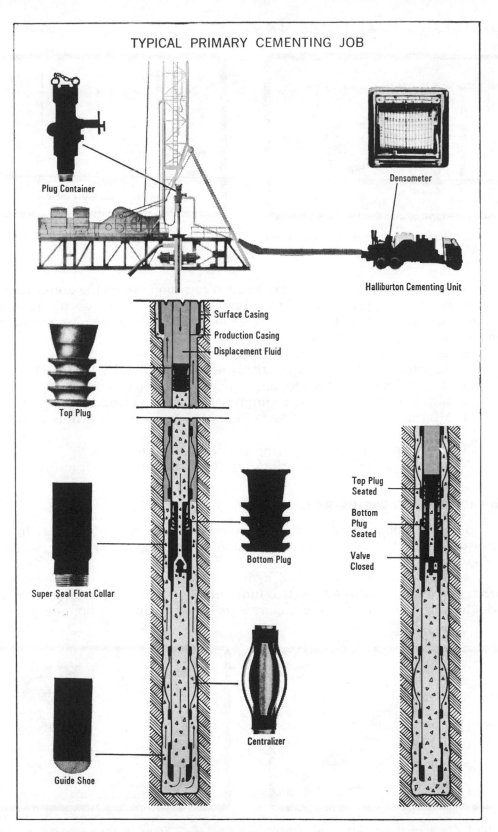

FIG. 16–1. *Typical primary cementing job*

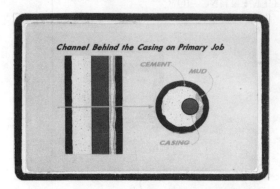

FIG. 16–2. *Channel behind the casing on primary job*

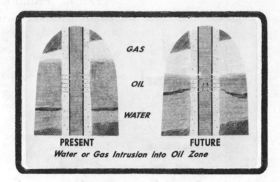

FIG. 16–3. *Water or gas intrusion into oil zone*

the most important use is to segregate hydrocarbon producing zones from those formations producing other fluids. The key factor on a squeeze cementing job is that of placing the cement at the desired point or points necessary to accomplish the purpose.

Squeeze cementing is also employed to:

1. Supplement or repair a faulty primary cementing job (Figure 16–2).
2. Reduce the gas-oil, water-oil, or water-gas ratio (Figure 16–3).
3. Repair defective casing or improperly placed perforations.
4. Minimize the danger of lost circulation in open hole while drilling deeper.
5. Abandon permanently a nonproductive or depleted zone.
6. Isolate a zone prior to perforating for production or to fracture.

MANUFACTURE AND CHEMISTRY OF CEMENT

The development of portland cements started when man calcined gypsum and later limestone to produce a bonding material for stone[2] (Figure 16–4).

As a result of efforts to find hydraulic cements which could be used under water, it was discovered that limes produced from impure limestones yielded mortars which were superior to those produced from the purer

FIG. 16–4. *Steps in the evolution of cement*

FIG. 16–5. *Steps in the evolution of cement*

limestones. Such discoveries led to the burning of blends of calcareous and argillaceous materials and to the granting of a patent by the government of Great Britain in 1824 to Joseph Aspdin for the manufacture of a cement for the construction of a lighthouse.[3] It was called "portland cement" because concrete produced from it resembled stone quarried on the Isle of Portland off the coast of England.

In order to understand the nature of the cement-hydration process, it is necessary to explain briefly the chemical reactions which occur in the cement kiln[4] (Figure 16.5).

The kiln feed consists of a blend of finely ground calcareous and argillaceous materials. These materials are primarily limestone or other materials high in calcium carbonate content, clay or shale, and some iron and calcium chlorides, if they are not present in sufficient quantity in the clay or shale.[5,6] These dry materials are finely ground and mixed thoroughly in the correct proportions either in the dry condition (dry process) or mixed with water (wet process). (Figure 16.6).

This raw mixture is then fed into the upper end of a sloping, rotary kiln, at a uniform rate, and slowly travels to the lower end. The operating temperature at the fired end of the kiln is between 2,600° and 2,800°F. During its travel through the kiln, the material is converted to cement clinker which varies in size from dust to about 2 inches in diameter. This clinker generally passes from the kiln to either a grate or rotary cooler where it is quenched with air.

In the hot zone of the kiln, about 25% of the clinker is in the form of a liquid. Some of this fails to crystallize during the quenching process and is present in the cooled clinker in the form of supercooled liquid or glass. Well-quenched clinkers will contain as much as 10 to 15% glass as determined by microscopic methods.

The clinker contains four compounds which are believed to be the principal cementing materials, i.e., those which hydrate to form or aid in the formation of a rigid structure (Figures 16–5 and 16–7). They are believed to serve the following functions in the cement:

1. Tricalcium aluminate ($3CaO \cdot Al_2O_3$) is the compound that promotes rapid hydration and is the constituent which controls the initial set and the thickening time of the cement. It is also responsible for the susceptibility of cement to sulfate attack, and to be classified as a high sulfate-resistant cement, the cement must have 3% or less ($3CaO \cdot Al_2O_3$).

2. Tetracalcium aluminoferrite ($4CaO \cdot Al_2O_3 \cdot Fe_2O_3$) is a low heat of hydration compound in the cement. The addition of an excess of iron oxide will increase the amount of tetracalcium aluminoferrite and decrease the amount of tricalcium aluminate.

3. Dicalcium silicate ($2CaO \cdot SiO_2$) is the slow-hydration compound and accounts for the gradual gain in strength which occurs over an extended period of time.

4. Tricalcium silicate ($3CaO \cdot SiO_2$) is the most prevalent compound in cement and the principal strength producing material. It is responsible for the early strength ranging from 1 to 28 days. High-early-strength cements have a higher percentage of this compound than do portland or retarded cements.

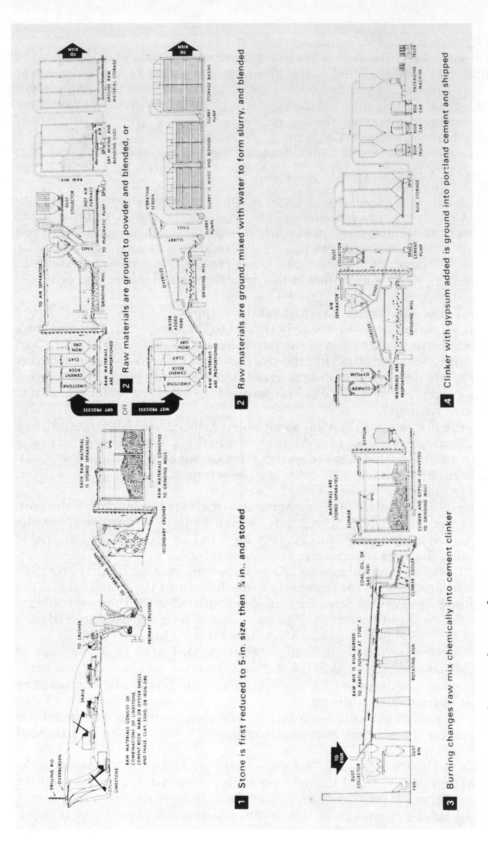

FIG. 16–6. *Steps in the manufacture of port-land cement*

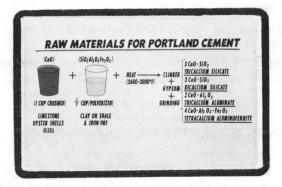

FIG. 16–7. *Raw materials for portland cement*

TABLE 16–1[4]
Typical Composition of Portland Cement

API Class	C_3S	Compounds—% C_2S	C_3A	C_4AF	Wagner Fineness Sq. Cm./Gram
A	53	24	8+	8	1600–1800
B	47	32	5–	12	1600–1800
C	58	16	8	8	1800–2200
D & E	26	54	2	12	1200–1500
G & H	50	30	5	12	1600–1800

Property	How Achieved
High early strength	Increasing the C_3S content: finer grinding
Better retardation	Control C_3S, C_3A Contents and grind coarser
Low heat of hydration	Limiting the C_3S and C_3A content
Resistance to Sulphate attack	Limiting the C_3A content

After a period of storage, the clinker is ground with gypsum to form the final cement. Gypsum is used to control the rate of setting and hardening of the cement paste. It is used in amounts of about 1.5 to 3.0% by weight of the cement (Table 16–1 and Figure 16–8).

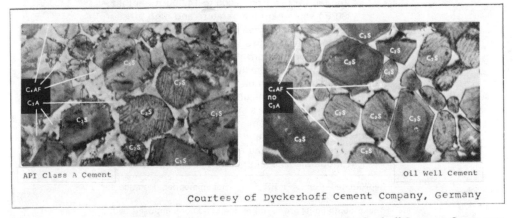

FIG. 16–8. *API Class A cement Oil well cement (Courtesy of Dyckerhoff Cement Company, Germany)*

CLASSIFICATIONS OF CEMENT

The portland cements for oil-well cementing carry the following API classifications (Tables 16-2 and 16-3).[7]

The usable depth of API cements is a function of temperature and pressure. In areas of subnormal temperatures, API cements can be used at

TABLE 16-2
API Cement Classification

API Class	Mixing Water Gals/Sk	Slurry Wt. Lbs/Gal	Well Depth (a) Ft	Static Temp-°F
A (Portland)	5.2	15.6	0-6000	80-170
B (Portland)	5.2	15.6	0-6000	80-170
C (High Early)	6.3	14.8	0-6000	80-170
D (Retarded)	4.3	16.4	6-10,000	170-230
E (Retarded)	4.3	16.4	6-14,000	170-290
F (Retarded)	4.3	16.4	10-16,000	230-320
G (Basic-Calif)	5.0	15.8	0-8000	80-200
H (Basic-Gulf Coast)	4.3	16.4	0-8000	80-200

(a) - Depth based on API casing well simulation schedule.

TABLE 16-3
API Classifications for Oil-well Cements*

Portland

Class A: Intended for use to 6,000-ft depth†—(170°F.) when special properties are not required. Similar to ASTM C 150, Type I Cement.

Portland

Class B: Intended for use to 6,000 ft depth†—(170°F.) when moderate sulfate resistance is required. Similar to ASTM C 150, Type II Cement.

Class C: Intended for use to 6,000-ft depth†—(170°F.) when high early strength is required. Similar to ASTM C 150, Type III Cement. Available in regular and high-sulfate resistant types.

Class D: Intended for use from 6,000 to 10,000-ft depths†—(230°F.) when moderately high temperature and high pressure are encountered. Available in regular and high-sulfate-resistant types.

Class E: Intended for use from 6,000 to 14,000-ft depths†—(290°F.) when high temperature and high pressure are encountered. Available in regular and high-sulfate-resistant types.

Class F: Intended for use from 10,000 to 16,000-ft depths†—(320°F.) when extremely high temperatures and extremely high pressures are encountered. Available in regular and high-sulfate-resistant types.

Class G and H: Intended for use as a basic cement from surface to 8,000-ft depth† as manufactured, or can be used with accelerators and retarders to cover a wide range of well depths and temperatures. Available in moderate and high-sulfate-resistant types.

* Reproduced from API RP 10B, "Recommended Practice for Testing Oil-Well Cements and Cement Additives."

† These depth limits are based on the conditions imposed by the casing well-simulation tests and should be considered as approximate values.

TABLE 16-4
Basis for API Well-Simulation Test Schedules

Well Depth Feet	Bottom Hole Temperature Static °F	Bottom Hole Temp. Circulating °F		
		Casing	Squeeze	Liner
2000	110	91(9)*	98(4)*	91(4)*
6000	170	113(20)	136(10)	113(10)
8000	200	125(28)	159(15)	125(15)
12,000	260	172(44)	213(24)	172(24)
16,000	320	248(60)	271(34)	248(34)
20,000	380	340(75)	———	———

* Time to Reach BHC Temperature.

greater depths; whereas, in areas of abnormally high temperatures, they may be limited to shallower depths. Normal API temperature gradient is considered to be 1.5°F/100 ft of depth[8] (Table 16–4).

Field Handling and Storage of Cement

A large percentage of the world's cementing jobs utilize bulk systems rather than manual handling in sacks. Such systems enable the preparation and supply of tailor-made compositions to suit any well-condition requirement. The cementing service companies offering bulk materials obtain the basic cement from manufacturers and transport it in hopper-bottom railway cars, trucks, or by barges to their field bulk-storage stations.

At the bulk blending stations, cement is unloaded by pneumatic sys-

FIG. 16–9. *Bulk rail transports*

FIG. 16–10. *Land based bulk storage plant for marine loading and transport*

tems operated under 30 to 40 psi air pressure into weather-tight bins or tanks of various designs. Land or marine transports are designed to handle cement at the wellsite at controllable feed rates.

Water-borne service vessels utilizing the pneumatic pressure system are usually equipped with their own weighing and blending plants or may obtain weighed and blended materials from a support vessel or shore stations within their operating areas (Figures 16–9 thru 16–14).

FIG. 16–11. *Land based bulk storage and blending plant*

FIG. 16–13. *Bulk cement and/or mud system mounted on offshore drilling platform*

FIG. 16–12. *Bulk transport for land operations*

FIG. 16–14. *In remote arctic areas cement is transported by air craft*

Bulk materials supplied under these conditions have contributed much to progress in well cementing, both technically and economically. Bulk cement blending stations provide many specific kinds of cement compositions which otherwise would not be practical. Jobs requiring a very large number of sacks would be most difficult to complete because of excessive time required for handling sacked cement.

For convenience and storage space, especially for high volume jobs, several types of field storage bins are used. The bins may be located at the well site, either on land or water, and filled in advance of the cement operation and used as needed. This limits the transport equipment on the job site and conserves storage at the time of cementing.

CEMENT ADDITIVES AND THEIR EFFECT ON CEMENTS

Bulk handling and the manufacture of basic cements has led to the development of "tailored cementing compositions" for specific well conditions. This is accomplished by the use of additives with API Classes A, B, G, or H cements. Additives are used to (1) reduce slurry density and increase slurry volume, (2) increase thickening time and retard setting, (3) reduce waiting-on-cement time and increase early strength, (4) reduce water loss and help protect sensitive formations; also help prevent premature dehydration, (5) increase slurry density to restrain pressure, and (6) restore circulation with increased fillup in annulus and reduce cost.

Accelerators

The average well drilled in the U.S. is less than 5,000 feet. Conductor and surface casing cements have lower temperatures requiring an accel-

CEMENT ACCELERATORS

➤ HA-5

➤ Calcium Chloride

➤ Sodium Chloride

➤ Diacel A

➤ Cal-Seal

FIG. 16–15. Cement accelerators

erator to promote the setting of cement and reduce excessive waiting times for cement to set in cold weather.

Early work by Farris[9] demonstrated that a compressive strength of 100 psi or 8 psi tensile strength will support casing and prevent communication of formation fluids. Bearden and Lane[10] suggested that, under ideal conditions, a 4 psi tensile strength is sufficient.

Industry generally accepts a compressive strength of 500 psi as being adequate for resuming drilling operations, which is a considerable safety factor based on the other investigations. Accelerators may be added to the cement or mixing water to shorten the waiting time on cement, thereby producing a saving in time and money. The most common cement accelerators are shown in Figure 16–15 and Table 16–5.

Retarders

For deep wells, cement retarders help extend the pumpability of the cement.

The primary factor that governs the use of additional retarder is the

TABLE 16–5
Pressure-Temperature Thickening-Time Data
API Common Cement
Hr:Min
API Casing Tests

Calcium chloride %	0% Bentonite		2% Bentonite		4% Bentonite	
	2,000 ft	4,000 ft	2,000 ft	4,000 ft	2,000 ft	4,000 ft
0	3:36	2:25	3:20	2:25	3:45	2:34
2	1:30	1:04	2:00	1:30	2:41	2:03
4	0:47	0:41	0:56	1:10	1:52	2:00

Compressive strength

Calcium chloride %	Curing temperature °F. and curing pressure					
	40 0 psi	60 0 psi	80 0 psi	95 800 psi	110 1,600 psi	140 3,000 psi
			8 hr			
0	N.S.	45	265	445	730	2,890
2	10	300	1,230	1,250	1,750	3,380
4	N.S.	450	1,490	1,650	2,350	2,950
			12 hr			
0	N.S.	80	560	800	1,120	2,170
2	64	555	1,675	1,310	1,680	2,545
4	15	705	2,010	2,500	3,725	4,060
			24 hr			
0	30	615	1,905	2,085	2,925	5,050
2	415	1,450	3,125	3,750	5,015	6,110
4	400	1,695	3,080	4,375	4,600	5,410

N.S.—Not set.

temperatures of the well. As the temperature increases, the chemical reaction between the cement and water is accelerated, which in turn reduces the thickening time or pumpability

Pressure will have some effect, but an increase of 20°F in temperature may mean the difference between an unsuccessful and a successful cementing job. Temperature is a critical factor in deep wells because of the requirement for longer displacement time. When cementing in deep, high-temperature wells, it is always recommended that materials be selected carefully and pilot tests be conducted using actual job materials under temperature and pressure conditions similar to those in the well.

In high-temperature environments cement retrogression may become a problem with all API classes of cement. For this reason, it is recommended that 35 to 40% silica flour be added to overcome strength retrogression.[11] High temperatures also increase cement permeability. Any additive having a high water requirement, such as bentonite, is not recommended with cement for use in wells where bottomhole temperatures exceed 230°F for reasons of progressive strength loss over prolonged periods of time.

Materials commonly used to retard or increase the setting time of ce-

TABLE 16–6
Pressure-Temperature Thickening-Time Data
API Class G or H Cement

Well Depth-Ft.	Temperature-°F		Lignin Retarder Percent	Approx. Thickening Time:Hr.
	Static	Circulation		
Casing-Cementing (Primary)				
0 to 6,000	80–170	80–113	0.0	2–4
6,000 to 10,000	170–230	113–144	0.0–0.5	2–4
10,000 to 14,000	230–290	144–206	0.5–0.7	3–4
14,000 to 18,000	290–350	206–300	0.7–1.5	*
Squeeze Cementing				
0 to 4,000	80–140	80–116	0.0	2–4
4,000 to 8,000	140–200	116–159	0.0–0.5	2–4
8,000 to 12,000	200–260	159–213	0.5–0.7	3–4
Below 12,000	260 plus	213 plus	0.7–1.5	*

* Require special laboratory tests. Other retarders may be recommended for these conditions.

ment include calcium lignosulfonate, sodiumcarboxymethylhydroxyethyl-cellulose derivatives, and blends of lignin materials with organic acids (Table 16–6).

Lightweight Additives

Additives used to reduce slurry density include bentonite, pozzolans, diatomaceous earth, expanded perlite and gilsonite.[12] Of these, bentonite is more widely used in cement slurries. Normal concentrations range up to 12 percent by weight of cement, but 4 per cent represents the average concentration.

Recommended water-cement ratios for given percentages of bentonite for API Class A, B, & G cements are shown in Table 16–7.

TABLE 16–7
Recommended Water-Cement Ratio for Bentonite Cements

Bentonite Per cent	Water Gal/Sk	Slurry Weight lbs/gal	Slurry Volume cu ft/sk
0	5.2	15.60	1.18
2	6.5	14.70	1.36
4	7.8	14.10	1.53
6	9.1	13.65	1.69
8	10.4	13.10	1.92
*10	11.1	12.95	2.02
*12	12.3	12.60	2.19

* With Dispersant.

Heavyweight Additives

These additives are normally added to cement when abnormally high pressures are to be encountered during cementing. There are many materials available to increase slurry density; however, some of the additives affect other properties of the cement in deep wells and should not be used. The specific gravities of the most common heavyweight additives range from 2.65 to 6.98. The most common materials are hematite, barite, and sand. However, the use of friction reducing additives in combination with hematite represents the most common technique of increasing cement slurry density to 22 pounds per gallon.

Lost-Circulation Additives

During drilling, the problem of lost returns or circulation is fairly common. In most instances additives are used with the drilling mud; however,

FIG. 16–16. *API class A cement*

under certain circumstances it is necessary to use cement containing lost-circulation materials to maintain circulation. A large number of materials are available to help restore circulation or prevent lost-circulation difficulties. These materials are generally classified as fibrous, granular, lamellated, and semisolids.

Fibrous Materials—Fibrous additives include shredded wood, bark, sawdust, etc. However, some of these materials are not suitable for cement slurries, because they contain tannins which either retard or in some instances prevent the cement slurry from setting. Nylon fibers $\frac{1}{8}$ to $\frac{1}{4}$ pound per sack of cement are sometimes used as they provide a high impact resistant cement (Right) as compared to neat cement (Left) (Figure 16–16).

Granular Materials—The materials commonly used are gilsonite, nut shells, plastics and perlites. These materials are considered as bridging agents to help prevent loss of or restore circulation. Gilsonite is considered to be the best material. In addition to being inert, it also has a very low specific gravity and requires a very small amount of mixing water. Normally $12\frac{1}{2}$ to 25 pounds of gilsonite per sack of cement is all that is required to restore circulation.

Lamellated Materials—These are materials such as mica, cellophane flakes, and related products. These materials help restore circulation by forming a mat at the face of the formation or an obstruction in a fracture, but possess little strength.

Low-Fluid-Loss Additives

The application of low-water-loss additives in oil-well cements to reduce filtration rates is similar to that in drilling muds or fracturing fluids.[13,14] These materials reduce the possibilities of water and/or emulsion blocks, and blocks caused by bentonite clay swelling within formations when penetrated by filtrate from cement slurries. They also help protect water-

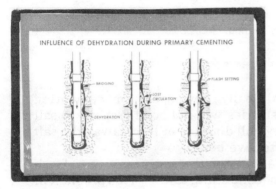

FIG. 16–17. *Influence of dehydration during primary cementing*

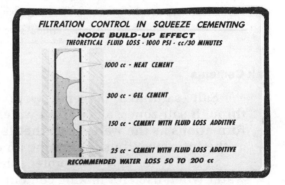

FIG. 16–18. *Filtration control in squeeze cementing*

sensitive shales and reduce the likelihood of premature dehydration of the cement slurry to the formation during high column cementing. Fluid-loss control tends to keep the slurry viscosity low and reduces the possibility of higher circulating pressures (Figure 16–17).

The ability of the low-fluid-loss additive to aid in reducing the filter-cake buildup during squeeze cementing is shown in Figure 16–18.

While low-water-loss additives were designed primarily for squeeze cementing, they are widely employed today in high column cementing, particularly on deep liners.

Friction Reducers

Additives or thinners reduce the apparent viscosity of the slurry. The lower viscosity slurries will go into turbulence at lower pumping rates reducing circulation rates and allowing cement to be pumped in turbulent flow at less than formation breakdown pressures.[15,16]

Additives currently in use with cementing slurries for promoting turbulent flow at low displacement rates include organic dispersants, salt, different types and mixtures of calcium lignosulfonate, and the use of a high-molecular-weight cellulose material in gel cement (Table 16–8).

TABLE 16-8
Effectiveness of Friction-Reducing Additive
to Obtain Turbulent Flow
API Class H Cement
Friction-Reducing Additive-Organic Dispersant
Hole Size—8¾ Inches
Casing Size—5½ Inches

Additive Per cent	Vc ann Ft/Sec	Flow Rate bpm	Frictional Pressure - psi 1,000 Ft.	
			Casing	Annulus
0	9.798	26.45	260	136
0.5	6.827	18.43	126	66
0.75	3.241	8.75	28	14
1.00	2.029	5.48	11	5

Note: Vc ann = critical velocity in annulus.

Salt Cements

Salt saturated cements were originally developed for cementing through salt zones, since fresh water slurries will not bond properly to salt formations as the water from the slurry will dissolve or leach away the salt at the interface, thus preventing an effective bond.

Salt slurries also help protect shale sections that are sensitive to fresh water when used for mixing cement not containing salt.[17,18] This problem in its most noticeable form shows up when sloughing or heaving occurs while pumping the slurry past particularly sensitive formations, resulting in some of the following possible conditions:

1. Excessive washouts and channeling behind the pipe;
2. Lost circulation into the weakened shale structure;
3. Annular bridging which may prevent slurry circulation.

Even shales which are apparently competent in the presence of fresh water may be weakened by continued exposure to fresh water slurry which contacts the formation before the cement sets (Figure 16–19).

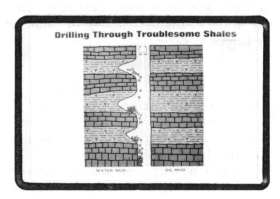

FIG.   16–19. *Drilling   through   troublesome shales*

SUMMARY

Cementing Composition Check Sheet

APPLICATION	ADDITIVE	TYPES OF CEMENTS	BENEFITS	OTHER CONSIDERATIONS
REDUCE W.O.C. TIME (Accelerators)	HA-5	Pozmix Cements and all API Cements	High Early Strength	Surface pipe, shallow wells and plugs
	Calcium chloride			
	CFR-2			Plugs for whipstocks - densified cements
	Cal-Seal	API Class A, B, C and G Cements	Fast Setting	Lost Circulation
INCREASE THICKENING TIME (Retarders)	HR-4	Pozmix Cements and all API Cements	Extend thickening time	Retarder for moderate temperatures
	HR-7			Viscosity reducer-improves flow properties
	CFR-1			Viscosity reducer-cement densifier
	HR-12	Pozmix 140 and API Class D, E and G Cements		Viscosity reducer-improves flow properties—retarder for extreme temperatures
	Diacel LWL	API Cement-Diacel Cement systems		Low water loss
DECREASE SLURRY DENSITY	Halliburton Light Cement	Special high strength filler formulation	Low density - high strength - economical	Filler cement
	Bentonite	Pozmix Cements and all API Cements	Low density - economical	Filler cement additive - greater set volume
	Gilsonite		Low density - high strength	Combats lost circulation
	Diacel D		Low density - high yield	
	Perlite		Low density	
INCREASE SLURRY DENSITY	Hi-Dense No. 3	Pozmix Cements and all API Cements	Combat high pressure	Hard plugs for whipstocks Slurries up to 22 lb/gal
	Sand			Hard plugs for whipstocks Slurries up to 18 lb/gal
	Barite	All API Cements		Slurries up to 19 lb/gal
	CFR-1	Pozmix Cements and all API Cements	High slurry weight - less water—low slurry viscosity	Higher strength-faster setting by reduced water ratios
	CFR-2			Slurries up to 18 lb/gal
LOST CIRCULATION	Gilsonite	All Cementing compositions	Combat lost circulation	Light weight slurry-high fillup above weak zones—bridges fractures - usable in squeeze cementing
	Tuf-Plug			Usable in drill mud for lost returns—blocking agent and fluid column lightener
	Perlites	Pozmix Cements and all API Cements (Usually mixed with 4% bentonite)		
	Flocele	All Cementing compositions	Minimize lost circulation	Usable in drill mud for lost returns
	Cal-Seal Cement	API Cements	Fast setting for plugging lost circulation zones	Plug back, water shut-off and blow-outs
	Bentonite-Diesel Oil (gunk squeeze)	Use with or without Class A Cement (Not mixed with water)	Combat lost circulation, plug fractures, vugs and crevices	Used as spearhead in squeezing fractured zones

SUMMARY CONT'D

Cementing Composition Check Sheet

APPLICATION	ADDITIVE	TYPES OF CEMENTS	BENEFITS	OTHER CONSIDERATIONS
LOW FLUID LOSS SLURRIES	Halad®-9	Pozmix Cements and all API Cements	Reduces slurry dehydration	Reduces flow rate for turbulence in Bentonite cement slurries
	CFR-2		Reduces slurry dehydration - filtration control	Reduces flow rate for turbulence
	Diacel LWL		Filtration control	Acts as retarder
	Halad®-14	API Class Cements	Same as Halad-9 for temperature above 200°F.	Acts as retarder and dispersant
	LA-2 (Latex)	Pozmix Cements, API Class A, B, C, G and some Class E Cements	Fluid loss control on squeeze, liner and primary cementing	Good bonding and perforating qualities—resistant to acid and corrosive fluids
LOW FRICTION CEMENT SLURRIES	CFR-1	Pozmix Cements and all API Cements	Provides turbulence at lower displacement rates - reduces hydraulic horsepower requirements	Retards thickening time - use above 200°F.
	CFR-2			Dispersant used in densified cements
	Salt (NaCl)			Bonds to shales, bentonitic sands, and salt zones
	Halad®-9	Bentonite cements (6% and higher)		Provides low fluid loss properties
	HR-12	API Class D, E and G Cements		Retarder for extreme temperatures
	HR-7	Bentonite Cements (all percentages)		Retards thickening time
SPECIAL ADDITIVES	Salt (NaCl)	Pozmix Cements and all API Cements	Bonds to salt, shale and bentonitic formations - Improves flow properties	Accelerates in low concentrations—retards in high concentrations—more expansion than fresh water cements—increases slurry density
	Halliburton Red Plug Cement	Special high strength formulation	High strength, particularly when densified - low fluid loss - good flow properties - expansion upon setting - affords good placement time	Tolerates high percentage of mud contamination - low permeability - excellent bonding to clays and shales
	Silica Flour	Pozmix Cements and all API Cements	Reduces high temperature strength retrogression in cements	Reduces permeability of set cement - usable in oil, thermal recovery, steam injection and geothermal steam wells
	Mud-Kil® I	Pozmix Cements and API Class A, B, C and G and some E Cements	Combat effect of highly treated drilling muds on cement	Mud-Kil I neutralizes contamination effect of quebracho, starch, sodium carboxymethylcellulose, tannins etc. on cements
	Mud-Kil® II	Pozmix Cements and API Class A, B, C and G Cements	Combat effect of highly treated drilling muds on cements: Calcium ligno-sulfonate, chrome lignin, chrome lignite, etc.	Mud-Kil II neutralizes the effect of calcium lignosulfonate retarders used in some Class E Cements
	Radioactive Tracers-RAC-1 and RAC-3	All cementing compositions	Helps locate top of cement or cement after squeeze	Helps locate leaks in casing with gamma log
	Casing-Kote™	Applied to external surface of casing prior to running in hole	Improve bond at the pipe-cement interface	Increases shear bond and resistance to communication at pipe-cement interface
	DOC-3	API Class A Cement	Thief zone treatment - squeeze - moderate penetration	Unlimited setting time - selective to water - mixed with diesel oil—does not activate until in contact with formation water
	DOC-10	API Class A Cement	Thief zone treatment - squeeze - deeper penetration	

®—Registered U. S. Pat. Office
™—Trademark used by Halliburton

JOB CONSIDERATIONS

Mixing Equipment

The mixing system on any cementing operation functions to proportion and blend the dry cementing composition with the carrier fluid. When this is achieved, a cementing slurry with predictable properties can be supplied to the wellhead.

The most widely used mixing method is the jet-type mixer (Figure 16–20). A stream of water mixes with cement by passing through the mixer bowl creating a vacuum which pulls the dry cement into the bowl from the hopper immediately above. As the cement enters the jet stream of water, it is thoroughly mixed by the turbulent flow that occurs in the discharge pipe. Mixers of this type, when supplied with sufficient water under the optimum mixing pressure and adequate feed rate of cement, are capable of producing a normal slurry at a rate of up to 50 cubic feet per minute.

Control of mixing speed is regulated by the volume of water forced through the jet, and amount of cement in hopper while mixing. A bypass line can supply extra water into the bowl discharge line for lowering of slurry weight by increase of water-cement ratio.

Modified jet mixers, called recirculating mixers, have been designed

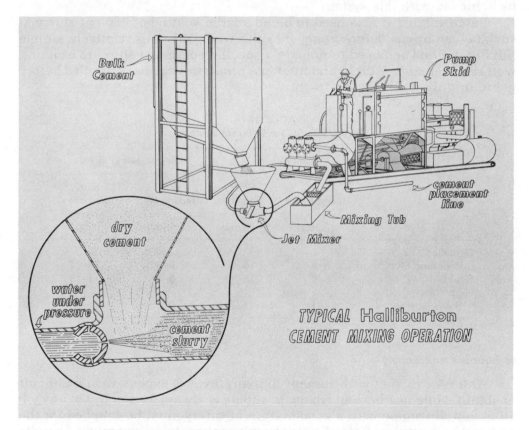

FIG. 16–20. *Typical Haliburton cement mixing operation*

FIG. 16–21. *Modified jet mixers*   FIG. 16–22. *Pneu-multi-matic mixer*

for mixing heavy densified slurries (Figure 16–21). A recirculating system is basically a pressurized jet type mixer with a large tub capacity. It uses recirculated slurry and mixing water to partially mix and discharge the slurry into the tub. Additional shear is provided by the recirulating pump, agitation jets, and an in-tub eductor adds additional energy and improves mixing. As a result, more uniform cement slurries having densities as high as 22 pounds per gallon and pumped as slow as 0.5 barrels per minute can be achieved with this system.

Batch mixing is also used to blend a large volume of cement slurry at surface conditions before going into the well. This is a relatively simple mixing method and used to prepare a specific volume of slurry to exacting well requirements. Weight and fluid loss properties can be controlled before going into the well (Figure 16–22).

TABLE 16–9
Mixing Ranges for Different Systems

	Jet Mixer		Recirculating Mixer	
	Mixing Rates	Slurry Densities	Mixing Rates	Slurry Densities
	(BPM)	(lb/gal)	(BPM)	(lb/gal)
Densified and Weighted Slurries	0–5	16–20	0–6	16–22
Neat Slurries	0–8	15–17	0–8	14–18
High Water Ratio Slurries	0–14	11–15	0–12	11–15

Water Sources and Supply

Water for mixing with cement in many areas is expensive and difficult to obtain while in others an adequate supply is always available. On any job an essential volume with a good margin of safety must be supplied for the cementing operation. Rate of water supply to the mixing and pumping units will affect and often may control rate of mixing of the cementing slurry. When this occurs, the time of pumping may encroach on the time the slurry is beginning to set.

The water supply normally available at a drilling well may be anything from reasonably clean and free from soluble chemicals to badly contaminated with varying quantities of silt, organic matter, alkali, salts, or other impurities.

Water from wells, streams, or ponds in some localities often contain sulfates, chlorides, or other dissolved mineral salts which may be detrimental both to drilling muds and cements. Potable water is always recommended where available. Unless a supply of clear water has a noticeable saline or brackish taste, it is usually suitable for use without further questions, although saline waters are usable if conditions are pretested and understood. In the event of questionable water, the purest available supply should be used for mixing with cement.

Water temperature during warm weather mixing is important. High-temperature mixing water (110°F) may result in a viscous slurry with a shorter pumping time.

Winter conditions must also be considered, primarily because of line freeze-up and because very low water temperatures may significantly increase the setting time of a slurry.

There must be no conflict with other water supply requirements for the drilling rig and possible emergency operations. It would not be wise to use a water supply line for cementing which would in any way interfere with water in reserve for blowout prevention.

Slurry Density

Cement systems can be designed to cover a density range from 10.8 to 22 pounds per gallon. Slurry density is directly related to the amount of mixing water and additives in the cement and the amount of slurry contamination from drilling mud or other foreign material. In critical low pressure fracture-gradient wells where lost circulation zones restrict the density of mud and cement, a slurry must be light enough to be supported by the weakest incompetent zone. Where high pressure gas or salt water flows exist, the cement slurry must be heavy enough to control the pressure and prevent a blowout.

Cement slurries should always be slightly heavier than the mud density, but the mechanisms for achieving light or heavy weight slurries are quite variable.

In field operations, slurry control is customarily maintained by measuring the density with the standard mud balance. For accuracy, samples should be selectively taken from the tub and vibrated to remove the finely entrapped air from the jet mixer. Automated weighing devices which fit into the discharge line between the mixing unit and wellhead, give a more uniform weight record and are being more widely used.

A radioactive weighing device is sometimes fitted into the slurry discharge line and provides a strip chart record of density measurements (Figure 16–23).

The Fluid Density Balance is a newer instrument which measures the cement slurry under sufficient pressure to compress the entrained air to a negligible volume. This compression of the slurry under approximately 250 psi allows a more accurate density measurement of the cement when sampled directly from the tub during the mixing process (Figure 16–24).

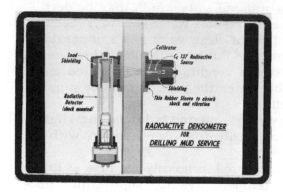

FIG. 16–23. *Radioactive densometer for drilling mud service*

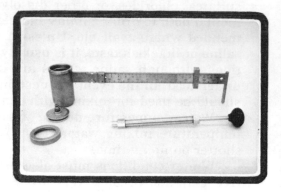

FIG. 16–24. *Tub cement sampler*

Effects of Drilling Fluids and Additives

A major problem in the cementing of wells is the effective removal of drilling fluids during cementing operations. Cementing systems are influenced by:

1. Mud Contamination
2. Mud Dilution
3. Mud Chemicals
4. Bonding to Filter Cake

Drilling mud additives, organic and inorganic, have varied effects on the setting of cement. Organic mud additives generally retard the set of cement; whereas, inorganic additives will accelerate the setting of cement (Table 16–10). The degree of retardation or acceleration will depend upon the concentration of chemicals present in a given system. With some types of cements, mud additives may not conform to a fixed pattern but cause an erratic gelation, or a viscous consistency resembling a set.

Some degree of mud-cement contamination occurs during most jobs, but probably the greatest problem is encountered when spotting cement plugs in highly chemical-treated mud systems. The volume of cement in relation to the volume of mud is small and the degree of mud contamination is always unknown. Contamination is evident when soft cement is observed in drilling out the plug.

Oil or inverted emulsion muds having powerful surfactants and sodium or calcium chloride in their formulation have been found to produce a rapid set of cement slurries under high temperatures and pressures. For such system, a generous volume of treated or untreated Diesel Oil flush is recommended as a spacer and flush to prevent interfacial gelation.

The most satisfactory means of combating the effect of drilling mud additives on oil well cement has been to reduce to a minimum the contamination of the cement with treated drilling fluids during a cement operation. This has been accomplished by the use of wiper plugs ahead of the cement in casings and by the use of buffer solutions or washes ahead of the cement, or both (Figure 16–25).

The wiper plug prevents any contact and resulting contamination between the drilling fluid and the cement slurry inside the casing,[19] and a buffer wash helps eliminate contamination in the annular space between the outside of the casing and the formation.

TABLE 16–10
Effects of Mud Additives
on Cement

Organic Chemicals Retard

CMC	Lignosulfonates
Starch	Organic Acids
Lignites	Wood Bark
Quebracho	

Inorganic Chemicals Accelerate

Hydroxides
Carbonates
Bicarbonates
Chlorides
Sulfates

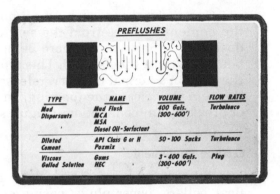

FIG. 16–25. *Preflushes*

Volume of Cement to be Considered

API Cements, based on depth and temperature requirements, may be purchased in most oil producing areas of the world. While much emphasis is placed on deep, hot wells the larger volume of cement used in wells is at depths less than 8,000 feet. API Class G and Class H Cement have been designed to fit these conditions where bottomhole static temperatures are less than 200°F. For deeper wells, other classes of API cements may be purchased or Class G or H Cements can be modified with additives to fit individual well conditions.

The volume of cement required for a specific fillup on a casing job should be based on field experience and regulatory requirements. In the absence of specific guides, a volume equal to 1.5 times the caliper survey volume should be used.

Where a good mud program is used, it is customary to assume that the hole size is the same as the bit size, yet allowances may be made to take care of hole irregularities. Caliper measurements generally show that such allowances are usually inadequate. Caliper logs may be necessary to determine the proper location of centralizers or scratchers to obtain the maximum efficiency, however, model studies indicate that scratchers, centralizers, or any casing device, whether in a gauge or washed-out section, will aid in inducing turbulence, provide better mud removal and consequently better fillup.

While regulatory rules and hole conditions may dictate the fillup necessary for a given cementing operation, it is desirable to have a minimum of 300 to 500 feet fillup behind the intermediate and/or production string of casing. The use of too much cement rather than too little is always advisable especially where there exists the possibility of contamination or dilution by the mud system.

CASING EQUIPMENT

Cementing Heads

Cementing heads or plug containers provide the union for connecting the cementing lines from the service unit to the casing. These heads are designed to hold one or more cementing plugs with the dual-plug heads being the most widely used. This type of head makes it possible to circulate the mud in a normal manner, release the bottom plug, mix and pump down the cement, release the top plug, and displace the cement without making or breaking any connections (Figures 16–1 and 16–26).

Subsurface Equipment

Subsurface equipment commonly used include: casing guide shoe, float collars, wiper plugs, casing centralizers, scratchers and stage tools. Although there are many different types of subsurface equipment, the following examples are representative of this type of equipment.

The function of the plain open ended guide shoe is to guide the casing past any obstruction in the hole; whereas, conventional combination float guide shoe and the float collar provide: (1) a means of floating or partially floating the casing into the hole, (2) a back-pressure valve to help prevent the backflow of cement after it has been placed, and (3) a safety valve to help prevent a blowout when going into the hole.

The essential element of this equipment is a check valve. Flow of fluids can be directed down the casing; however, the valve prevents flow into and up the casing (Figures 16–1, 16–27, and 16–28).

Fillup Floating Equipment

Because of lost returns, trouble frequently experienced while cementing, plus the extra time required to fill casing, modified types of float collars have been introduced. This type of float permits the casing to partially fill from the bottom as the casing is run. Fluid entry into the casing is controlled by a differential valve or an orifice (Figure 16–29).

The backpressure valve effect is put into effect by pump pressure, so that the same advantages of preventing cement backflow are realized with this equipment as with conventional floats. However, there is no protection against blowouts.

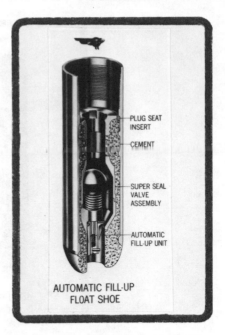

FIG. 16–27. *Automatic fill-up float shoe*

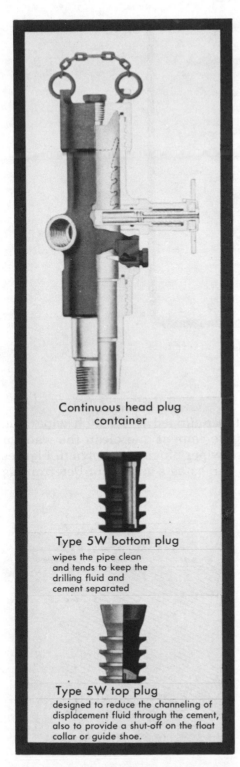

Continuous head plug container

Type 5W bottom plug

wipes the pipe clean and tends to keep the drilling fluid and cement separated

Type 5W top plug

designed to reduce the channeling of displacement fluid through the cement, also to provide a shut-off on the float collar or guide shoe.

FIG. 16–26. *Types of plugs*

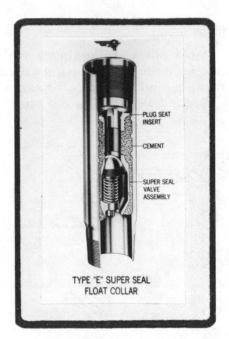

FIG. 16–28. *Type "E" super seal float collar*

FIG. 16–29. *Self-fill differential collar*

Wiper Plugs

Wiper plugs are equipped with rubber-cupped fins which wipe mud from the walls of the casing ahead of the cement and clean the walls of cement behind the slurry.[19] Examples of wiper plugs are shown in Figures 16–29A and 16–30. The top plug also serves as a means of determining

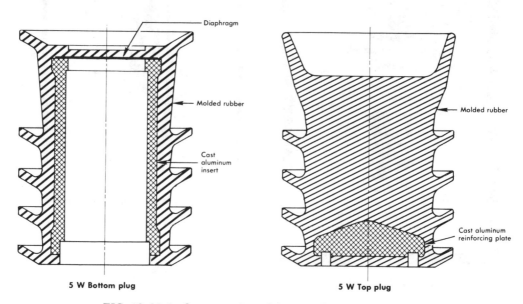

FIG. 16–29-A. *Cross section of improved cementing plugs*

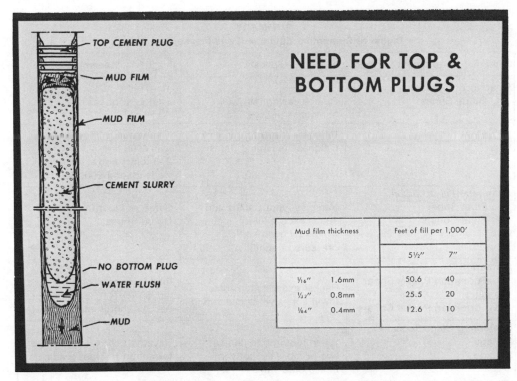

FIG. 16-30. *Need for top and bottom plugs*

when cement has been completely displaced. When this plug contacts the bottom plug or hits the plug seat in a float collar, surface pressure increases instantly.

It is also good practice to count pump strokes in displacing cement, since it is possible to encounter an obstruction before reaching the float collar. Also pumps should be slowed down before the plug hits the float collar to prevent the buildup of an excess pressure inside the casing. Other casing equipment is described in Table 16–11.

PRIMARY CEMENTING—THE CRITICAL PERIOD

The cementing of casing is one of the most critical operations during the drilling and completion of an oil well. The preparation of the hole, the assembly of the necessary surface and subsurface equipment, rigging up, and running the casing are all preliminary to the most important stage, which is that period between the running of the last few joints of casing and the final displacement of the cement.

During this critical period, the success or failure of the entire operation is likely to be determined. A successful casing cementing job is one in which the casing is landed at the exact specific depth and the annular space between the casing and formation is sealed with set cement so that any one fluid or gas may be produced to the exclusion of any other fluid or gas.

Most primary cement jobs are performed by pumping the slurry down the casing and up the annulus; however, there are modified techniques

TABLE 16–11
SUMMARY
Digest of Cementing Equipment and Mechanical Aids

Cementing Equipment & Types	Application	Placement
Floating Equipment 1. Guide Shoes	Guides casing into well Minimizes derrick strain	First joint of casing
2. Float Collars	Prevents cement flow back Catches cementing plugs	1 joint above shoe in wells less 6000 feet 2–3 joints above shoe in wells greater 6000 feet

Automatic Fill-up Equip.
  1. Float Shoes
  2. Float Collars
	Same as Float Collars and
	shoes except fill-up is con-
	trolled by hydrostatic
	pressure in annulus
		Same as Float Collars or
		Guide Shoes

Formation Packer Tools
  1. Formation Packer Shoes
 	
  2. Formation Packer Collars
	Packer expands to protect
	lower zones while cementing
		First joint of casing
		
		As hole requirements dictate

Cementing Stage Tools 2 Stage 3 Stage	When required to cement two or more sections in in separate stages	Based on critical zones and formation fracture gradients

Plug Containers
  1. Quick Opening
  2. Continuous Cementing
    Heads
	To hold cementing plugs in
	casing string until released
		Top joint of casing at surface
		of well

Cementing Plugs
  1. Top & Bottom Wiper Plugs
  2. Ball Plugs
  3. Latch Down Plugs
	Mechanical Spacer between
	Mud and cement (bottom
	plug) and cement and
	displacement fluid (top plug)
		Between well fluids and
		cement

Casing Centralizers Variable Types	Center casing in hole or pro- vide minimum standoff to improve distribution of cement in annulus Prevent differential sticking	Straight hole – 1 per joint through and 200 feet above and below pay zones; 1 per 3 joints in open hole to be cemented Crooked hole – Variable with deviation
Scratchers or Wall Cleaners 1. Rotating	Remove mud cake and circu- latable mud from wellbore.	Place through producing for- mations and 50 to 100 feet above. Rotate pipe 15 to 20 RPM
2. Reciprocating	Aid in creating turbulence Improve cement bond	Same as Rotating, Reciprocate pipe 10 to 15 feet off bottom

TABLE 16-11

Cementing Equipment & Types	Application	Placement
Squeeze Cementing Tools 1. Drillable	Where pressure is required to be held on cement after squeeze job	Purchased and left in well as permanent plug or drilled out
2. Retrievable	Perform multiple squeeze jobs and then retrieve	Can be moved up or down hole to squeeze as many zones as may be required
Liner Hanger Tools 1. Mechanical	To support liner in casing while cementing and remove after operation	Set at specific location in last casing string. (Usually 200 feet overlap in last casing) Set by disengaging J slot and reducing casing weight
2. Hydraulic	Same as mechanical	Same as mechanical except set by differential pressure
Bridge Plugs 1. Wire Line 2. Tubing	For permanent or temporary plugging in open hole or casing	May be placed in well on wire line, on tubing or below retrievable squeeze packers
Cementing Baskets and External Packers	For casing or liner where mechanical support is necessary until the cement column sets	Below stage tools or where weak formations occur downhole

for special situations (Figures 16–31 through 16–37). The various placement methods include:

1. Cementing Through Casing (normal displacement technique)
2. Stage Cementing (for wells having critical fracture gradients)
3. Inner String Cementing Through Drill Pipe (for large diameter pipe)
4. Outside or Annulus Cementing Through Tubing (surface pipe or large casing)
5. Multiple String Cement (for small diameter tubing)
6. Reverse Circulation (critical formations)
7. Delayed Setting (critical formations and to improve placement)

Suggested procedures to be followed to obtain a good primary cement job:

1. Condition hole by reaming, if necessary, to remove tight places.
2. Condition mud. Circulate over screen until mud is mostly free of cuttings. Keep viscosity and gel strength low. Select a water loss comparable to that used to drill the lower portion of the hole.
3. Install a guide shoe and float collar. The float collar should be about 30 ft. above the guide shoe. This helps prevent overdisplacement of cement and obtain a good cement slurry around casing shoe.
4. Install scratchers spaced according to location of permeable zones. Use only those needed to obtain adequate coverage.
5. Use casing centralizers. Utilize logs to determine location as well as routine spacing of 60 to 90 ft. apart.

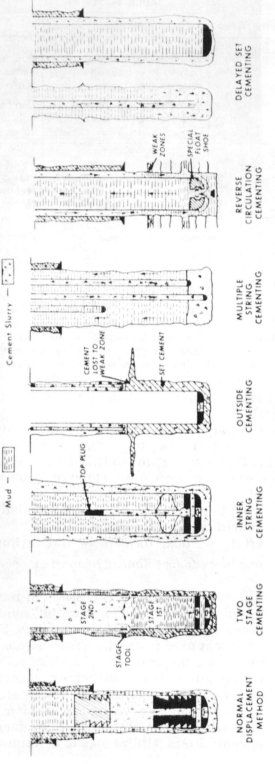

FIGS. 16–31 through 16–37. "Placement techniques for primary cement"

6. Use a cement slurry that is as heavy as or heavier than the drilling fluid. This will help prevent overdisplacement. Cement selected will depend on temperature, pressure and hole conditions.

7. If caliper log is available, use it to calculate cement volumes. If none is available, increase volumes according to knowledge of area. Generally, volumes are increased by 50 to 100%. Remember that volume varies as the square of the hole diameter.

8. Use a top and bottom cementing plug.

9. Rotate or reciprocate casing until the top wiper plug has hit the float collar. Premature cessation of movement will permit the filter cake to reform.

10. After cement is in place watch the annulus fluid level. Keep the annulus full of fluid.

11. Maintain tension in the entire casing string while cement is setting. Setting time can be altered to fit almost any condition as noted in the previous section.

12. Before drilling the cement plug or perforating for production, pressure test the cement job. Use a maximum pressure of 80% of the minimum yield of the weakest point.

To determine the height of the cement column in the annulus, it is common practice to run a temperature log from 12 to 24 hours after placement of the cement. Because cement generates heat when it hydrates it is possible to locate the top of the cement by the anomaly in the temperature log. A good approximation of the location of the cement top may be obtained using Equation No. 1:

$$H = \frac{P_s - P_f}{0.052(W_c - W_m)} \tag{1}$$

Where:
H = height of cement, ft
W_c = weight of cement slurry, lb/gal
W_m = weight of mud, lb/gal
P_s = surface displacement pressure, psi
P_f = friction pressure, psi

To use this formula, the pump should be slowed down until it is just barely moving the fluid. At this rate, friction losses can probably be neglected. If friction is neglected it should be remembered that the height calculated is probably above the actual cement column.

CONSIDERATIONS DURING CEMENTING

Cement Mixing

Job success is correlated with the quality of the cement mixing operation. Slurry density is used to control the relative amounts of water and cement used and should be monitored and recorded to insure maintenance of the correct water-solids ratio. To avoid the effect of aeration, samples for weighing should be obtained from a special manifold on the discharge side of the displacement pump rather than from the mixing tub (Figure 16–38).

This figure shows cement slurry sampling taken near wellhead for density and rheological measurements. Special manifold reduces shearing and provides representative well sample.

The last volume of cement mixed will be placed around the shoe; thus, particular care should be taken to insure that it meets desired weight specifications. In most instances cement slurries are mixed with a lower water requirement toward the end of the job to provide better strength around the shoe joint.

Where densified or heavy weight slurries are used (17.0 lbs/gal plus) and pumped or displaced at rates of less than five barrels per minute, a

FIG. 16–38. *Cement slurry sampling*

recirculating type mixer improves uniformity of slurry over the standard jet mixer.

For critical cementing jobs, *Batch Mixing* provides much greater slurry precision, and permits measuring and adjusting rheological properties, fluid loss control or other specific properties of the slurry before pumping into the well.

Displacement

The most predominant cause of cementing failure appears to be channels of gelled mud remaining in the annulus after the cement is in place[20,21,22] (Table 16–12). If mud channels are eliminated a variety of cementing compositions will provide an effective seal. Should mud channels remain after cementing, no matter what the quality of the cement, there will not be an effective seal between formations. Under dynamic conditions of pressure and temperature water base muds and gel deposit a thick filter cake making removal more difficult. Model research supported by field practices indicate the following forces aid mud displacement in the annulus.[23,24,25]

TABLE 16–12
Factors Which Contribute to Cementing Failures[19A]

Factors	Results
Improper Water Ratio Improper Temperature Assumption Mechanical Failures Wrong Cement or Additives Hot Mixing Water Static too Long Before Movement Improper Mud-Cement Spacers Contaminates in Mixing Water Dehydration of Cement Insufficient Retarder	Flash Setting of Cement
Plug Did Not Leave Head Did Not Allow for Compression	Failure to Bump Plug
Mechanical Failure Insufficient Water or Pressure Failure of Offshore Bulk Systems	Could Not Finish Mixing
Pipe Lying Against the Formation Poor Mud Properties (High YP & PV) Failure to Move Pipe Low Displacement Rates	Channeling
Insufficient Hydrostatic Head Cement Mud Gelation Cement Did Not Cover Gas Sands	Gas Leakage in Annulus

1. Cement exhibits drag stresses upon mud to differences in flow rates and flow properties. A well conditioned mud is much more effectively removed than a thick viscous mud using a thin cement in turbulent flow.
2. Drag stress of pipe upon mud and cement due to pipe motion—either rotation or reciprocation. Any pipe movement will minimize static mud pockets in the annulus.
3. Buoyant forces due to density difference between mud and cement aid mud removal.

In evaluating factors affecting displacement of mud, it is necessary to consider the flow pattern in an eccentric annulus condition with the pipe closer to one side of the hole than the other. Flow velocity in an eccentric annulus is not uniform. Highest flow rate occurs in the side of the hole with the largest clearance as shown in Figure 16–39.

Field studies have shown that the length of time cement moves past a point in the annulus in turbulent flow is important.[23] Thus, if a mud channel is put in motion, even though its velocity is much lower than cement flowing on the wide side of the annulus, given enough time the "mud channel" may move above the critical productive zone. Contact time is not an important factor with the cement in laminar flow conditions where apparently the cement does not exert sufficient drag stress on the mud to start the mud channel moving.

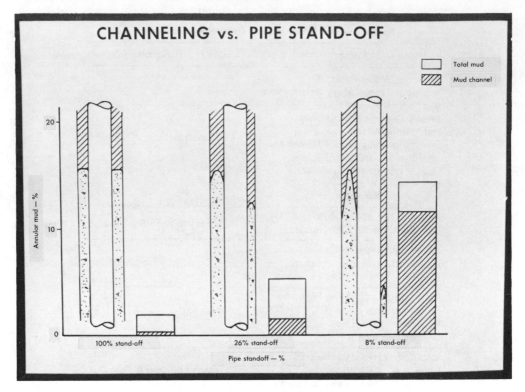

FIG. 16–39. *Channeling vs. pipe stand-off*

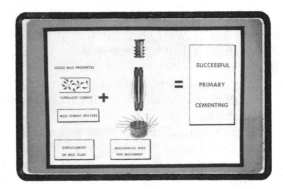

FIG. 16–40. *Elements in primary cementing*

At a given displacement rate, contact time is directly proportional to the volume of cement.

It is significant that essentially all the laboratory and field work performed in recent years indicate the same factors as contributing to the success of primary cementing during this critical period. Each of the following factors are aimed at removing mud from the annular cross section[26] (Figure 16–40).

1. *Pipe centralization* significantly aids mud displacement.

2. *Pipe movement,* either rotation or reciprocation, is a major driving

force for mud removal. Pipe motion with scratchers substantially improves mud displacement in areas of hole enlargement.

3. *A well conditioned mud* (low PV and YP) greatly increases mud displacement efficiency.
4. *High displacement rates* promote mud removal. At equal displacement rates a thin cement slurry in turbulent flow is more effective than a thick slurry in laminar flow.
5. *Contact time* (cement volume) aids in mud removal if cement is in turbulent flow in some part of the annulus.
6. *Buoyant force* due to density difference between cement and mud is a relatively minor factor in mud removal.

Gas Leakage

In gas well completions the leakage problem has become more critical, particularly in deep holes, causing a pressure buildup in the annuli of the production and intermediate casings or liners.[27,28]

The following parameters have been found directly related to gas migration in a wellbore:
1. Ineffective hydrostatic head
2. Borehole mud removal
3. Cement filtration control

The density of the cement slurry and other wellbore fluids determine the hydrostatic pressure exerted at any particular depth. When completing gas wells the density of the drilling mud, preflush and cementing composition, either separately or in combination, must exceed the formation gas pressure to prevent it from entering the annulus.

One cubic foot of gas migrating from 12,000 feet to the surface under a given set of conditions will expand to 316 cubic feet.

	Conditions at 12,000 Feet	*Conditions at Surface*
Pressure	5366 psi gauge (5380 Abs.) (8.6#/gal.)	14.7 psi
Temp.	160°F (620°R)	70°F (530°R)
Z	.99	1

$$V_2 = \frac{V_1 \, Z_2 \, T_2 \, P_1}{Z_1 \, T_1 \, P_2}$$

Where V = Volume of gas
Z_1 = Compressibility factor under pressure in formation — dimentionless
T_1 = Temperature of formation
P_1 = Pressure of formation
Z_2 = Compressibility factor — surface
T_2 = Temperature at surface
P_2 = Pressure at surface
V_2 = Volume of gas at surface

TABLE 16–13
Summary—Primary Cementing

	Type Job and General Conditions		
	Conductor	Surface	Intermediate and/or Production String
Casing Size—Inches	20–30	7–20	7–11 $\frac{3}{4}$
Setting Depth—Ft.	30–1500	40–4500	1000–15,000
Hole Conditions	Probably Enlarged	Probably Enlarged	Probably Enlarged (particularly in salt)
Muds	Native	Native	Native, Water-Base or Oil Emulsion
Properties	Viscous-Thick Cake	Viscous-Thick Cake	Controlled Viscosity and Fluid Loss
Cemented To	Surface	Surface	May or May Not Be Circulated to Surface Pipe
Cement—API Class	A-G-H	A-G-H With	A-C-G-H With
Additives	2–3% CaCl	Bentonite or Pozzolan	High gel, filter or pozzolan bentonite
			Dispersant + retarder if needed
			Salt saturated through salt sections for bonding
Tail-In Slurry	Same-Densified (Ready-Mix concrete may be dumped in annulus)	Densified for High Strength (Deep well may use high strength slurry for entire job)	Densified for high strength over lower 500–1000 ft.
Technique	Through D.P. using small plugs & sealing sleeve or	Same as Conductor	Down casing with plugs (top and bottom) or
	Down casing with large plugs or		Stage Cementing depending on frac gradient
	Cement pumped or dumped in annulus		String very heavy, may be set on bottom & cement through ports
	Float Collar may or may not be used	Same as Conductor	Float Collar—Guide Show required
Placement Time	Generally short—less than 30 min.	Usually short—less than 45 min.	Variable depending on cement volume 45 min.—2 $\frac{1}{2}$ hrs.
	Low or high displacement rates employed.	High displacement rates	High displacement rates, 6–12 bbl/min.

TABLE 16-13
Summary Cont'd

	Type Job and General Conditions		
	Conductor	Surface	Intermediate and/or Production String
W.O.C. Time	6–8 Hours*	6–12 Hours*	6–12 Hours*
Mud-Cement Spacers	Plugs and Water Flush	Plugs and Water or Thin Cement Spacer	Plugs and Thin Cement or Spacer Compatible With Mud
Cementing Hazzards	Casing can be pumped out of hole.	Should have competent casing seat.	May cover both weak and high pressure zones requiring variable weight cement slurries.
	Cement may fall back downhole after circulating to surface.	Same as Conductor.	Prolonged drilling may damage casing.
		Lower joints sometimes lost downhole with deeper drilling.	Bottomhole temperature necessary for design in hot wells.
		Casing can be easily stuck.	Centralize in critical areas.
		Centralize lower casing.	

* See Regulatory Rules

Prior to cementing a gas well should be circulated for an excessive period of time to condition the mud and remove any trapped gas bubbles located in the annulus. After circulation and prior to cementing the pumps should be shut down for a short period and then started again to help remove any microscopic gas bubbles that may be trapped in the washout areas or that may be adhering to the walls of the hole.

This gas is considered to be a potential second gas kick and, if not removed prior to cement placement, may result in a lowering of the fluid column density. During the cementing process, the cement slurry should be maintained as heavy as possible to restrain any gas from cutting the slurry.

The following factors have been found beneficial in reducing gas migration into an annulus during and after primary cementing:[27]

1. Centralization of casing string in hole.
2. Increased flow rates during displacement.
3. Pipe movement during displacement.
4. Scratchers employed across washout sections.

5. Mud or cement density (positive hydrostatic head).
6. Cement filtration control.
7. Cement setting or changing from a slurry to a solid in a minimum time after placement.

Deep Well Cementing

The procedures used in cementing deep casing or liner strings are generally the same as those used for shallower depths yet the physical factors encountered in shallow wells are of minor importance when com-

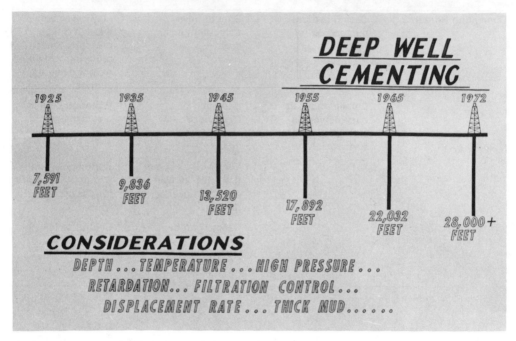

FIG. 16–41. *Deep well cementing*

pared to those in deep holes[29,30] (Figure 16–41). The magnitude of these factors increases with depth and include such conditions as:

1. Higher temperatures and pressures with corrosive waters or H_2S gas zones in some areas.
2. The increased length and reduced clearances of the casing-casing annulus.
3. Greater mechanical loads imposed on the casing string and drilling rig.
4. Longer time intervals of pulling the bit off bottom and the running of casing prior to cementing.
5. Heavily treated mud systems having high densities.

Deep holes demand considerable planning to achieve adequate zonal isolation with any cementing composition selected for deep hot hole cementing (Figure 16–42). Such plans include:

1. Design of slurry to provide adequate rheological properties and pumping time. Pretesting should always be done with rig water and cements to be used on specific jobs.
2. The best technique for displacement of mud by cement slurry. Field mud should be evaluated in the selection of mud spacer to precede the cement.
3. Attainment of desired slurry properties (weight, viscosity, fluid loss control, etc.) during the mixing process to effectively remove drilling mud.
4. Control setting cements to resist gas leakage, strength retrogression and corrosive environments encountered at elevated temperatures.

The casing-to-hole clearances in deep wells are effectively smaller because of greater lengths of open hole involved. This results in greater lengths of permeable zones and consequently thicker filter-cake. The higher temperatures, higher pressures, and exposure time each contribute to gelation and filter cake deposition across these permeable sections.[31,32] This factor alone complicates mud removal as cement will have a tendency to channel through gelled mud and follow a path of least resistance (Figure 16–43). This figure shows cement channeling through Ferrochrome mud under dynamic test conditions[25] without pipe movement.

Lengthy open hole exposure will cause shales to loosen and accumulate unless oil or invert mud systems are used. Higher mud densities also cause increased hydrostatic pressure differentials over some of the intervening formation pressures, tending to crowd the casing further into the mud cake on the contact side of the hole and result in differential sticking or frictional drag should the casing remain stationary for any period of time.

Where the geometry of the pipe and borehole prohibit turbulent rates, the next most effective mud removal system is to place the cement in plug flow. Plug flow depends on the same criteria as turbulence with the exception that a different Reynolds number is used.[33] The pertinent information is derived from the cement slurry by the use of a Fann VG Meter in the laboratory.

These data are then utilized in a graphical log-log plot to derive a flow behavior index (n') and a consistency factor (K') for a particular slurry. These parameters, along with a Reynolds number of 100, are utilized in the equations shown below to calculate the plug flow velocity. This readily converts to a maximum pump rate in order to remain in plug flow. In general, these rates will correspond well with the 90 feet/minute rule-of-thumb normally used for nominal annuli.

1. *Velocity at Some Specific Reynolds Number*
   For generalized calculations:

   N_{Re} for Plug Flow = 100 (maximum)
   N_{Re} for beginning of Turbulence = 2100 (minimum)

$$V^{2-n'} = \frac{N_{Re} \; K' \; (96/D)^{n'}}{1.86\rho} \qquad\qquad V = \left(\frac{N_{Re} \; K' \; (96/D)^{n'}}{1.86\rho} \right)^{\frac{1}{2-n'}}$$

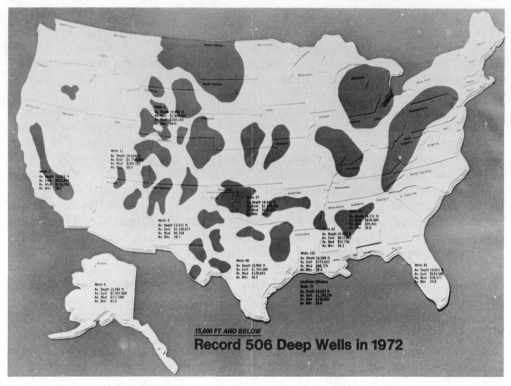

FIG. 16–42. *Casing programs used in deep wells*[29]

where, V = velocity, feet per second
K′ = consistency index, pound-seconds per square foot
n′ = flow behavior index, dimensionless
ρ = slurry density, pounds per gallon
D = inside diameter of pipe, inches
N_{Re} = specified Reynolds Number, dimensionless
For annulus, $D = D_o - D_I$
where, D_o = outer pipe inside diameter or hole size, inches
D_I = inner pipe outside diameter, inches

2. Displacement Rate $Q_b = \dfrac{VD^2}{17.15}$ or $Q_{cf} = \dfrac{VD^2}{3.057}$

where, V = velocity, feet per second
Q_b = pumping rate, barrels per minute
Q_{cf} = pumping rate, cubic feet per minute
D = inside diameter of pipe, inches
For annulus, $D^2 = D_o^2 - D_I^2$
where, D_o = outer pipe inside diameter or hole size, inches
D_I = inner pipe outside diameter, inches

The desirable cement flow characteristics are a PV and YP in excess of the mud system. If possible, cement Fann readings in the range of twice

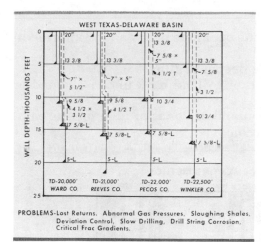

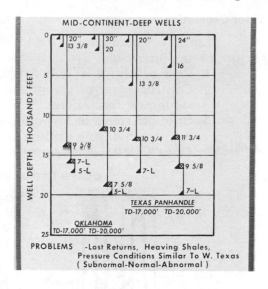

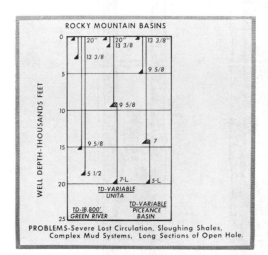

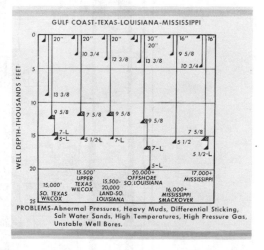

those of the mud are preferable. Quite often at higher densities this is not practical. The mud system must meet bore hole requirements which may make it impractical to design for this ratio, in which case the two systems should approach the PV and YP of the mud system.

FIG. 16–43. *Simulated liner cementing*[25]

Liner Cementing

Liners are commonly used to seal the openhole below a long intermediate casing string to:

1. Case off the open hole to enable deeper drilling.
2. Control water or gas production or to hold back unconsolidated or sloughing formations.
3. Case off zones of lost circulation and/or zones of high pressure encountered during drilling operations.

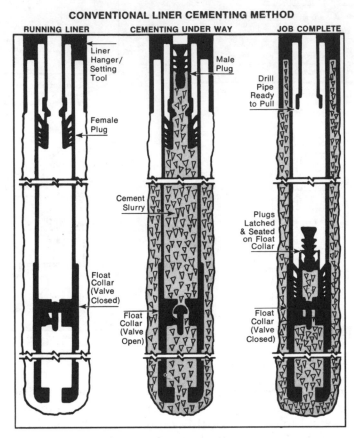

FIG. 16–44. *Conventional liner cementing method*

Once the liner has been set, cement is circulated down the drill pipe, out the liner shoe and up the outside of the liner. Plugs are used to prevent mud cement mixing in the liner just as they are in a full casing string during primary cementing. At the completion of the cementing operation, the excess cement is reversed out, or allowed to set above the liner and drilled out after setting (Figure 16–44).[32]

Tie-back liners and/or shorter scab liners are run and cemented to protect the intermediate casing against corrosion under high pressures.[32] They also serve the function of:

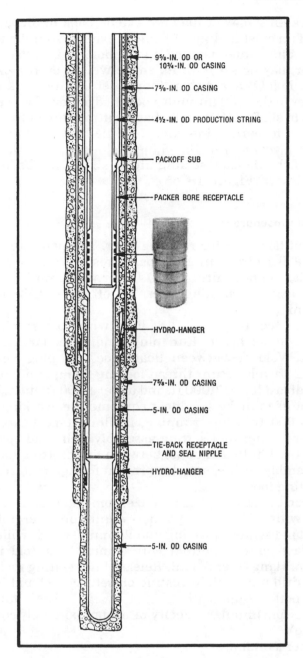

9⅝-IN. OD OR
10¾-IN. OD CASING

7⅝-IN. OD CASING

4½-IN. OD PRODUCTION STRING

PACKOFF SUB

PACKER BORE RECEPTACLE

HYDRO-HANGER

7⅝-IN. OD CASING

5-IN. OD CASING

TIE-BACK RECEPTACLE
AND SEAL NIPPLE

HYDRO-HANGER

5-IN. OD CASING

FIG. 16–45. *Tie back liner showing sealing nipple*

1. Reinforcing the intermediate casing worn by drilling.
2. Providing greater resistance to collapse stress from abnormal pressures.
3. Providing corrosion protection.
4. Sealing an existing liner which may be leaking gas.

The tie-back liner may or may not extend back to the surface but does cover the top of an existing liner. Should it extend from the top of a production liner to a point some distance up the hole, it is called a "stub" liner. This short liner may be set with its entire weight on the production liner or hung up the hole. When either liner is set as a tie-back section, it connects to the liner in the well through a polished or honed receptacle having a length of three to six feet. It usually has a setting thread for holding and releasing the liner to permit insertion of a tie-back sealing nipple before, during, or after placement of the cement slurry around the tie-back liner. The diameter of the tie-back sealing nipple should not be less than that of the liner set below it (Figure 16–45).

Liner Operational Procedure[32]

1. Trip to condition hole for running liner. Temperature subs should be used where BH circulating temperatures are unknown. Drop hollow drift (rabbit) to check drill pipe I.D. for pump down plug. Strap drill pipe to be used for running liner. Tie off other drill pipe on opposite side of board.
2. Run _____ feet of _____ liner with float shoe and float collar spaced ____ joints apart. Run plug landing collar ____ joints above float collar. Volume between float shoe and plug landing collar is _____ barrels. Run thread-locking compound on bottom 5–8 joints. Sandblast lower 1,000 ft. and upper 1,000 ft. liner. Pump through first few joints to make sure float equipment is working.
3. Fill each 1,000 ft. while running, if fill-up type floats are not used.
4. Install liner hanger and setting assembly. Fill dead space (if pack-off bushing is used in lieu of cups) between liner setting tool and the liner hanger assembly with inert gel to prevent foreign material from settling around setting tool.
5. Run liner on _____ (size, joint) _____ (grade) drill pipe with _____ pounds minimum over-pull rating. Run 1–2 min./stand while in casing and 2/3 min./stand while in open hole. Circulate last joint to bottom with cement manifold installed. Shut pump down. Hang liner 5' O.B. Release liner setting tool and leave 10,-000 lbs. of drill pipe weight resting on setting tool and liner top.
6. Circulate bottoms-up with _____ BPM rate to achieve _____ fpm annular velocity rate (approximately equal to previous drilling rate).
7. Cement liner as follows:
8. If unable to continue circulation or cementing due to plugging or bridging in liner-open hole annulus, pump on annulus between drill pipe and casing to maximum _____ psi and attempt to remove bridge. Do not overpressure and break down formation. If unable to break circulation, pull out of liner and reverse out any cement remaining in drill pipe.
9. Slow down pumps just before pump down plug reaches the liner wiper

plug. _____ bbls. is drill pipe capacity. Watch for plug shear, recalculate or correct cement displacement and continue plug displacement plus _____ bbls. maximum over-displacement.

10. If no indication of plug shearing, pump calculated displacement plus _____ bbls. (100% + 1 − 3%).

11. Pull out 8–10 stands or above cement, whichever is greatest, and hold pressure on top of cement until cement hardens to prevent gas migration.

12. Trip out of hole.

13. W.O.C. _____ hours.

14. Run _____ in. O.D. bit, drill cement to top of liner. Test liner overlap with differential test, if possible. Trip out.

15. Run _____ in. O.D. bit or mill, drill out cement inside liner as necessary. Displace hole for further drilling, spot perforating fluid (if in production liner) or other conditioning procedures as desired.

Regulatory Rules[34,35]

Practically all rules (State and Federal) cover the setting of specific sections of casing and methods for plugging wells. Conductor pipe is specificially emphasized in offshore federal regulations since it is used to start the hole in a vertical direction and to seal off badly sloughing surface material or water. Conductor pipe also serves as a connection for blowout preventer equipment when drilling shallow gas zones. Although not always done, cement should be circulated to the ocean or Gulf floor level in setting the conductor string.

Most state rules are very specific in the setting of surface casing to protect fresh water sands from contamination, and to form a good, solid anchor for blowout-preventer equipment while drilling the well. States having mineral deposits require cementing of surface pipe below such deposits to prevent any migration of fluids from the lower oil and gas producing horizons into the mineral deposits should they be mined at some later date.

Thirty years ago, regulations were almost nonexistent or applicable rules frequently were not enforced. In the absence of cementing regulations, most operators have casing cementing practices today which usually exceed the minimum requirements of state or federal regulatory bodies.

Regulatory rules were not written by the same groups nor were they prepared during the same time period. For these reasons they are expressed in a variety of ways.

Many states have appointed industry representatives to assist in writing laws applying to the drilling and cementing of wells. Kansas, New Mexico, Oklahoma, Texas and Wyoming represent a few states which have followed this practice. Much of the reference literature guiding these committees dates back to Farris[9] and others who have published data on the minimum compressive or bond strength required to adequately support pipe.

Technology changes and operating environments can often be attributed to the wide divergence in state regulations, particularly in regard to the volume of cement which should be used and the time allowed for strength development. There are also variations in cementing practices in fields of comparable characteristics in the same state.

ACOUSTIC LOGGING

With the introduction of acoustic logging to the oil industry as a means of evaluating a primary cementing job, more emphasis has been placed on using good cementing practices. Most of the available bond logging instruments record only a portion of the acoustic signal transmitted. Therefore, only a small portion of the available information is received, and these measurements may be influenced by factors which are not related to bonding. For example, the speed of sound is faster in certain formations than in the steel casing.

The most recent technique of bond logging is the use of a variable density film recording of the entire acoustic signal.[36] Therefore, it is not subjected to many of the interpretation limitations of the single curve "bond log."[37] Interpretations of the variable-density film recording provides qualitative information of the condition of the cement sheath surrounding the pipe.[38]

When logging using the variable-density film recording method, the first arriving pipe-borne portion of the acoustic signal will have high amplitude if the pipe is not cemented, as shown in Fig. 16–46(A). When the ce-

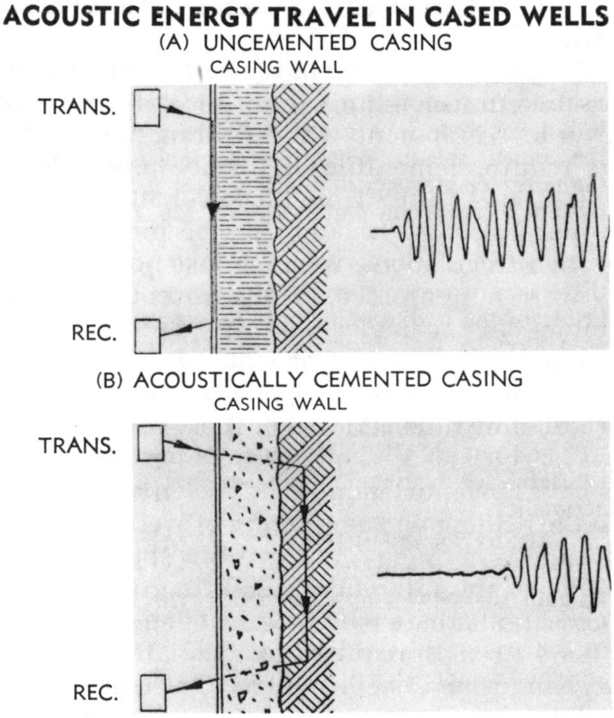

ACOUSTIC ENERGY TRAVEL IN CASED WELLS
(A) UNCEMENTED CASING

(B) ACOUSTICALLY CEMENTED CASING

FIG. 16–46. *Acoustic energy travel in cased wells*

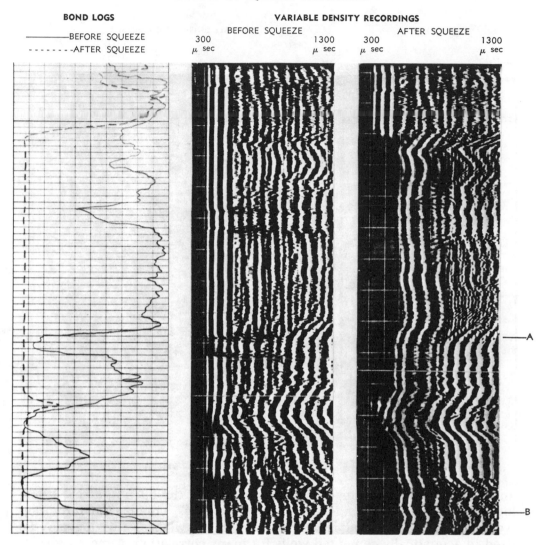

EFFECTS OF SQUEEZE CEMENTING

BOND LOGS

VARIABLE DENSITY RECORDINGS

BEFORE SQUEEZE

AFTER SQUEEZE

————BEFORE SQUEEZE

- - - - - - AFTER SQUEEZE

300 μ sec   1300 μ sec   300 μ sec   1300 μ sec

A

B

FIG. 16–47. *Effects of squeeze cementing*

ment is acoustically bonded to both the pipe and formation the wave will travel through the formation without appearance of a pipe signal. This is illustrated in Fig. 16–46(B). Where channeling exists in the cement sheath both the casing and formation paths may be evident. This is illustrated in Fig. 16–47 titled "Before Squeezing and After Squeezing." As illustrated in this figure, the cement became acoustically bonded after squeezing. Points A and B are locations of perforations where cement was squeezed into the annulus.

Fig. 16–48 shows a section of a log made in the open hole and the same section made in the cemented casing in the same well. Acoustic energy transmission through the formation is shown by the absence of pipe arrivals

ENERGY TRANSMISSION COMPARISON

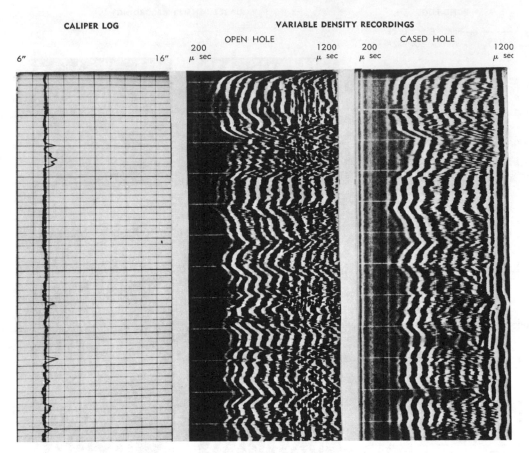

FIG. 16–48. *Energy transmission comparison*

and the correlation of the formation arrivals on the open and cased hole logs. This correlation could not exist if the pipe, cement, and formation were not adequately coupled. This example shows that this log may be used in acoustically cemented wells for many correlation purposes.

REFERENCES

1. Union Oil Co. of California: *On Tour*, Nov-Dec (1952).
2. Bogue, R. H., "A Digest of the Literature on the Constituents of Portland Cement": Concrete, July, 1926 to February 1927.
3. Aspdin, Joseph, "An Improvement in the Modes of Producing Artificial Stone": British Patent No. 5022(1824)
4. Portland Cement Association: "Design and Control of Concrete Mixtures," July 1968.
5. Lea, F. M. and Desch, C. H., "Chemistry of Cement and Concrete": published by Arnold & Co., London, 1935 (reprinted, 1937, 1940).
6. "Recommended Practice for Testing Oil-Well Cements and Cement Additives": API RP 10B, thirteenth edition, March 1964.

7. API Specification for Oil-Well Cements and Cement Additives, American Petroleum Institute, New York, N.Y., 1972.

8. API Recommended Practice for Testing Oil-Well Cements and Cement Additives, American Petroleum Institute, New York, N.Y., 1972.

9. Farris, R. Floyd, "Method for Determining Minimum Waiting-on-Cement Time": Petroleum Trans., AIME v 165, 1946, p 178–188.

10. Bearden, W. G. and Lane, R. D. "Engineered Cementing Operations to Eliminate WOC Time": Mid-Continent API District Meeting, April 5–7, 1961, Tulsa, Okla.

11. Carter, Greg, and Smith, Dwight K., "Properties of Cementing Compositions at Elevated Temperatures and Pressures": AIME Paper 892-G, Thirty-Second Annual Fall Meeting, Dallas, Tex., October 6–9, 1957.

12. Murphy, W. C. and Smith, Dwight, "A Critique of Filler Cements": SPE Preprint, March 1967.

13. Morgan, B. E. and Dumbauld, G. K.: "A Modified Low-Strength Cement, *J. Pet. Tech.* v 3 n 6, June 1951, p 165–70.

14. Beach, H. J.; O'Brien, T. B.; and Goins, W. C., Jr.: "The Role of Filtration in Cement Squeezing," Spring Meeting of API Southern District, Div. of Prod. Shreveport, 1961.

15. Parker, P. N., Ladd, B. J. and Wahl, W. W.: "An Evaluation of a Primary Cementing Technique Using Low Displacement Rates: API 1234, Presented at SPE 40th Annual Fall Meeting of AIME, Denver, Colo. (Oct. 3–6, 1965).

16. Slagle, Knox A.: "Rheological Design of Cementing Operations," *J. Pet. Tech.* (March, 1962) 323–328.

17. Slagle, K. A. and Smith, D. K.: "Salt Cement for Shale and Bentonitic Sands," *Journal of Petroleum Technology* (Feb., 1963) 187–194.

18. Hewitt, Charles H.: "Analytical Techniques for Recognizing Water-Sensitive Reservoir Rocks," *Jour. Pet. Tech.* (Aug. 1963) 813–818.

19. Owsley, W. D.: "Improved Casing Cementing Practices in the United States," Paper presented at AIME Meeting, San Antonio, Texas, Oct. 6, 1949: and Los Angeles, Calif., Oct. 20, 1949; *Oil and Gas J.* v 48 n 32, Dec. 15, 1949, p 76, 78.

19A. Clark, E. H. and Murray, A. S.: "A Study of Primary Cementing," API Spring Meeting, Los Angeles, California, May 22–23, 1958.

20. Jones, P. H. and Berdine, D.: "Oil-Well Cementing – Factors Influencing Bond Between Cement and Formation," *Drill. & Prod. Prac.*, API (1940) 45.

21. Teplitz, A. J. and Hassebroek, W. E.: "An Investigation of Oil-Well Cementing," *Drill. & Prod. Prac.*, API (1946) 76.

22. Piercy, N. A. V., Hooper, M. S. and Winney, H. F.: "Viscous Flow Through Pipes with Cores," *Phil. Mag.* (1933) 15, No. 99, 674.

23. Brice, J. W. Jr., and Holmes, R. C.: "Engineering Casing Cementing Programs Using Turbulent Flow Techniques" *JPT*, May 1964, p 503.

24. McLean, R. H., Manry, C. W., Whitaker, W. W.: "Displacement Mechanics in Primary Cementing," SPE Santa Barbara (1966) SPE 1488.

25. Clark, Charles R., and Carter, Greg L.; "Mud Displacement with Cement Slurries," (October 1972), San Antonio, Texas Annual SPE Meeting, SPE Paper 4090.

26. Graham, Harold L.; "Rheology-Balanced Cementing Improves Primary Success," *O&GJ*, December 18, 1972.

27. Carter, Greg L., and Cook, Clyde: "Cementing Research in Directional Gas Well Completions," London SPE (April 1973), SPE Paper 4313.

28. Carter, Greg, and Slagle, Knox A.: "A Study of Completion Practices to Minimize Gas Communication," Amarillo, Texas Regional SPE Meeting (November 1970), SPE Paper 3164.

29. *World Oil*, "A Look at Deep Drilling," May 1968, p 57.

30. Kirk, W. L.: "Deep Drilling Practices in Mississippi," *Journal of Petroleum Technology*, June 1972, p 633.

31. Gibbs, Max A.: "Delaware Basin Cementing – Problems and Solutions," *Journal of Petroleum Technology*, October 1966.

32. Lindsey, H. Ed Jr., and Bateman, S. J.: "Liner Cementing in High-Pressure Gas Zones," 1973.

33. Slagle, Knos A.: "Rheological Design of Cementing Operations," *Journal of Petr. Tech.* (March 1962).
34. McRee, Boyd C., "Cementing Regulations Applied to Oil and Gas Wells," *Oil Well Cementing Practices in the United States*, Chapter 20.
35. Regulations Pertaining to Mineral Leasing, Operations and Pipelines on the Outer Continental Shelf, *Code of Federal Regulations*, United States, Department of Interior, April 1971.
36. Anderson, T. O., Winn, R. H. and Walker, Terry: "A Qualitative Cement-Bond Evaluation Method," API Paper 875-18-A, Spring Meeting of the Rocky Mountain District, Division of Production, Billings, Montana, April 20–22, 1964.
37. Pickett, G. R., "Acoustic Character Logs and Their Applications in Formation Evaluation," *Journal of Petr. Technology*, June 1963, Vol 15, No. 6.
38. Fertl, W. H., Pilkington, P. E. and Scott, J. B.: "A Look at Cement Bond Logs," *Journal of Petroleum Technology*, June 1974.